MECHANICAL ENGINEERING
LIBRARY
IMPERIAL COLLEGE

TYRE MODELS FOR
VEHICLE DYNAMIC ANALYSIS

TYRE MODELS FOR VEHICLE DYNAMIC ANALYSIS

Edited by

F. Böhm and H.-P. Willumeit

*Proceedings of the 2nd International Colloquium
on Tyre Models for Vehicle Dynamic Analysis,
held at the Technical University of Berlin, Germany,
February 20-21, 1997*

*Supplement to
Vehicle System Dynamics, Volume 27*

SWETS & ZEITLINGER PUBLISHERS
LISSE   ABINGDON   EXTON (PA)   TOKYO

Library of Congress Cataloging-in-Publication Data

Applied for

Printed in the netherlands by Grafisch Produktiebedrijf Gorter, Steenwijk

Copyright © 1996 Swets & Zeitlinger B.V., Lisse, the Netherlands

All rights reserved. No part of this publication may be reproduced, stored in a retrieval system, or transmitted in any form or by any means, electronic, mechanical, photocopying, recording, or otherwise, without the prior written permission of the publisher.

ISBN 90 265 1488 3

CONTENTS

Introduction 1

Lectures

G. Mastinu, S. Gaiazzi, F. Montanaro and D. Pirola
A Semi-Analytical Tyre Model for Steady- and Transient-State Simulations   2

G. Leister
New Procedures for Tyre Characteristic Measurement   22

H. Oldenettel and H.-J. Köster
Test Procedure for the Quantification of Rolling Tire Belt Vibrations   37

A. Higuchi and H.B. Pacejka
The Relaxation Length Concept at Large Wheel Slip and Camber   50

P.W.A. Zegelaar and H.B. Pacejka
Dynamic Tyre Responses to Brake Torque Variations   65

K. Guo and Q. Liu
Modelling and Simulation of Non-Steady State Cornering Properties and
Identification of Structure Parameters of Tyres   80

M. Gipser, R. Hofer and P. Lugner
Dynamical Tire Forces Response to Road Unevennesses   94

M. Eichler
A Ride Comfort Tyre Model for Vibration Analysis in Full Vehicle
Simulations   109

F. Böhm
Elastodynamics of Cars and Tires   123

S. Bruni, F. Cheli and F. Resta
On the Identification in Time Domain of the Parameters of a Tyre Model
for the Study of In-Plane Dynamics   136

D.J. Allison and R.S. Sharp
On the Low Frequency In-Plane Forced Vibrations of Pneumatic Tyre /
Wheel / Suspension Assemblies   151

D. Zachow
3D Membrane Shell Model in Application of a Tractor and PKW Tyre 163

F.R. Fassbender, C.W. Fervers and C. Harnisch
Approaches to Predict the Vehicle Dynamics on Soft Soil 173

Son-Joo Kim and A.R. Savkoor
The Contact Problem of In-Plane Rolling of Tires on a Flat Road 189

P. Fancher, J. Bernard, C. Clover and C. Winkler
Representing Truck Tire Characteristics in Simulations of Braking and
Braking-in-a-turn Maneuvers 207

F. Negrus, G. Anghelache and A. Stanescu
Finite Element Analysis and Experimental Analysis of Natural Frequencies
and Mode Shapes for a Non-Rotating Tyre 221

S. Yamazaki, O. Furukawa and T. Suzuki
Study on Real Time Estimation of Tire to Road Friction 225

Invited lectures

H.B. Pacejka and I.J.M. Besselink
Magic Formula Tyre Model with Transient Properties 234

A.E. Belkin, B.L. Bukhin, O.N. Mukhin and N.L. Narskaya
Some Models and Methods of Pneumatic Tire Mechanics 250

J.J.M. van Oosten, H.-J. Unrau, G. Riedel and E. Bakker
TYDEX Workshop: Standardisation of Data Exchange in Tyre Testing
and Tyre Modelling 272

Ch. Oertel
On Modelling Contact and Friction: Calculation of Tyre Response on
Uneven Roads 289

F. Böhm
On the Roots of Tire Mechanics 303

Poster

R.W. Allen, J.P. Chrstos and T.J. Rosenthal
A Tire Model for use with Vehicle Dynamics Simulations on Pavement
and Off-Road Surfaces 318

E.M. Negrus and M. Cocosila
Flexible Model to Simulate Wheel Pass Over Singular Road Obstacles 322

S. Detalle, J. Flament and F. Gailliegue
A Tyre Model for Interactive Driving Simulators 326

Ch. von Holst and H. Göhlich
The System Tractor - Tire under the Influence of Tractor Development 330

W.E. Krabacher
A Comparison of the Moreland and Von Schlippe-Dietrich Landing Gear
Tire Shimmy Models 335

J.P. Maurice and H.B. Pacejka
Relaxation Length Behaviour of Tyres 339

List of addresses 343

INTRODUCTION

The 2nd International Colloquium on Tyre Models for Vehicle Dynamic Analysis was held on 20./21. February 1997 at the Technical University of Berlin. Since the 1st International Colloquium held at Technical University of Delft mainly dealt with the stationary behaviour of tyres, the subject of the 2nd Colloquium was „transient tyre behaviour". Based on the results of the Sonderforschungsbereich 181 (special research project 181) „High frequency rolling contact of wheels" (1986 - 1994), which was financed by Deutsche Forschungsgemeinschaft (DFG), the need for this Colloquium arose. The task of the project was to investigate the high frequency mechanisms of the tyre - road surface contact.

The industry implemented the results in many of their applications. This was also confirmed by the presentations at the 2nd Colloquium, and simultaneously it was shown that this research field is in further and fast development. Mechanics of tyres is complicated not only because of the rubber and cord nonlinearities, but also because of contact and friction. In order to comprehend the complex processes, suitable test and modelling methods are necessary. The modelling methods of the driving behaviour have to take the high frequency rolling processes of the tyre into account.

This 2nd Colloquium was similar to the 1st one held under the patronage of the International Association for Vehicle System Dynamics (IAVSD). The Scientific Committee of this Colloquium mainly consisted of the following members of IAVSD: R. Wade Allen, Prof. Dr. F. Böhm, Prof. Dr. R. Gnadler, Dr. M. Gipser, Prof. Dr. K. Guo, Prof. Dr. P. Lugner, Prof. Dr. Ch. Oertel, J. J. M. van Oosten, Prof. Dr. H. B. Pacejka, Prof. Dr. L. Palkovics, Dr. A. R. Savkoor, Prof. R. S. Sharp, Prof. Dr. H.-P. Willumeit.

Members of the Local Organizing Committee were Prof. Dr. F. Böhm, Prof. Dr. Ch. Oertel and Prof. Dr. H.-P. Willumeit. 18 lectures, 5 invited lectures and 9 posters/presentations were held in the new building of the Physical Institute of the Technical University of Berlin.

In the evening of 20. February 1997 a dinner took place at „Zitadelle Spandau" (Fortress Spandau, historical site of Berlin), as well as the visit of research facilities for dynamic tyre testing at the Institute of Mechanics and the Institute of Machinery Construction for testing agricultural tyres were organized on 22. February 1997.

The Organizing Committee acknowledges very special thanks to the Daimler-Benz AG, Stuttgart for sponsoring the proceedings for all participants of the Colloquium.

For the Local Organizing Comittee:
Prof. Dr. F. BÖHM, Prof. Dr. H.-P. WILLUMEIT

A Semi-Analytical Tyre Model for Steady- and Transient-State Simulations

G. MASTINU, S. GAIAZZI, F. MONTANARO and D. PIROLA

ABSTRACT

On the basis of previous research papers, an enhanced tyre model is presented which could be useful both for vehicle dynamics analysis and for tyre design. The physical phenomena involved in tyre-ground force generation are described mathematically in a semi-analytical form. Given the main mechanical parameters of the tyre, the friction coefficients and the running conditions, the model returns the values of the forces generated at the tyre-ground interface either at steady- or at transient-state condition. Comparisons between experimental and computed data have been performed with satisfactory results.

INTRODUCTION

A tyre model is presented which could be useful both for vehicle dynamics analysis and for tyre design. The proposed model has been developed in order to be both as simple as possible (to provide basic insight to the problems of tyre-ground force generation) and as accurate as possible (to allow reliable simulations of non-existing tyres). In-depth reviews on tyre models for steady- and transient-state simulations can be found in [8] and [12].

The tyre model presented here has been developed in previous studies which have been published in [1] and [2]. According to the definition given in [12], the model is a "physical" one, i.e. it is based on the mathematical description of actual physical phenomena involved in tyre mechanics. Particularly, the mathematical description is "semi-analytical", i.e. both formulae in analytical form and numerical procedures are used to compute the longitudinal force F_x, the lateral force F_y and the self-aligning torque M_z. Wide ranges of longitudinal and lateral slip ($-1 < s_x < 1$; -20 deg $< \alpha < 20$ deg) can be taken into consideration.

In [1] the main features of the tyre model presented here were introduced. The pneumatic tyre was described by a belt connected to the rigid wheel centre via radial and lateral non-linear elastic elements. The tread pattern was modelled by longitudinal and lateral elastic elements. The belt lateral deformation in the contact area was described by a parabola whose parameters were proportional to the lateral force F_y and to the self-aligning torque M_z. These parameters were estimated by a relatively simple FEM sub-model. In [2] a validation of the model was performed which showed that, at steady-state, the longitudinal force F_x, the lateral force F_y and the self-aligning torque M_z could be predicted with reasonable accuracy, provided that a preliminary parameter identification had been undertaken.

By this early formulation some features were not adequately refined. In particular, the contact area shape and pressure distibution were calculated roughly. In the present formulation a better computation of these variables is proposed.

At transient-state, the lateral force F_y and the self-aligning torque M_z are computed. The basic procedure on how doing it was described in [1]. Here some significant enhancements have been added. In particular, the gyroscopic effects have been introduced.

1. CONTACT AREA

1.1 Contact area length

In Fig. 1 the portion of the pneumatic tyre in contact with the ground is represented. The belt is modelled as an Euler beam. The lower continously distributed stiffness represents the tread elasticity.

The stiffness per unit length k_L is

$$k_L = \frac{E_r \cdot b_b}{\tau} \cdot \Lambda \qquad (1.1)$$

(Λ is a coefficient accounting for actual tread pattern - the tread is not a continuous strip but there are voids between single tread elements).

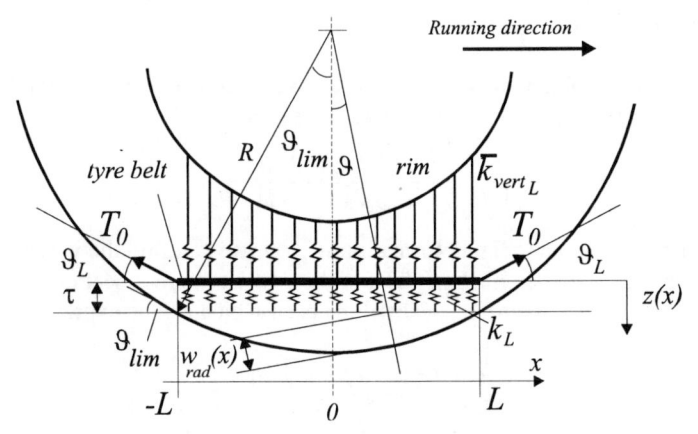

Fig.1. The tyre belt in longitudinal direction as a beam in between two continuously distributed stiffnesses.

The upper continuously distributed stiffness in Fig.1 represents the effect of the inflated inner volume of the carcass. Fig. 2 shows the cross section of the carcass at a generical longitudinal position x.

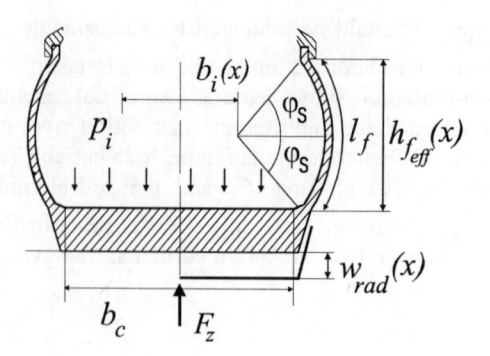

Fig. 2. Carcass cross section at generical longitudinal position x (see Fig.1).

From the observation of Fig. 2 the following relation between the geometrical parameters of the carcass can be derived

$$\frac{sin(\varphi_s(x))}{\varphi_s(x)} = \frac{h_{f_{eff}}(x)}{l_f} \qquad (1.2)$$

where $h_{f\,eff}(x) = h_f - w_{rad}(x)$ represents the effective heigth of the sidewalls at x.

Developing Eq.(1.2) in Mc Laurin's Series up to the third order and solving with respect to φ_S we obtain

$$\varphi_s(x) = \sqrt{6 \cdot \left(1 - \left(\frac{h_{f_{eff}}(x)}{l_f}\right)\right)} \qquad (1.3)$$

Appling the membrane theory [1,7], it is possible to calculate the radial force for unit (longitudinal) length for the generical cross section

$$f_r(x) = p_i \cdot b_i(x) = p_i \cdot \left[b_c - \left(\frac{l_f \cdot cos(\varphi_s(x))}{\varphi_s(x)}\right)\right] \qquad (1.4)$$

Dividing $f_r(x)$ by the radial deformation $w_{rad}(x)$, we can find the vertical stiffness per unit (longitudinal) length $k_{vert\,L}(x)$ for the generical cross section. For sake of simplicity, the mean value of $k_{vert\,L}(x)$ will be used which can be expressed as

$$\bar{k}_{vert_L} = p_i + k_{f_L} = p_i + \frac{1}{L}\int_{-L}^{L}(k_{vert_L}(x) - p_i) \cdot dx \qquad (1.5)$$

In Fig. 1 T_0 represents the effect of the draught into the belt due to the inflation pressure and its value is approximated here by Mariotte's formula

$$T_0 = p_i \cdot R \cdot b_b \qquad (1.6)$$

The belt deformation in x direction can be obtained by considering the equilibrium of an infinitesimal beam element, as shown in Fig. 3.

Appling Euler theory to the beam represented in Fig. 1 and Fig. 3, the following differential equation can be derived

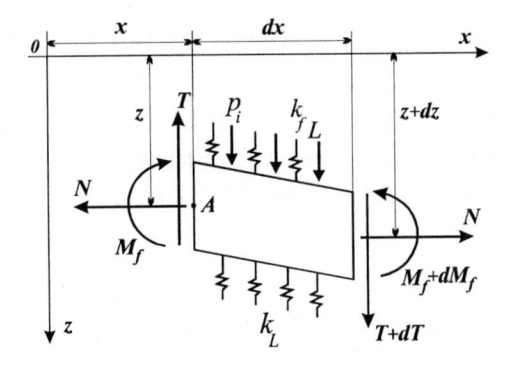

Fig. 3. Moment and forces acting on an infinitesimal beam element representing a longitudinal portion of the belt (see Fig.1).

$$EJ_L \cdot z^{IV}(x) - T_0 \cdot \left(\cos(\vartheta_L)\right) \cdot z^{II}(x) + \left(k_L + k_{f_L}\right) \cdot z(x) = p_i \cdot b_b \tag{1.7}$$

The solution is

$$z(x) = \left(A_L e^{\alpha_L x} + B_L e^{-\alpha_L x}\right) \cos\left(\beta_L x\right) + \\ + \left(C_L e^{\alpha_L x} + D_L e^{-\alpha_L x}\right) \sin\left(\beta_L x\right) + \frac{p_i \cdot b_b}{k_L + k_{f_L}} \tag{1.8}$$

where

$$\alpha_L = \sqrt[4]{\frac{k_L + k_{f_L}}{EJ_L}} \cdot \left\{\cos\left[\frac{1}{2} \cdot \arccos\left(\frac{T_0 \cdot \cos(\vartheta_L)}{\sqrt{4\left(k_L + k_{f_L}\right) \cdot EJ_L}}\right)\right]\right\} \tag{1.9}$$

$$\beta_L = \sqrt[4]{\frac{k_L + k_{f_L}}{EJ_L}} \cdot \left\{\sin\left[\frac{1}{2} \cdot \arccos\left(\frac{T_0 \cdot \cos(\vartheta_L)}{\sqrt{4\left(k_L + k_{f_L}\right) \cdot EJ_L}}\right)\right]\right\} \tag{1.10}$$

To calculate the values of the four constants A_L, B_L, C_L, D_L in the Eq. (1.8), the following boundary conditions are imposed

$$z(-L) = 0 \qquad z(L) = 0 \qquad z'(-L) = \vartheta_L \qquad z'(L) = -\vartheta_L \tag{1.11}$$

It is useful to express ϑ_L as function of ϑ_{lim} (see Fig. 1). To do this Eq. (1.8) can be rewritten by setting $x = R\vartheta$, then the function $z(\vartheta)$ can be derived with respect to ϑ in order to obtain $\vartheta_L = \frac{dz}{d\vartheta}(\vartheta_{lim})$. One finds

$$\vartheta_L = \sum_{i=1}^{4} X_i \cdot e^{(\lambda_i \cdot \vartheta_{lim})} \qquad (1.12)$$

As the parameters X_i and λ_i are not known, they have to be identified.

1.2 Contact width

The same approach used to compute the deformation of the belt in longitudinal direction is applied in lateral direction too (see Fig.4).

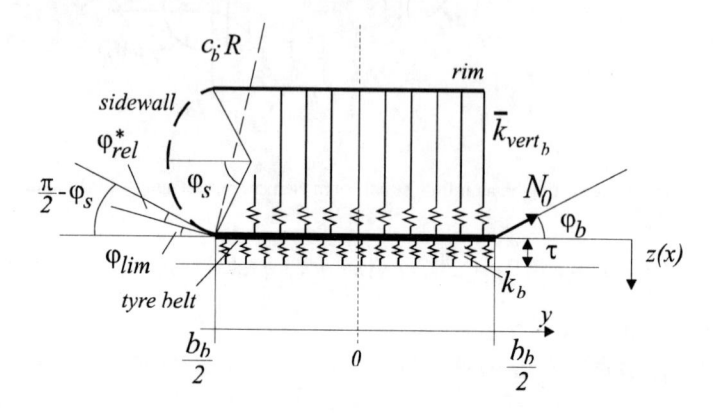

Fig. 4. The tyre belt in lateral (cross) direction as a beam in between two continuously distributed stiffnesses.

The vertical stiffnesses per unit length k_b and $\bar{k}_{vert_b}$ are

$$k_b = \frac{E_r \cdot 2 \cdot L}{\tau} \cdot \Lambda \qquad (1.13)$$

$$\bar{k}_{vert_b} = p_i + k_{f_b} = p_i + \frac{k_{f_L} \cdot 2 \cdot L}{b_b} \qquad (1.14)$$

The draught in the beam is given by

$$N_0 \cdot \cos(\varphi_b) = N = \frac{p_i \cdot 2 \cdot L \cdot h_{f_{eff}}(0)}{2} = p_i \cdot L \cdot h_{f_{eff}}(0) \qquad (1.15)$$

The equation for the deformation of the belt in lateral direction is

$$z(y) = \left(A_b \, e^{\alpha_b y} + B_b \, e^{-\alpha_b y}\right) \cos(\beta_b y) + \\ + \left(C_b \, e^{\alpha_b y} + D_b \, e^{-\alpha_b y}\right) \sin(\beta_b y) + \frac{p_i \cdot 2L}{k_b + k_{f_b}} \qquad (1.16)$$

where

$$\alpha_b = \sqrt[4]{\frac{k_b + k_{f_b}}{EJ_b}} \cdot \left\{ \cos\left[\frac{1}{2} \cdot arccos\left(\frac{N}{\sqrt{4(k_b + k_{f_b}) \cdot EJ_b}}\right)\right]\right\} \quad (1.17)$$

$$\beta_b = \sqrt[4]{\frac{k_b + k_{f_b}}{EJ_b}} \cdot \left\{ \sin\left[\frac{1}{2} \cdot arccos\left(\frac{N}{\sqrt{4(k_b + k_{f_b}) \cdot EJ_b}}\right)\right]\right\} \quad (1.18)$$

The boundary conditions are

$$z(-b_b/2) = 0 \quad z(b_b/2) = 0 \quad z'(-b_b/2) = \varphi_b \quad z'(b_b/2) = -\varphi_b \quad (1.19)$$

with

$$\varphi_b = \sum_{i=1}^{4} Y_i \cdot e^{(\xi_i \cdot \varphi_{rel}^*)} \quad (1.20)$$

$$\varphi_{rel}^* = \varphi_{lim} - \left(\frac{\pi}{2} - \varphi_s(0)\right) = arcsin\left(\frac{b_b}{2 \cdot c_b \cdot R}\right) - \left(\frac{\pi}{2} - \varphi_s(0)\right) \quad (1.21)$$

The parameters Y_i and ξ_i have to be identified. c_b is set equal to 2 which means that the lateral belt profile is assumed to belong to a sphere whose diameter is $c_b R$.

1.3 Computation of the bending stiffnesses

To solve equations (1.8) and (1.16), the belt bending stiffnesses EJ_L and EJ_b are needed.

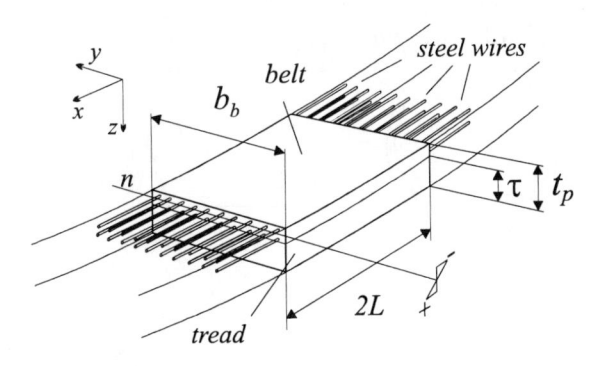

Fig. 5. Belt section in longitudinal and lateral directions.

A satisfactory approximation of EJ_L seems

$$EJ_L = E_{steel} \cdot \sum_{i=1}^{n_w} \left(\frac{1}{4}\pi \cdot r_i^4 + \pi \cdot r_i^2 \cdot d_i^2\right) \quad (1.22)$$

The greatest contribution to the bending stiffness EJ_L is due to the steel wires constituting the belt. As indicated in [17], the longitudinal wires have no influence in lateral direction. It is possible to write the analytical expression of the belt lateral bending stiffness by considering a partial contribution of radial wires

$$EJ_b = 3 \cdot E_r \cdot \frac{1}{12}(t_p - \tau)^3 \cdot 2L \qquad (1.23)$$

1.4 Trapezoidal contact area

As shown in [5], [6] and [8], when a shear lateral force is developed by the running tyre, the shape of the contact area becomes approximatively trapezoidal. The deformation from the static nearly rectangular shape to the quasi-static nearly trapezoidal shape is related to lateral force and to tyre vertical deflection. Experimental results suggest that, in absence of camber, the half contact length L_s at $y=0$ (see Fig. 6) can be considered equal to that in static condition. We can define

$$\gamma_c = \arctan\left(\frac{y_{tot}}{R - w_{rad}(0)}\right) \qquad (1.24)$$

where y_{tot} [1,2] is given by the sum of the lateral deformation of the carcass $y_c(0)$ (see Eq. 2.1) and that of tread y_{tr}

$$y_{tot} = y_c(0) + y_{tr} = \frac{F_y}{K_c} + \frac{F_y}{2 \cdot L \cdot k_y} \qquad (1.25)$$

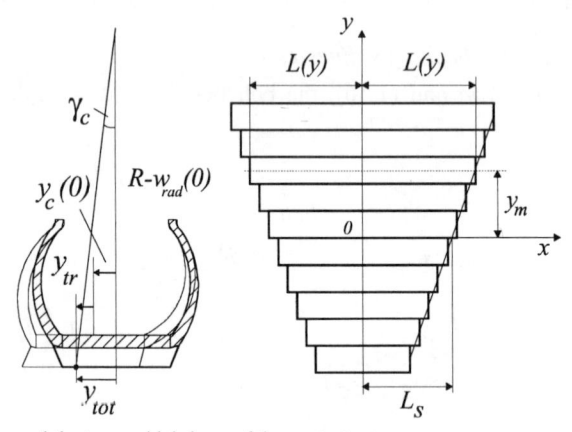

Fig. 6. The angle γ_c and the trapezoidal shape of the contact area.

We can write

$$L(y) = L_s + K \cdot \gamma_c \cdot y \qquad (1.26)$$

where K is a constant typical for each tyre. $L(y)$ gives the half contact length for each of the n_s longitudinal stripes the contact area is divided into. By imposing the proper rotation ϑ_L at each stripe end, it is possible to calculate the longitudinal pressure distribution along each stripe (see Fig. 14).

1.5 Contact area size and pressure distribution computation

The contact area size is defined by the values of $2L$ and b_b. The values of $2L$ and b_b are found when the two integrals of system

$$\begin{cases} F_z = \int_{-L}^{L} z(x) \cdot k_L \cdot dx \\ F_z = \int_{-b_b/2}^{b_b/2} z(y) \cdot k_b \cdot dy \end{cases} \quad (1.27)$$

are equal to F_z. Because $z(x)$ is actually $z(x,L,b_b)$ and $z(y)$ is actually $z(y,L,b_b)$, a numerical iterative method is used.

The bi-dimensional pressure distribution on the ground is obtained as

$$p(x,y) = \frac{p(x) \cdot p(y)}{F_z} = \frac{(z(x) \cdot k_L) \cdot (z(y) \cdot k_b)}{F_z} \quad (1.28)$$

At running conditions, additional rotations at beam ends are assumed in order to consider pressure distribution distortion both in longitudinal and lateral direction. New boundary conditions are

$$z'(-L) = \theta_L + \eta_x \cdot V_0 \qquad z'(L) = -\theta_L + \eta_x \cdot V_0 \quad (1.29)$$

$$z'(-b_b/2) = \varphi_b + \eta_y \cdot \alpha \qquad z'(b_b/2) = -\varphi_b + \eta_y \cdot \alpha \quad (1.30)$$

1.6 Contact area size computation and measurement

Fig. 7 shows a comparison beetween experimental and computed contact area size. From the examination of the results the capability of the model to compute the contact area size can be judged. Even though the mechanical model of the tyre is rather simple, experimental data are fitted very well by computed results.

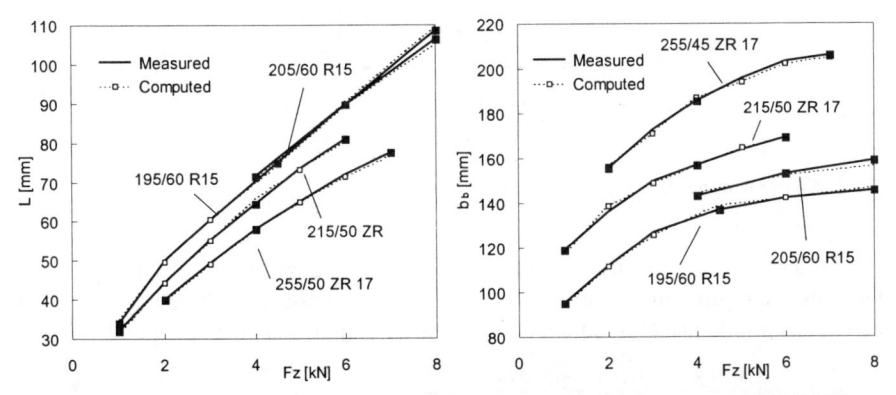

Fig. 7. Measured and computed contact area size (L half contact length ; b_b contact width): 195/60 p_i = 1.97 atm ; 205/60 p_i = 2.2 atm ; 215/50 p_i = 2.44 atm ; 255/45 p_i = 2.64 atm. The black boxes indicate the combinations of F_z-L or F_z-b_b used to identify the parameters X_i, λ_i in Eq.(1.12) or Y_i, ξ_i in Eq.(1.20). Complete data set in Appendix.

2. STEADY-STATE SHEAR FORCE COMPUTATION

2.1 Mechanical model of the tyre

Given the main mechanical parameters of the tyre, the friction coefficients and the running conditions, the model returns the values of the forces generated at the tyre-ground interface at steady-state condition.

Referring to [1] and [7], the carcass is assumed to be composed by the sidewalls and the belt. The belt is a circular beam inextensible in tangential direction and flexible in lateral and radial directions. The beam is connected to the wheel hub through a distributed stiffness both in lateral and radial directions.

The continuous lateral distributed stiffness varies along the circumference. It depends mainly on inner tyre pressure and on radial deflection: the higher the pressure the higher the stiffness, the higher the radial deflection the lower the stiffness. In the contact region, where the radial deflections are sensible, the lateral stiffness varies considerably [1].

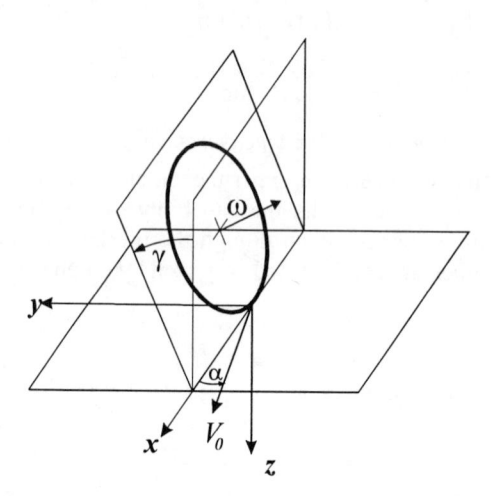

Fig. 8. Conventions for describing tyre mechanics.

Taking into account the varying lateral stiffness along the tyre circumference, the computer program computes by F.E.M. the carcass lateral deformation. The carcass lateral deformation in the contact region is approximated by a parabola

$$y_c(x) = \left(\frac{sin(\gamma)}{2R} - \frac{F_y}{K_W}\right) \cdot x^2 + \frac{M_z}{K_\psi} \cdot x + \frac{F_y}{K_c} \qquad (2.1)$$

Camber effect is roughly taken into account.

To calculate the shear force, it is important to compute the adhesion area end points for each strip in which the contact area is subdivided (see Fig. 6). In each of n_s strip the adhesion area ends at abscissa x_s where the horizontal force reaches the limit given by adhesion friction coefficients and contact pressure

$$\left(\frac{f_{x_a}(x_s)}{\mu_x}\right)^2 + \left(\frac{f_{y_a}(x_s)}{\mu_y}\right)^2 = \left(p(x_s, y_m) \cdot \frac{b_b}{n_s}\right)^2 \qquad (2.2)$$

The forces per unit length in the adhesion area are
$$f_{x_a} = k_x \cdot u \qquad f_{y_a} = k_y \cdot v \qquad (2.3)$$
while the forces for unit surface in the sliding region are

$$f_{x_s}(x,y) = \frac{-s_x}{\sqrt{\left(\frac{s_x}{\mu_{x_s}}\right)^2 + \left(\frac{\tan\alpha}{\mu_{y_s}}\right)^2}} \cdot p(x,y)$$

$$f_{y_s}(x,y) = \frac{\tan\alpha}{\sqrt{\left(\frac{s_x}{\mu_{x_s}}\right)^2 + \left(\frac{\tan\alpha}{\mu_{y_s}}\right)^2}} \cdot p(x,y) \qquad (2.4)$$

Integrating Eq.(2.3) and (2.4) and writing the equations for self-aligning torque as shown in [2], an implicit system of non linear algebraic equations is obtained

$$\begin{cases} F_x = F_{x_a} + F_{x_s} = F_x(\alpha, s_x, \bar{x}_s) \\ F_y = F_{y_a} + F_{y_s} = F_y(\alpha, s_x, \gamma, \bar{x}_s, F_y, M_z) \\ M_z = M_{z_a} + M_{z_s} = M_z(\alpha, s_x, \gamma, \bar{x}_s, F_y, M_z) \\ \bar{x}_s = \bar{x}_s(\alpha, s_x, \gamma, \bar{x}_s, F_y, M_z) \end{cases} \qquad (2.5)$$

The generalized forces F_x, F_y and M_z are the solutions. The *analytical* expressions of the first three equations of system (2.5) are reported in [2].

2.2 Computer program

The computer program that has been implemented has two working modes: a *calculation mode* (the generalized forces F_x, F_y and M_z are computed given V_0, α, γ, s_x, F_z, p_i, R, τ, h_f, l_f, EJ, k_x, k_y, X_i, λ_i, Y_i, ξ_i, μ_x, μ_y, μ_{xs}, μ_{ys}, η_x, η_y) and an *identification mode* (there are 3 ways of operations : I) the parameters X_i, λ_i, Y_i, ξ_i are computed given 3 sampled sets of F_z, L and b_h; II) the parameters k_x, μ_x, μ_{xs}, η_x are computed given 5 sampled sets of F_x, s_x; III) the parameters EJ, μ_y, μ_{ys}, η_x, η_y are computed given 5 sampled sets of F_y, M_z, α). Fig. 9 shows a flow-chart of the calculation mode.

2.3 Shear force computation and measurement
Fig. 10, 11, 12, and 13 show the comparison between computed and measured shear force referring to four different pneumatic tyres. A satisfactory agreement beetwen

computed and experimental results seems to have been reached. With respect to previous results presented in [2], an enhancement is evident due to the more accurate computation of the contact area size and pressure distribution (see Fig. 14). Slight variations of the identified parameters are found as F_z varies. Further studies are needed to investigate this occurence.

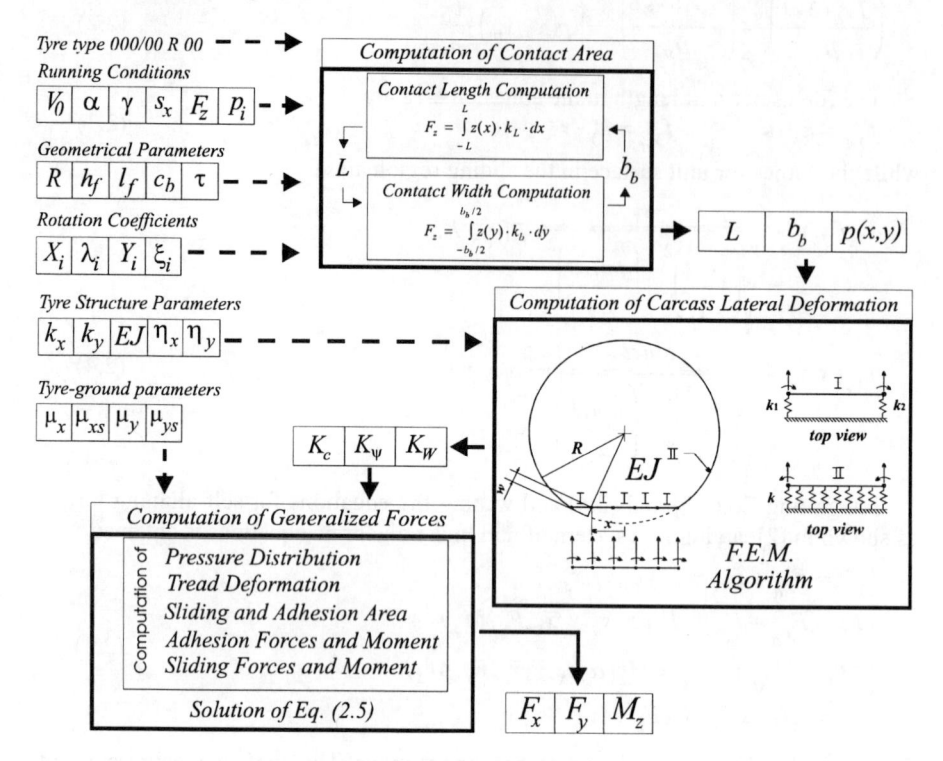

Fig. 9. Flow-chart of the computer program for steady-state simulations.

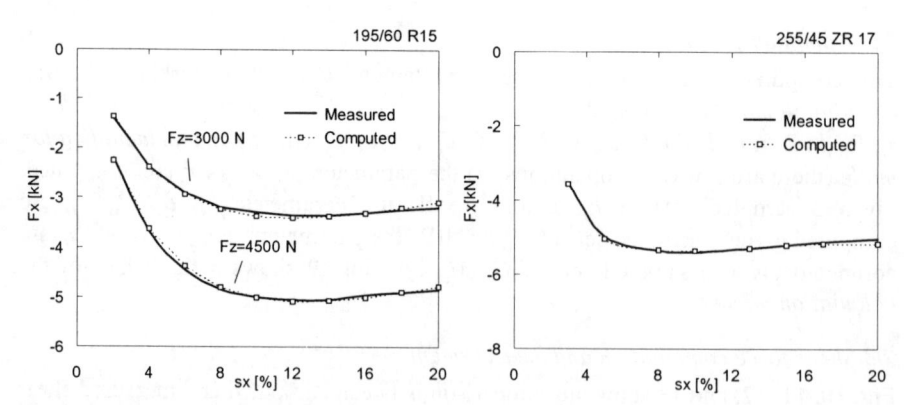

Fig. 10. Pure longitudinal slip: 195/60 R 15, p_i = 1.97 atm; 255/45 ZR17, p_i = 2.56 atm. Complete data set in Appendix.

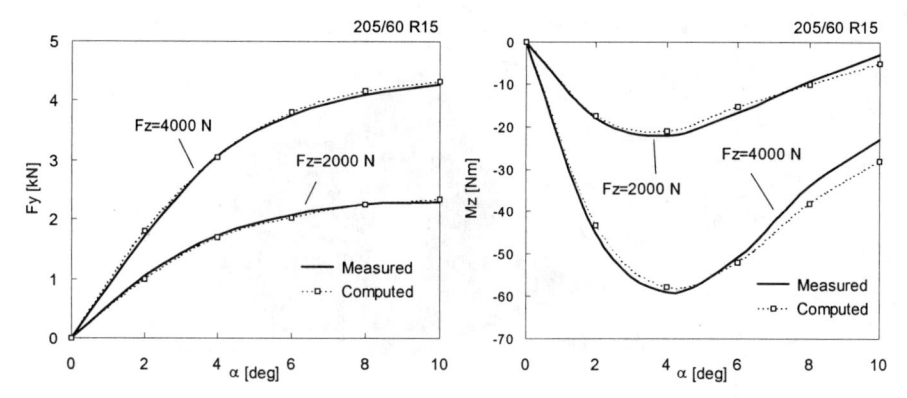

Fig. 11. Pure lateral slip: tyre size 205/60 R15, p_i = 2.2 atm. Complete data set in Appendix.

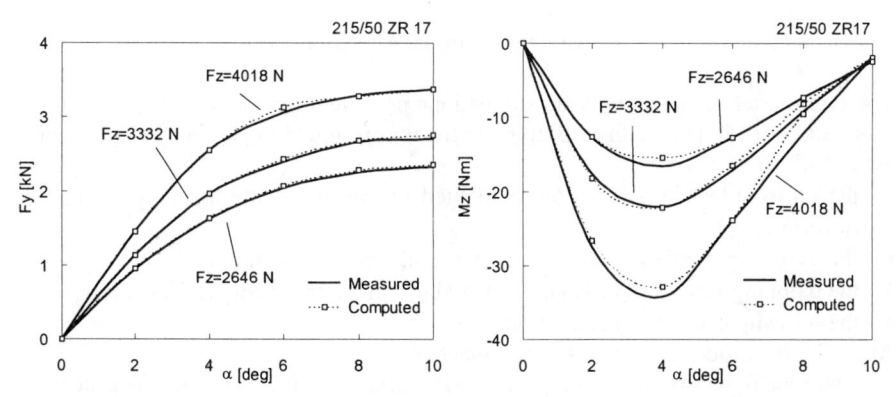

Fig. 12. Pure lateral slip: tyre size 215/50 ZR17, p_i = 2.45 atm. Complete data set in Appendix.

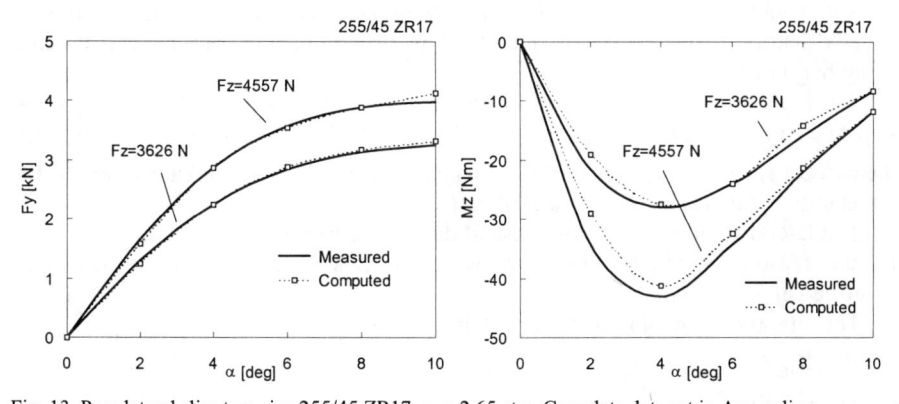

Fig. 13. Pure lateral slip: tyre size 255/45 ZR17, p_i = 2.65 atm. Complete data set in Appendix.

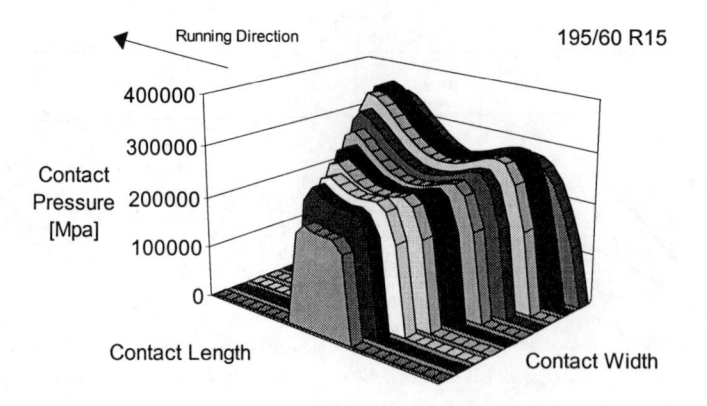

Fig. 14. Example of a computed tyre-ground contact pressure distribution; $F_z = 4500$ N.

3. TYRE MODEL FOR TRANSIENT STATE

The tyre model for transient state simulations presented here is basically that which was introduced in [1]. With respect to the theory presented in [1], the enhancements are as follows:
- the half contact length is now computed by means of the theory presented in Section 1;
- the gyroscopic effects have been now introduced as shown in [7];
- the damping effects have been included both in the tread and in the carcass;
- the turn slip effect has been introduced.

As in [1], the model has a vanishing contact area width.

The model is able to reconstruct at each time step the deformation history of every tread element and of the carcass given the maximum amplitude Y_{max} of a sinusoidal lateral displacement $Y_{imp}(t) = Y_{max} \cdot sin(\omega_s \cdot t)$ of the centre of the rim.

Referring to Fig. 15, the tread in contact with the ground can be divided into $n+1$ equally distributed nodes as to assume the tyre structure made from the carcass and a number n of tread elements.

Starting the simulation at time $t = 0 + \Delta t$ with integration time step $\Delta t = \Delta x / V_x = (2L/n)/V_x$, it is possible to calculate the lateral deformation of the carcass $y_c(x)$ by means of Eq. (2.1) where F_y and M_z, being unknown, are to be set as initial guesses, i.e. $F_y = F_{yI}$ and $M_z = M_{zI}$.

The lateral stiffness K_c is not constant during the lateral deformation of the tyre: for this reason a medium value of K_c which is 15% lower than the one which is identified at pure lateral slip condition is used.

The lateral deformation of each node is (see Fig. 15)

$$v_i = Y_{imp}(t) - y_c(x_i) - y_i \tag{3.1}$$

The lateral force per unit longitudinal length at the i-th node is

$$\Delta F_{y_i} = v_i \cdot k_y \tag{3.2}$$

The deformation velocity for each tread element is calculated in order to consider the damping of the rubber

$$\Delta F_{y_{d_i}} = c_s \cdot \frac{v_i(t) - v_i(t - \Delta t)}{\Delta t} \qquad (3.3)$$

ΔF_{ydi} represents the lateral damping force per unit longitudinal length.

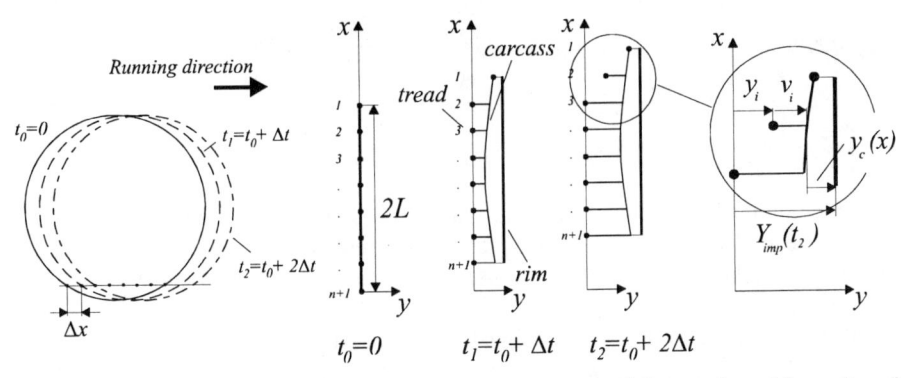

Fig. 15. Tyre model for transient state simulations. Left: lateral view. Right: top view of the portion of the tyre in contact with the ground.

Aderence conditions are checked by

$$\left(\frac{\Delta F_{x_i}}{\mu_x}\right)^2 + \left(\frac{\Delta F_{y_i} + \Delta F_{y_{d_i}}}{\mu_y}\right)^2 \leq p(x)^2 \qquad (3.4)$$

with $\Delta F_{x_i} = k_x \cdot u_i$. If Eq. (3.4) is not satisfied, the element slides and the transmitted force is given by the contact pressure multiplied by the sliding coefficient in the direction of skidding as shown in [1].

The tread lateral force F_{yt} and self aligning torque M_{zt} are

$$\begin{cases} F_{yt} = \sum_{i=1}^{n} \left[\frac{(\Delta F_{y_i} + \Delta F_{y_{d_i}}) + (\Delta F_{y_{i+1}} + \Delta F_{y_{d_{i+1}}})}{2} \right] \cdot \Delta x \\ M_{z_t} = \sum_{i=1}^{n} \left[\frac{(\Delta F_{y_i} + \Delta F_{y_{d_i}}) + (\Delta F_{y_{i+1}} + \Delta F_{y_{d_{i+1}}})}{2} \right] \cdot \Delta x \cdot b_{p_i} \end{cases} \qquad (3.5)$$

where $b_{p\,i}$ is the distance of the tread element from the centre of the contact area.

The motion of the tyre carcass generates damping reactions both for lateral force and for self aligning torque

$$F_{y_{d_c}} = c_{d_c} \cdot 2 \cdot L \cdot \frac{y_c(0,t) - y_c(0, t - \Delta t)}{\Delta t} \qquad (3.6)$$

$$M_{z_{d_c}} = c_{d_{tc}} \cdot \frac{\alpha(t) - \alpha(t - \Delta t)}{\Delta t} \qquad (3.7)$$

M_{zdc} is related to the damping effects due to carcass rotation around a vertical axis.
Turn slip [10] and gyroscopic effects [7] are computed as

$$F_{y_{ts}} = c_{ts} \cdot \frac{1}{V_x} \cdot \frac{\alpha(t) - \alpha(t-\Delta t)}{\Delta t} \tag{3.8}$$

$$M_{z_{gyr}} = -C_{gyr} \cdot V \cdot \frac{F_y(t) - F_y(t-\Delta t)}{\Delta t} \tag{3.9}$$

Finally we have

$$\begin{cases} F_y = F_{y_t} + F_{y_{d_c}} + F_{y_{ts}} \\ M_z = M_{z_t} + M_{z_{d_c}} + M_{z_{gyr}} \end{cases} \tag{3.10}$$

To compute F_y and M_z an iteration is needed. Actually (as explained before) F_y and M_z depend on $y_c(x)$ which depends on F_y and M_z (see Eq. 2.1). The initial guesses F_{yI} and M_{zI} are changed by a bisection method until convergence occurs, i.e. $F_{yI} \cong F_y$ and $M_{zI} \cong M_z$.

The positions of node contact points to be used with $t = 0 + 2\Delta t$ are

$$y_i = Y_{imp}(t) - y_c(x_i) - \frac{\Delta F_{y_i}}{k_y} \tag{3.11}$$

The model can set $t = 0 + 2 \cdot \Delta t$ and repeat the previous computation at each time t. In this way a time varing behaviour of the lateral force and the self aligning torque is obtained.

Fig. 16 and Fig. 17 show comparisons of computed and measured results for time varying lateral displacement of the rim. The parameters μ_y, μ_{ys}, K_c, K_ψ, K_W were identified (see Fig. 11, $F_z = 4000$ N) while the parameters c_s, c_{dc}, c_{dtc}, c_{ts} were estimated. The *same* parameters used for steady state simulations are used for transient state simulations too.

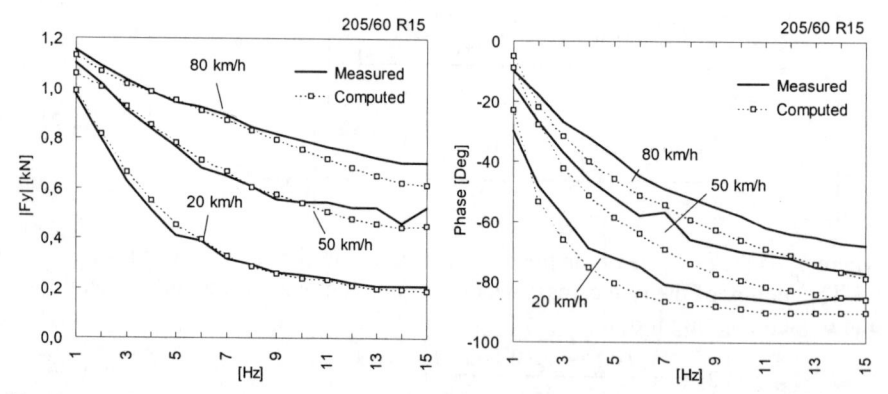

Fig. 16. Simulations at given rim lateral slip angle variation : $p_i = 2.2$ atm , $\alpha_{max} = 1.2°$. 20 km/h $F_z = 3200$ N ; 50 km/h $F_z = 3300$ N ; 80 km/h $F_z = 3370$ N. Measurements after [16]. Complete data set in Appendix.

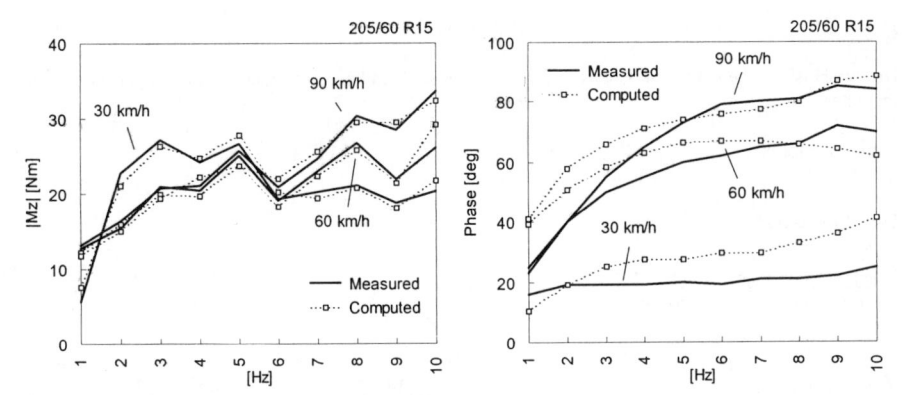

Fig. 17. Simulation at given rim lateral displacement variation: F_z = 4000 N ; 1-3 Hz Y_{max} = 0.0095 m, 4-5 Hz Y_{max} = 0.008 m, 6-8 Hz Y_{max} = 0.0055 m, 9-10 Hz Y_{max} = 0.0045 m. Measurements after [3]. Complete data set in Appendix.

CONCLUSIONS

A physical tyre model with semi-analytical formulation has been presented. Given the mechanical parameters of the tyre, the friction coefficients and the running conditions, the model returns the contact area size, the lateral deformation of the carcass and finally the values of the forces generated at the tyre-ground interface, i.e. the longitudinal force F_x, the lateral force F_y and the self-aligning torque M_z. Some of the parameters of the model have to be identified, precisely some of the parameters of the flexibility of the belt, the stiffnesses of tread and the friction coefficients. At steady-state, the agreement between computed and measured contact area size and between computed and measured F_x, F_y and M_z appears to be satisfactory.

At transient-state, according to the present formulation, the model allows the computation of F_y and M_z up to 10-15 Hz. The simulations seem to give a satisfactory representation of the actual tyre mechanics.

The time needed for the computation of tyre generalized forces is reasonable both for time domain vehicle dynamics simulations and for optimal tyre design purposes.

Acknowledgements

The Authors wish to acknowledge Prof. H.B. Pacejka for his help given by having hosted at TU-Delft Mr. A. Bersellini and Mr. R. Pagani who produced some of the experimental data presented in the paper.

APPENDIX

The symbols and the numerical values of the parameters which have been identified are reported by bold characters.

Common parameter values for all tyres: $E_r = 2 \cdot 10^6 \text{ N/m}^2$; $E_{steel} = 2 \cdot 10^{11} \text{ N/m}^2$, $c_b = 2$

Tyre 195/60 R15

R [m]	hf [m]	lf [m]	τ [m]	ri [m]	tp [m]	Λ	K
0,308	0,089	0,098	0,014	6E-04	2,6	0,7	6

Fig. N°	Fz [N]	pi [atm]	Vo [km/h]	μx	μxs	μy	μys	kx [N/m²]	ky [N/m²]	EJ [Nm²]	ηx	ηy
10	3000	1,97	60	**1,98**	**0,95**	-	-	**4,36E+07**	-	-	**0,093**	-
10	4500	1,97	60	**1,90**	**0,98**	-	-	**4,43E+07**	-	-	**0,101**	-

X1	X2	X3	X4	λ1	λ2	λ3	λ4
0.01	**0.19**	**-0.11**	**0**	**-7.26**	**-2.35**	**-5.32**	**0**
Y1	Y2	Y3	Y4	ξ1	ξ2	ξ3	ξ4
8.25	**4.32**	**-3.52**	**0**	**-7.29**	**-1.35**	**-2.37**	**0**

Tyre 205/60 R15

$K_c = 108667 \text{ N/m}, K_\psi = 7764 \text{ Nm}, K_w = 5387 \text{ Nm}$

R [m]	hf [m]	lf [m]	τ [m]	ri [m]	tp [m]	Λ	K	Cs [Ns/m²]	Cdc [Ns/m²]	Cdtc [Nms]	Cts [Nm]	Cgyr
0,3105	0,09	0,099	0,014	6E-04	2,6	0,7	6	150	70	12	1870	1,07E-05

Fig. N°	Fz [N]	pi [atm]	Vo [km/h]	μx	μxs	μy	μys	kx [N/m²]	ky [N/m²]	EJ [Nm²]	ηx	ηy
11	2000	2,20	60	-	-	**1,425**	**0,982**	-	**1,09E+07**	**203**	**0,095**	**0,204**
11	4000	2,20	60	-	-	**1,352**	**0,952**	-	**1,13E+07**	**207**	**0,101**	**0,215**

X1	X2	X3	X4	λ1	λ2	λ3	λ4
0.02	**0.24**	**-0.15**	**0**	**-6.25**	**-3.91**	**-4.25**	**0**
Y1	Y2	Y3	Y4	ξ1	ξ2	ξ3	ξ4
7.26	**3.52**	**-4.21**	**0**	**-5.81**	**-1.25**	**-3.65**	**0**

Tyre 215/50 ZR 17

R [m]	hf [m]	lf [m]	τ [m]	ri [m]	tp [m]	Λ	K
0,323	0,091	0,098	0,014	6E-04	2,6	0,7	5

Fig. N°	Fz [N]	pi [atm]	Vo [km/h]	μx	μxs	μy	μys	kx [N/m²]	ky [N/m²]	EJ [Nm²]	ηx	ηy
12	2646	2,45	30	-	-	**1,477**	**0,868**	-	**1,08E+07**	**212**	**0,121**	**0,156**
12	3332	2,45	30	-	-	**1,243**	**0,831**	-	**1,76E+07**	**213**	**0,113**	**0,105**
12	4018	2,45	30	-	-	**1,530**	**0,815**	-	**1,23E+07**	**203**	**0,136**	**0,112**

X1	X2	X3	X4	λ1	λ2	λ3	λ4
0.03	**0.22**	**-0.202**	**0**	**-8.28**	**-3.39**	**-4.94**	**0**
Y1	Y2	Y3	Y4	ξ1	ξ2	ξ3	ξ4
10.26	**5.449**	**-6.51**	**0**	**-8.81**	**-4.66**	**-5.14**	**0**

A SEMI-ANALYTICAL TYRE MODEL

Tyre 255/45 ZR 17

R [m]	h f [m]	l f [m]	τ [m]	r i [m]	t p [m]	Λ	K
0,323	0,091	0,096	0,014	6E-04	2,6	0,7	4

Fig. N°	Fz [N]	p i [atm]	V 0 [km/h]	μx	μxs	μy	μys	k x [N/m²]	k y [N/m²]	EJ [Nm²]	ηx	ηy
10	4655	2,65	60	1,71	1,11	-	-	2,38E+07	-	-	0,112	-
13	3626	2,65	30	-	-	1,477	0,868	-	1,08E+07	357	0,128	0,117
13	4557	2,65	30	-	-	1,243	0,831	-	1,76E+03	369	0,142	0,129

X₁	X₂	X₃	X₄	λ₁	λ₂	λ₃	λ₄
0.06	0.18	-0.19	0	-10.58	-5.04	-8.3	0

Y₁	Y₂	Y₃	Y₄	ξ₁	ξ₂	ξ₃	ξ₄
10.45	1.71	-9.23	0	-6.31	-2.75	-4.82	0

NOMENCLATURE

A_L, B_L, C_L, D_L	coefficients for longitudinal beam deformation;
A_b, B_b, C_b, D_b	coefficients for lateral beam deformation;
b_b	contact area width (see Fig.4);
$b_i(x)$	(see Fig. 2);
b_c	(see Fig. 2);
b_{pi}	distance of the generical tread element from the centre of the contact area;
c_b	coefficient to describe the lateral undeformed belt profile;
C_{gyr}	coefficient for gyroscopic effect;
c_{dc}	damping coefficient for carcass lateral displacement;
c_{dtc}	damping coefficient for carcass rotation around the vertical axis;
c_s	damping coefficient for tread element;
c_{ts}	coefficient for turn slip effect;
d_i	steel wires distance from cross section neutral axis;
d_x	(see Fig.3);
E_r	rubber elastic modulus;
E_{steel}	steel wires elastic modulus;
EJ	carcass bending stiffness on x-y plane;
EJ_L	carcass bending stiffness on x-z plane;
EJ_B	carcass bending stiffness on y-z plane;
$f_r(x)$	radial force per unit length from road to tyre;
$f_{xa}(x), f_{ya}(x)$	adhesion area longitudinal and lateral force per unit length;
$f_{xs}(x,y), f_{ys}(x,y)$	sliding area longitudinal and lateral force per unit surface;
F_x	total longitudinal force;
F_y	total lateral force;
F_{yI}, M_{zI}	initial guesses for transient state simulations;
F_z	vertical load on the tyre;
F_{xa}, F_{ya}, M_{za}	generalized forces in adhesion area;
F_{xs}, F_{ys}, M_{zs}	generalized forces in sliding area;
F_{ydc}, M_{zdc}	damping lateral force and self aligning torque (carcass deformation);
F_{yt}, M_{zt}	lateral force and self aligning torque (tread elements);
F_{yts}	lateral force due to turn slip;
h_f	sidewall heigth at undeformed conditions;
$h_{f\,eff}(x)$	sidewall heigth at deformed conditions;
k_L	tread vertical stiffness per unit length (longitudinal direction);
k_b	tread vertical stiffness per unit length (lateral direction);
$k_{vert\,L}(x)$	vertical carcass stiffness per unit length in longitudinal direction;
$\overline{k_{vert\,L}}$	mean carcass stiffness per unit length in longitudinal direction;
$\overline{k_{vert\,b}}$	mean carcass stiffness per unit length in lateral direction;

k_{fL}	longitudinal stiffness (membrane effect) per unit length;
k_{fb}	lateral stiffness (membrane effect) per unit length;
K_f	radial carcass stiffness;
k_x	tread longitudinal stiffness per unit length;
k_y	tread lateral stiffness per unit length;
K	patch deformation parameter;
K_c, K_ψ, K_W	parameters of carcass deformation;
$L(y)$	generical half contact length;
L, L_s	static half contact length;
l_f	sidewall length;
M_z	total self aligning torque;
M_{zgyr}	gyroscopic torque;
N	(see Eq. 1.15);
N_0	lateral draugth on the belt;
n	number of tread element in transient state simulation;
n_w	number of longitudinal steel wires in the belt;
n_s	number of longitudinal stripes;
$p(x)$	contact pressure along contact length;
$p(y)$	contact pressure along contact width;
$p(x,y)$	local contact pressure;
p_i	inflation pressure;
R	tyre external radius;
r_i	steel wires radius;
s_x	longitudinal slip ($s_x > 0$: braking);
t_p	thickness of belt + tread;
T_0	longitudinal draugth on the belt;
u	longitudinal deformation of tread element;
u_i	longitudinal deformation of tread node;
V_0	tyre velocity relative to the road;
V_x	tyre longitudinal velocity relative to the road;
v	lateral deformation of tread element;
v_i	lateral deformation of tread node;
$w_{rad}(x)$	radial deformation of carcass section;
x_i	abscissa of the contact point of the node;
x_s	abscissa of the point where the adhesion area ends;
Y_{max}	amplitutude of the sinusoidal lateral displacement of the rim;
$Y_{imp}(t)$	lateral displacement of the rim;
$y_c(x)$	lateral caracass displacement;
y_i	node contact point position with respect to $y=0$;
y_m	(see Fig. 6);
y_{tot}	total lateral displacement of contact area centre with respect to the rim;
y_{tr}	lateral tread displacement;
$z(x), z(y)$	belt longitudinal and lateral deformation;
α	lateral slip angle;
α_L, β_L	(see Eq.(1.9) and (1.10));
α_b, β_b	(see Eq.(1.17) and (1.18));
α_{max}	amplitude of the sinusoidal lateral slip angle variation;
γ	camber angle;
γ_c	(see Fig. 6);
$\Delta F_{yi}, \Delta F_{ydi}$	elastic and damping forces per unit length on the tread element;
ΔF_{xi}	longitudinal force for unit length;
Δt	integration step time;
Δx	(see Fig. 15);
η_x	constant for the longitudinal pressure distribution distortion;
η_y	constant for the lateral pressure distribution distorsion;
φ_b	rotation at beam ends (see Fig.4);
φ_{lim}	(see Fig. 4);

φ_s	(see Fig. 2);
φ^*_{rel}	(see Fig. 4);
ϑ	polar reference coordinate (see Fig. 1);
ϑ_L	rotation at longitudinal beam ends;
ϑ_{lim}	(see Fig. 1);
Λ	tread pattern coefficient;
μ_x, μ_y	longitudinal and lateral adhesion friction coefficient;
μ_{xs}, μ_{ys}	longitudinal and lateral sliding friction coefficient;
τ	tread thickness;
λ_i, X_i	coefficients for the computation of ϑ_L;
ξ_i, Y_i	coefficients for the computation of φ_b;
ω	tyre angular velocity;
ω_s	lateral sinusoidal displacement pulsation;

REFERENCES

[1] Mastinu G. and Fainello M. - Study of the Pneumatic Tyre Behaviour on Dry and Rigid Road by Finite Element Method - Vehicle System Dynamics - Vol. 21, 1992, pp.143-165
[2] Mastinu G. and Pairana E. - Parameter Identification and Validation of a Pneumatic Tyre Model - 1-st International Colloquium on Tyre Model for Vehicle Dynamics Analysis - Delft, 1991
[3] Bersellini A. and Pagani R.E.G. - Determination of Pneumatic Tyre Response during Transient - State Running Conditions and Simulation by Means of Computer Model - Dissertation - TU-Delft, 1993
[4] Guerciotti L. - Identificazione dei Parametri di un Modello di Pneumatico con un Metodo di Ottimizzazione Vincolato - Progetto di Laurea - Politecnico di Milano, A.A.1991-1992
[5] Wang Y.Q., Gnadler R. and Schieschke R. - Two-Dimensional Contact Area of a Pneumatic Tire Subjected to a Lateral Force - Vehicle System Dynamics, Vol.23, 1994 , pp. 149-163
[6] Drews V.R. and Schindler H. - Meßtechnische Einrichtungen zur Bestimmung der radialen und lateralen Reifenverformung - ATZ, No. 93 , 1993
[7] Clark S.K.- Mechanics of Pneumatic Tires - US Dep. Of Transportation, NHTSA, Washington DC, 1982
[8] Willumeit H.-P. and Böhm - Wheel Vibrations and Transient Tire Forces - Vehicle System Dynamics, 24 (1995), pp. 525 - 550
[9] Pottinger M.G. and Fairlie A.M. - Characteristics of Tyre Force and Moment Data - Tyre Science and Technology , Vol. 17, No. 1, January - March 1989
[10] Pacejka H.B. and Takahashi T. - Pure Slip Characteristics of Tyre on Flat and on Undulated Road Surface - AVEC '92, 1992
[11] Pacejka H.B., Bakker E. and Lidner L. - A New Tyre Model with an Application in Vehicle Dynamics Studies - SAE Paper n. 890087 - 4 th Autotechnologies Conference, Montecarlo
[12] Pacejka H.B. and Sharp R.S. - Shear Force Development by Pneumatic Tyres in Seady-State Conditions. A Review of Modelling Aspects - Vehicle System Dynamics, Vol. 20, No. 3-4, pp.121 - 176
[13] Ishiara H. - Development of a Three Dimensional Membrane Element for the Finite Element Analysis of Tires - Tire Science and Technology, Vol. 13, No. 2, April - June 1985
[14] Rothert H., Idelberger H., Jacobi W. And Laging G. - On the Contact Problems of Tires, including Friction - Tire Science and Technology, Vol. 13, No. 2, April-June 1985
[15] Gauβ F. and Willmerding G. - Die Normalkraftverteilung des Gürtelreifens - ATZ, No. 79, 1977
[16] Berzeri M. - Investigation into the Non Stationary Behaviour of a Brush Type Tyre Model - Tesi di Laurea - Politecnico di Milano - 1994/1995
[17] Assaad M.C. - Mechanics of The Dynamic Flex Test - Tyre Science and Technology - Vol. 19, 4, Oct.-Dec. 1991
[18] Colorni A. - Ricerca Operativa - CLUP - Milano 1992
[19] Gotusso L. - Calcolo Numerico - CLUP - Milano 1990
[20] Baldissera C., Ceri S. and Colorni A. - Metodi di Ottimizzazione e Programmi di Calcolo - CLUP, Milano, 1981
[21] Himmelblau D.L. - Applied Non-Linear Programming - Mc Graw Hill, 1972

New Procedures For Tyre Characteristic Measurement

G. Leister

ABSTRACT

The properties of tyres mainly influence the dynamic behaviour of vehicles in reality and in simulation. The slip-angle characteristic and the characteristic of the aligning moment play an important role with respect to cornering manoeuvres. Regarding the conditions of a tyre on a vehicle and the typical conditions of a tyre on a test-bench there are large differences. This is one reason for bad correlation of vehicle dynamic's measurement and simulation using measured tyre data. Therefore, it is important to measure the properties of tyres „as realistic as possible". To realize this, new procedures are developed.

Tyre model oriented measurements: some tyre models are developed to describe the established measurements. On the other hand tyre models can be used to create measurement procedures.
Realistic measurements: tyre load, lateral force, driving velocity, camber angle is measured at a vehicle or simulated with a vehicle model. While tyre load, driving velocity and camber angle are variables to the test bench, slip angle is controlled by lateral force. This allows a reproduction of the real tyre behaviour on a car.
Measurements with constant sliding velocity: the tyre characteristic is usually measured with constant driving velocity. This results in unrealistic high slip velocities. To avoid this, driving velocity is computed in such a way, that the sliding velocity remains constant. This additionally leads to realistic tyre temperatures.
State space oriented measurements: considering the state space of wheel load and lateral force it can be shown, that only a few areas of the tyre characteristic measurement are covered during standard driving manoeuvres. The definition of procedures using a slip angle control measuring these important areas improves the measurements.

INTRODUCTION

The dynamic behaviour of vehicles is mainly influenced by the properties of tyres. Therefore, it is very important to be able to measure and characterize these properties. In the development phase of a vehicle the application of simulation tools is very important to save time and costs. The knowledge of the interaction forces and moments between road and tyre is necessary for the simulation. Measurements of the interaction forces and moments are used to identify the parameters of tyre models. These tyre models describe the non-linear behaviour of the longitudinal force, the lateral force and the aligning moment in dependence of the wheel load, the slip angle, the camber and the longitudinal slip. The lateral force and aligning moment characteristics plays the most important role in cornering which is regarded here.

The tyre characteristics usually are measured on special test benches, Fig. 1, varying one input. All other variables are kept constant. The variables for tyre measurement on a test bench are wheel load, slip angle, camber, longitudinal slip, and driving velocity.

Fig. 1: Universal test bench of Mercedes-Benz

The advantages and the reasons for doing this proceeding are:
o   All curves can be used without processing.
o   Interpolation during simulation can be done easily.
o   Mathematical tyre models describing these measurements exist.

These methods result in an unrealistic heating and wear of the tyre and may produce wrong results. Therefore, the development of stochastic procedures has been started e.g. Dori and Halbmann [1]. Here, procedures are presented, which avoid unrealistic heating and wearing by looking at tyre states like averaged sliding velocity and the real state space of a tyre.

1   STANDARD PROCEDURES FOR TYRE CHARACTERISTIC MEASUREMENT

For the determination of the cornering properties lateral force and aligning moment of a free rolling tyre are of main importance. The slip angle is varied continuously, while wheel load and camber are varied in steps.

	Slip angle measurement
Wheel load	e.g. 4 wheel loads
Camber	-5, 0, 5 degrees
Slip angle	-14 to +14 degrees
Slip	free rolling
Velocity	constant

The wheel loads are fixed in dependence of the load index of a tyre. The main outputs are the lateral force and the aligning moment. Typically, the curve of the slip angle has a triangular shape. This reduces the time the tyre remains at high slip angles in comparison with a sinusoidal shape. The maximum value of the slip angle depends on the capacity of the test-bench. At least the maximum of the lateral force should be reached for tyre characterization.

In a separate procedure cornering stiffness and pneumatic trail are determined. This is done at slip angles less than two degrees. The shape of the curve is either sinusoidal or triangular.

	Cornering stiffness, pneumatic trail
Wheel load	e.g. half axle load
Camber	0 degrees
Slip angle	-2 to +2 degrees
Slip	free rolling
Velocity	constant (left and right rotation)

For the determination of the camber properties of the tyre the lateral force and the aligning moment are measured.

	Camber stiffness, camber trail
Wheel load	half axle load
Camber	-6 to +6 degrees
Slip angle	0 degrees
Slip	free rolling
Velocity	constant

The wheel load sensitivity is determined by variation of the wheel load at a constant slip angle.

	Wheel load sensitivity
Wheel load	0 to axle load
Camber	0 degrees
Slip angle	e.g. -5,-2,-1,1,2,5 degrees
Slip	free rolling
Velocity	constant

In this manoeuvre the wheel load curve has the shape of a ramp. The maximum wheel load is determined either by the load index or the axle load. The result is the lateral force coefficient (lateral force divided by wheel load) versus wheel load.

The manoeuvres described above are developed for measuring the cornering properties. The properties of tyres w.r.t. the longitudinal direction by braking or driving the tyre are measured by a µ-slip measurement.

	µ-slip measurement
Wheel load	e.g. 4 wheel loads
Camber	0 degrees
Slip angle	0 degrees
Slip	-30 to +30 %
Velocity	constant

The main result is the longitudinal force versus slip. The range of slip changes in dependence of the purpose of the measurements up to 100%.

In addition to the consideration of pure longitudinal or lateral properties combined manoeuvres are interesting e.g. for braking in a corner. The decrease of longitudinal force due to slip angle is measured by a µ-slip measurement at different slip angles.

	µ-slip meas. combined with slip angle
Wheel load	e.g. 4 wheel loads
Camber	e.g. -5, 0, 5 degrees
Slip angle	e.g. -5,-2,-1,1,2,5 degrees
Slip	-30 to +30 %
Velocity	constant

The outputs of the μ-slip measurement are the longitudinal force, the lateral force, and the aligning moment.

2 DETERMINATION OF TYRE CHARACTERISTICS

For practical use characteristic values that are comparable to each other are necessary. These values are determined from measured data. A linear regression analysis of the lateral force and the aligning moment for the cornering stiffness and pneumatic trail measurement can be done in the range of one degree slip angle, see Eq. (1).

$$F_y(\alpha) \approx F_{y0} + c_{Fy,\alpha}\alpha$$
$$M_z(\alpha) \approx M_{z0} + c_{Mz,\alpha}\alpha \tag{1}$$

Lateral force and aligning moment at zero slip angle are described by F_{yo}, M_{zo}. It is necessary to change the rotation direction of the tyre to identify the rotation dependent ply steer force and the aligning moment as well as the rotation independent conicity force and moment. The slip angle for zero lateral force can easily be determined from Eq. (1) as well as the slip angle for aligning moment zero. These values are important for the straight on behaviour of a vehicle.

The cornering stiffness and the pneumatic trail are determined by $c_{Fy,\alpha}$ and $c_{Mz,\alpha}/c_{Fy,\alpha}$. The cornering stiffness is the derivative of the lateral force with respect to the slip angle at zero slip angle. Pneumatic trail is the partial derivative of the aligning moment with respect to the lateral force at zero slip angle. Cornering stiffness is of great influence for vehicle dynamics. The pneumatic trail is important for the axle design and influences the hand moment at the steering wheel. Due to the fact that the cornering stiffness is defined at zero slip angle the value at half of the axle load is of most importance.

The same procedure can be done with camber stiffness measurement

$$F_y(\gamma) \approx F_{yo} + c_{Fy,\gamma}\gamma$$
$$M_z(\gamma) \approx M_{zo} + c_{Mz,\gamma}\gamma \tag{2}$$

Camber stiffness is important to characterize the vehicle's behaviour on uneven roads.

From slip angle measurements it is possible to determine the maximum value of the lateral force and the corresponding angle. This angle is usually nearby the angle at which the aligning moment becomes zero. Especially the maximum, at a wheel load nearby the axle load (outer wheel at cornering), and at low wheel loads are important. The same measurement gives the lateral force at very high slip angles. This is important for the case of sliding. In this context, the lateral force at high and low wheel loads is important.

A method to describe the tyre characteristic is given by the magic formula developed by Pacejka [2]. There are some constants with a physical meaning that partially enable to describe tyre characteristics. One of the main problems using this approach is the nearly annual change and the resulting many different versions of this formula. The characterisation of the tyres (not valid for '96 version of magic formula [3]) can be done by the parameters a3 and a4. The value a3 describes the maximum of cornering stiffness. The parameter a4 describes the corresponding wheel load. The function $c_{Fy,\alpha}(F_z) = a3\sin(2\tan^{-1} F_z / a4)$ describes the change of cornering stiffness versus wheel load. The pneumatic trail can be expressed by $n_1(F_z) = (c3 F_z^2 + c4 F_z) / (a3 \sin(2 \tan^{-1} F_z / a4))$. The maximum value of lateral force is described by $a1 F_z^2 + a2 F_z$ and the maximum of the aligning moment is described by $c1 F_z^2 + c2 F_z$.

3 TYRE MODEL ORIENTED MEASUREMENT

For the identification of the parameters of a tyre model **different** measurements are necessary. The measurements that have to be carried out depend partially on the tyre model itself, due to the sensitivity if its parameters. The magic formula for example was developed for describing the usual tyre slip angle measurements. As a consequence this measurement is a very good procedure to determine the magic formula parameters.

Milliken and Milliken [4] describe a tyre model based on a similarity method. In this case,

- o measurements, at small slip angle, at several loads to obtain the cornering stiffness as well as the pneumatic trail,
- o measurements at small camber angles, but at zero slip angle to obtain the camber stiffness and the camber pneumatic trail (several loads),
- o measurements at one or two high slip angle values, at zero camber angle, and several loads to obtain the lateral friction coefficient,
- o measurement of the complete cornering curve with slip angle at one intermediate load, zero camber angle, and free rolling have to be done.

Considering the required measurements, the following procedure has been developed:

	wheel load	wheel load	wheel load	slip angle
wheel load	0 to axle load	0 to axle load	0 to axle load	0.5 * axle load
camber	0 Grad	-5, 5 Grad	0 Grad	0 degrees
slip angle	-0.5 0.5	0 Grad	-10, 10 Grad	-10 - +10 deg
slip	free rolling	free rolling	free rolling	free rolling

4 REAL PROCEDURES

In a real driving situation wheel load, camber slip angle, slip, and driving velocity are changing continuously. It is possible to give these quantities as input in the time domain to the test-bench. This is applicable for wear simulation, too [5]. In order to have a good correlation to the real driving conditions on a road it is better to look at the forces a tyre has to generate and not for the slip angle which is a result of the tyre and vehicle properties. Therefore, a PID controller was implemented changing the slip angle due the lateral force. The resulting slip angle can be understood as the necessary slip angle for generating the cornering force.

If in the concept phase of a vehicle no measurements are available, simulations have to be performed. Usually, standard vehicle dynamic's measurements of steering angle, velocity, longitudinal acceleration and lateral acceleration done with a similar vehicle are available. Supported by a computer model the tyre force that will appear can be computed with the help of simulation. The method used here is:

(1) Use measured scaled steering angle in the model
(2) Control longitudinal acceleration with driving pedal
(3) Control longitudinal velocity with brake
(4) Check lateral acceleration, if no correlation, change scale in (1)

(5) Simulate the vehicle with different loads to get the load dependency of the tyre

Here the vehicle dynamic software CASCaDE (Computer Aided Simulation of Car, Driver, and Environment) [8] together with the tyre model BRIT [6] was used.

The following figures show some results of a „normal driver". Fig. 2 shows the longitudinal acceleration versus the lateral acceleration in a simulation of 50 km course.

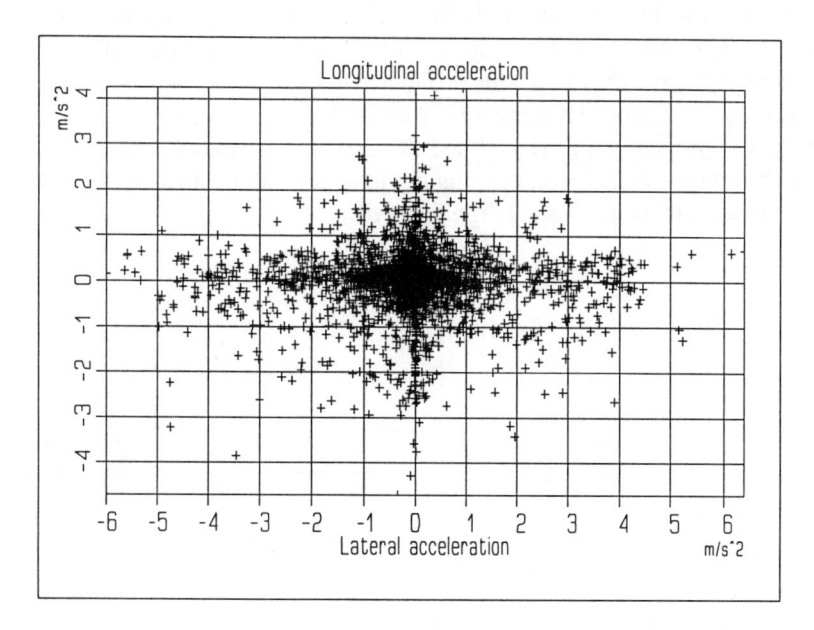

Fig. 2: Longitudinal acceleration versus lateral acceleration
(braking at front and rear)

The simulated time signals of wheel load, lateral force, camber and driving velocity can be delivered to the test bench together with wheel load and lateral force control. This allows an exact reproduction of the tyre behaviour. The results can be used to fit the parameters of a tyre model in an identification routine for the parameters.

	Real tyre measurement
Wheel load	input
Camber	input
Slip angle	**controlled**
Slip	**controlled**
Velocity	input
Lat. force	input
Long. force	input

With this method it is e.g. possible to use stochastic signals [1], too.

5 MEASUREMENT WITH CONSTANT SLIDING VELOCITY

Measurements of tyre characteristics usually are done at constant driving speed on the test benches. This results in large changes of sliding velocity of a tread rubber particle. The sliding velocity influences the local heating of the rubber. Measurements have shown that the difference in tread surface temperature during one slip angle sweep of 15 degrees is up to 50 degrees C. This temperature shock changes the properties of the tyre and the cornering stiffness decreases rapidly. Fig. 3 shows the lateral force, the aligning moment, and the temperature obtained from a measurement at constant driving speed.

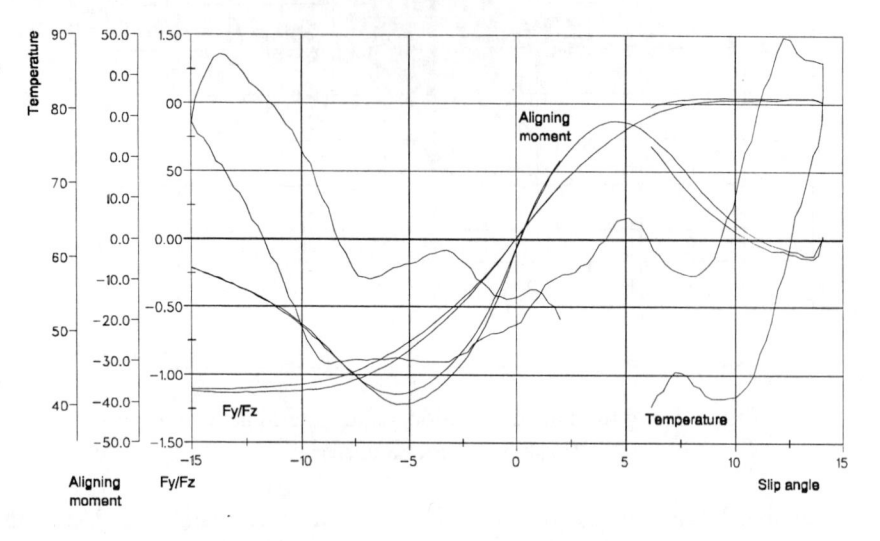

Fig. 3: Measurement with constant driving velocity

To avoid this effect and to solve the thermal problems there is one possibility: changing the driving velocity during measurement. This results in a new procedure that tends to keep the sliding velocity constant instead of the driving velocity. This has a positive influence to the rubber friction coefficient, too. Grosch [7] shows that the rubber friction coefficient is mainly a function of sliding velocity and temperature. The temperature of the contact area is partially a function of sliding velocity. Therefore, at constant wheel load, theoretically a constant sliding velocity result is constant temperature conditions. One possibility is e.g. max. driving velocity of 50 km/h and a sliding velocity of 1.745 km/h or 48,47 cm/s. This corresponds to a slip angle of 2 degrees at zero longitudinal slip with 50 km/h. With this method it is firstly possible to investigate separately the influence of temperature and driving speed, see [4]. In addition, it makes sense to restrict the maximal driving velocity to a certain value.

For the driving velocity this results in:

$$v = \min\left(\frac{v_{sliding}}{\sqrt{s^2 v^2 + \sin^2 \alpha}}, v_{max}\right), \text{ with } s = \frac{\omega r - \cos\alpha\, v}{v} \qquad (3)$$

Fig. 4 shows the averaged sliding velocity of the above described 50 km course.

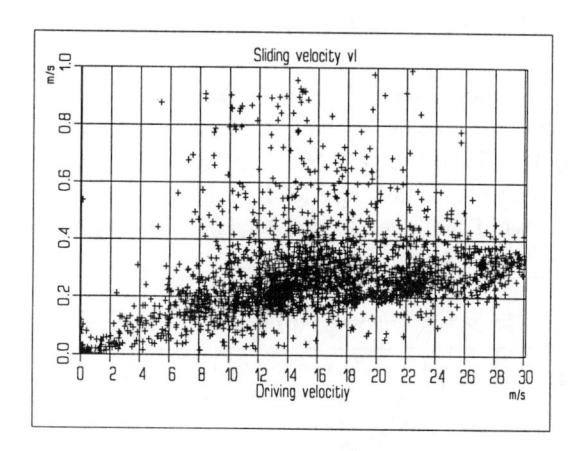

Fig. 4: Sliding velocity versus driving velocity at front left wheel

With this method surface temperature remains nearly constant in the area of normal heating, Fig 5 (different size as in Fig 4).

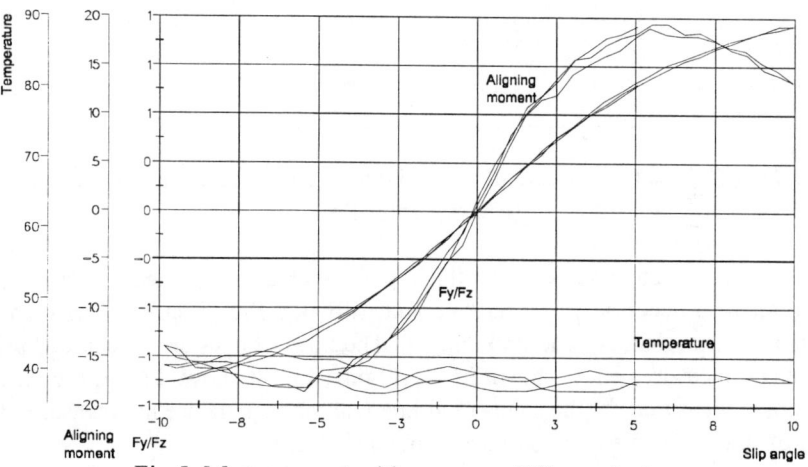

Fig 5: Measurement with constant sliding velocity

	Constant sliding velocity
Wheel load	e.g. 4 wheel loads
Camber	-5, 0, 5 degrees
Slip angle	-14 to +14 degrees
Slip	free rolling
velocity	$v = \min(v_{sliding} / \sin\alpha, v_{max})$

6 TYRE STATE SPACE ORIENTED MEASUREMENT

In the state space of wheel load and lateral force it can be shown, that only a very small region is used by a tyre on a vehicle. Dividing the lateral force by the wheel load leads nearly to a linear function. Fig. 6 shows simulation results at front left wheel generated with the almost described method. In this case braking is done only in the rear to avoid combined longitudinal and lateral forces.

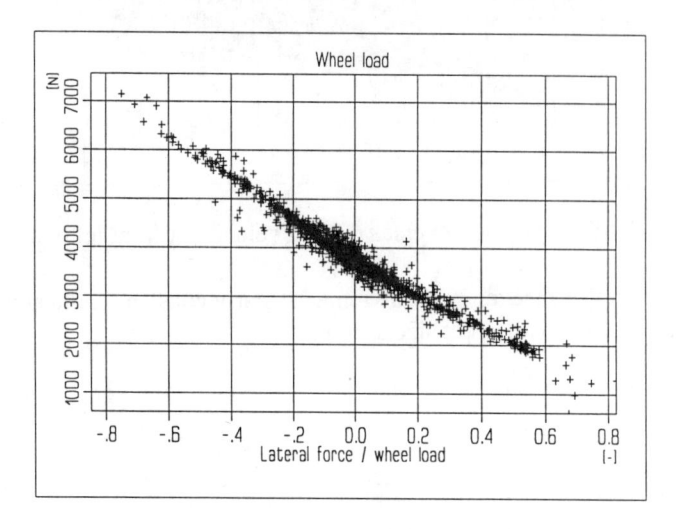

Fig. 6: Wheel load versus lateral force divided by wheel load at front left wheel (braking only at rear)

Looking even at an extreme driver this restricted area can be seen. For driving at the limit it is necessary to accelerate and to brake. Fig. 7 shows measurement results with accelerating and braking, too. The lines of constant wheel load using slip angle measurement is shown in this Figure, too. It shows that in the case of slip angle measurement the states used in a real manoeuvre are not measured and a lot of states which are measured are not reachable at all.

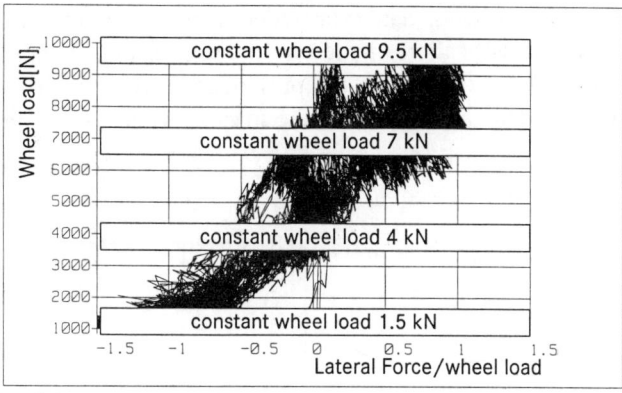

Fig. 7: Real force measurement front left of a rear driven car and measurement regions of slip angle measurement

The curve of wheel load versus lateral force divided by wheel load is roughly described by half of the axle load. The line begins at a minimum wheel load and ends at a maximum wheel load. Physically, these are the wheel loads of a vehicle in a stationary manoeuvre at maximum of the lateral acceleration. Along x-axis the line start at -µ and ends at µ. The value µ is the friction coefficient of wheel and road. This curve is a rough approximation tyre behaviour at a stationary manoeuvre. For measuring this line with a test bench a slip angle control has to be implemented. To get an array it is necessary to measure parallel lines in this diagram with a shift e.g. 1 kN. The measurement of camber influence can be done be measurements with camber of -5 and +5 degrees. Camber can also be changed linear to the wheel load.

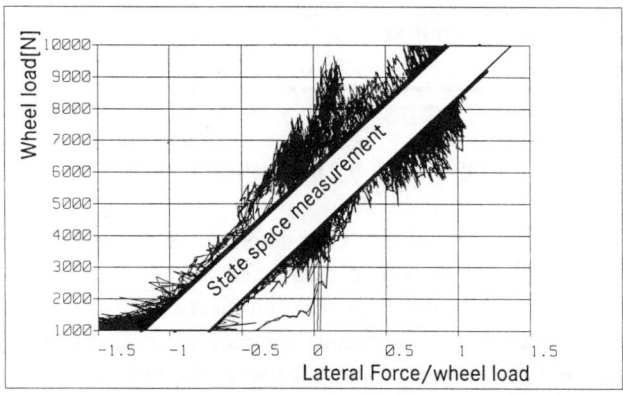

Fig. 8: Real force measurement front left of a rear driven car and state space measurement

This procedure results in better data with reduced measurement overhead, and additionally a direct classification of the tyre properties. Fig. 9 shows the slip angle, which is necessary to generate the prescribed forces. Fig. 10 shows the lateral force versus the aligning moment in the so called Gough diagram.

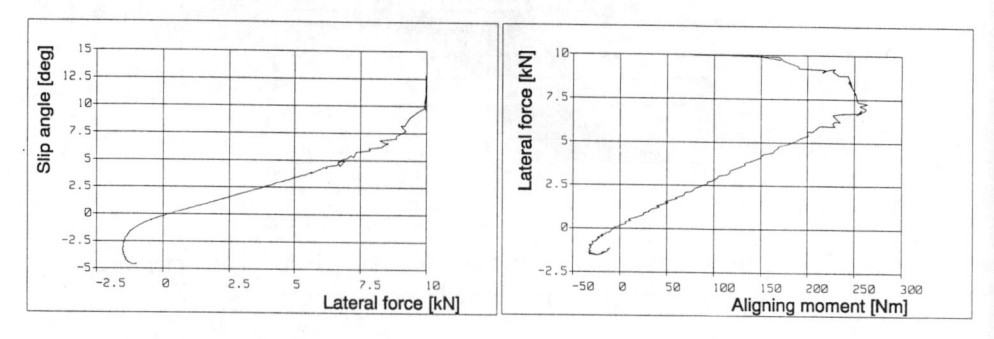

Fig. 9: Slip angle versus lateral force   Fig 10: Lat. force versus align. mom.

This enables to characterize the tyre properties in a large relevant region and the pneumatic trail during cornering.

	State space tyre measurement
Wheel load	linear from Fmin to axle load-2*Fmin
Camber	-5, 0, 5 Grad (or linear function of load)
Slip angle	controlled
Slip	free rolling
velocity	$v = \min(v_{sliding} / \sin\alpha, v_{max})$
Lat.force / wheel load	linear from $-\mu$ until $+\mu$

7 IDENTIFICATION OF TYRE MODEL PARAMETERS

All of these measurements can be used for determining tyre model parameters. The determination can be done with a mathematical optimization routine. The goal of an optimization in minimising a criterion function varying the parameters of the tyre model. The choice of the weight function is the first problem. A very easy method is using a least square method between the measurement points and the points generated through the model using the same input parameters as the test bench. In this case usually high values are weighted too high. Scaling the values for example dividing the lateral force by the wheel load and the aligning moment by the wheel load to the power of 1.5 solves this problem partially, but increases the influence of measurement errors at the very important area of small slip angles. An example of a the result of an identification of magic formula done by a tyre manufacturer shows the problem:

Wheel load	cornering stiffness measurement	cornering stiffness recomputed from magic formula	error
1 kN	285 N/o	259 N/o	9.0 %
3 kN	737 N/o	727 N/o	1.6 %
5 kN	819 N/o	906 N/o	10.6 %

To avoid the errors in the important areas due to identification for slip angle measurements a special weight function has been developed with a higher weight in the state space a tyre real is used. In the case of real tyre measurement and state space measurement the least square method is good due to the fact only realistic states are measured. Identification tests are done with two tyre models. A mathematical description using Pacejkas approach, and a physical description using BRIT, a brush and ring tyre model developed by Gipser [6]. The results of convergence tests with the different test procedures are:

	Magic formula	BRIT	
Slip angle measurement	++	+	special weight function
Milliken and Milliken	-	+	special weight function
State space	+	+	
Real measurement: normal driver	-	-	
Real measurement: racing driver	+	+	

For the identification of the parameters of both tyre models it is necessary to measure over the maximum of the lateral force. So it is difficult to identify with real data of a normal driver the parameters of a non-linear tyre model.

For a good convergence with mathematical models it is necessary to measure states that are not realistic. The optimal procedure is to have a separate measurement for each of the tyre model parameters, due to the fact that the identification is very sensitive to measurement procedure.

For physical models it was found out that the number of measurements can be reduced due to the fact that many effects are built in physically. The accuracy of identification is less than in a mathematical approach.

CONCLUSION

It has been shown, that it is helpful to develop tyre measurement procedures „as realistic as possible". Having a look at the process of rubber sliding, the state space of a tyre and the identification process of tyre model data. The result is:

Measurements with real functions in time domain are very important in special cases, but for measurement of the tyre characteristic. This procedure needs a lot of time and the measurement of camber influence is difficult.

Measurement only with constant sliding velocity helps to measure standard slip angle measurement avoiding unrealistic heating.

The state space oriented measurement is a very good procedure which very short measurement time. The heating process is realistic due to constant sliding velocity, realistic states are measured only. The influence of wheel load, camber, sliding velocity can be determined. An identification of tyre model parameters can be done easily.

REFERENCES

[1] Dori, C.; Halbmann, W.: Messung von Reifenkennfeldern auf dem Prüfstand mit realitätsnaher, stochastischer Belastung. In: Reifen, Fahrwerk, Fahrbahn. VDI-Berichte 778, 1989.

[2] Bakker, E.; Nyborg, L.; Pacejka, H.B.: Tyre Modelling for Use in Vehicle Dynamics Studies. SAE Technical Paper Series 870421, Detroit, Michigan, 1987.

[3] Pacejka, H.B.: The Tyre as Vehicle Component. XXVI FISITA Congress Prague, June 1996.

[4] Milliken, W.F; Milliken, D.L.: Race Car Vehicle Dynamics. ISBN 1-56091-526-9.

[5] Parekh, D.; Whittle, B.; Stalnaker, D.; Uhlir, E.: Laboratry Rire Wear Simulation Process Using ADAMS Vehicle Model. SAE Technical Paper Series 961001, 1996.

[6] Gipser, M.: Reifenmodellierung: BRIT Version 2.0. Stuttgart: Daimler-Benz F1M/SD CASCaDE- Dokumentation, 1995

[7] Grosch. K.A.: The Speed and Temperature Dependence of Rubber Friction and Ist Bearing on the Skid Resistance of Tires. In: Hays, D.F.; Browne, A.L. (eds.) The Physics of Tire Traction Theory and Experiments. New-York, London, Plenum Press, 1974.

[8] Rauh, J.: Fahrdynamiksimulation mit CASCaDE. In: VDI-Tagungsbericht Berechnung im Automobilbau, Würzburg 1990. Düsseldorf: VDI-Ber. 816, 1990, pp 599-608.

Test Procedure for the Quantification of Rolling Tire Belt Vibrations

H. OLDENETTEL and H. J. KÖSTER

ABSTRACT

The ride comfort on uneven roads is affected by vibrations of the tire which contribute to the audible noise inside the car. For any efficient development process, objective testing is essential to obtain a physical understanding of the vibrational behavior. The results of vehicle road testing show that tire modes have an influence on both the accelerations of the axles and on the interior sound pressure. Changes in tire design result in changes of frequency and damping of the belt modes.

Indoor testing was carried out with a measuring hub, while the tire was running on a cleated drum. Concerning the vertical forces, the vibrational behavior of the tire can be described by the enveloping process, measured at very low speeds, and the transfer behavior of the tire, defined by a transfer function. Based on this model, a test procedure is derived which allows the measurement of frequency and damping of the first vertical belt mode and of the torsional belt mode. The influence of the inflation pressure and of the tread compound's physical properties is discussed. Measurements on different test stands show, that belt vibrations can be influenced by insufficient stiffness of the test stand.

INTRODUCTION

The human perception of ride comfort on uneven roads is influenced not only by the mechanical vibrations felt by the driver, but also by the sound. Investigations on the relation between the subjective rating of test persons and objective measurements [1] have shown the importance of the interior noise level. The tire contributes not only to the mechanical but also to the acoustical ride comfort, because it has modes within the audible frequency range which are excited when the tire passes an obstacle [7]. As a consequence, the optimization of the vibrational behavior is one of the goals of the tire development process.

Considering the multitude of requirements a tire has to meet, an optimization based on subjective evaluations and a trial-and-error strategy would be very costly and time consuming. The optimization of the ride comfort may adversely affect other tire properties, and a tire design has to be found which gives the best possible compromise. For an effective development process, the subjective rating must be correlated to vehicle measurements. This is the key to the identification of the tire modes concerned. When the essential tire modes and their shapes are known, the influence of tire design parameters can be analyzed by indoor testing.

With the help of modal analysis, a quantification of tire vibrations can be made in terms of natural frequencies and damping ratios. The main disadvantage of the modal analysis is the fact that the vibrational behavior of the standing tire is quantified, and this differs considerably from the behavior of the rolling tire. This paper will discuss two different ways for the quantification of rolling tire vibrations, based on cleat impact testing.

VEHICLE TESTING

Vehicle road testing is the first step for an analysis of ride comfort. The vibrational behavior of the car with its full complexity is considered. This offers the advantage that the conditions of the measurement are identical to those of the subjective evaluation.

The selection of the test track is of great importance. In order to gain a good correlation between the measurements and the subjective rating, it is imperative to choose a test track where the differences between tires are most perceptible. Though this approach is useful to find correlations, it has two major shortcomings:

The first is that the precision which is needed to distinguish between different tires is difficult to reach. On most roads, even a small lateral deviation of the vehicle path causes differences in the vehicle vibrations which may be greater than the differences between tires. The other disadvantage is that the tire is continuously excited by the uneven road, and the decay of vibrations cannot be examined. This makes an evaluation in the time domain nearly impossible. With the help of the transformation into the frequency domain, an analysis of the involved frequencies and tire modes is possible, but the damping of the modes - which is often essential for the subjective rating - can hardly be extracted with sufficient accuracy.

For these reasons, an artificial obstacle, consisting of a rectangular cleat of 12 mm height and 120 mm width, was preferred. The vehicle passes the cleat at a speed of 80 kph; the data acquisition is triggered by a light barrier, and the interior noise and the accelerations at different points of the car are measured. By averaging several runs, the accuracy is improved.

Figure 1 shows the interior sound pressure level, measured near the driver's right ear, for a summer and a winter tire. Differences between the tires can be found mainly in the range from 80 to 100 Hz.

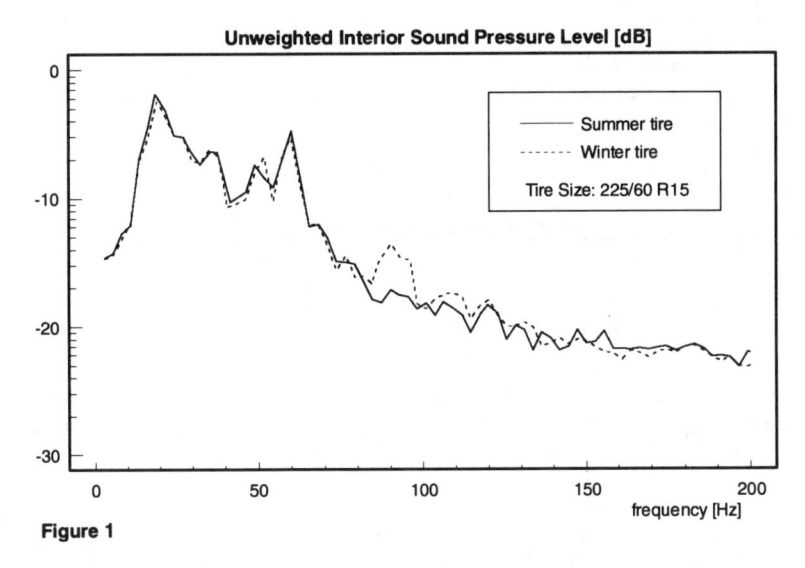

Figure 1

A comparison with the accelerations of the rear axle (Figures 2 and 3) exhibits peaks in the spectra of both the vertical and the fore-aft direction in this frequency range. These peaks are due to the second vertical and the third fore-aft mode of the tire-axle system [2]. The belt can mainly be considered as a stiff ring which moves in the vertical or fore-aft direction, and thus causes forces that act on the axle. Forces acting on the axle are of high importance to the interior noise, because most of the low-frequency noise is structure-borne.

Measurements in the frequency domain could have been made equally well on a "normal" uneven road. The advantage of the cleat becomes obvious when the accelerations are plotted in the time domain (Figure 4). The excitation of the vibration and the decay can be analyzed separately, and it is interesting to note that the first excitation which is caused when the tire hits the cleat is nearly the same for both tires. A comparison of figures 2 and 4 shows that the higher peak of the winter tire at 80 Hz is not caused by a stronger excitation, but by a lower damping.

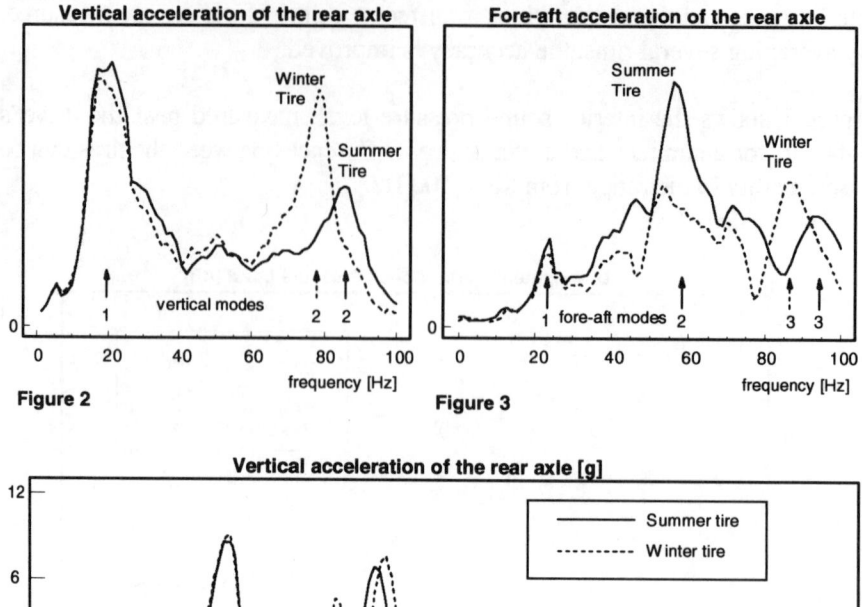

Figure 2: Vertical acceleration of the rear axle

Figure 3: Fore-aft acceleration of the rear axle

Figure 4: Vertical acceleration of the rear axle [g]

The difference between the damping behavior of the summer high performance tire and the winter tire is mainly due to differences in stiffness. With regard to other tire properties like handling on dry road or high speed performance, summer tires are designed stiffer than winter tires which need a soft tread for good winter performance. The high stiffness of the summer tire results in a greater amount of energy being dissipated by each vibration of the belt, and thus gives higher damping. The higher stiffness also explains the shift in frequency of the second vertical mode, which can be seen in figure 2.

In most cases, a strong increase of tire stiffness is not a realistic way of improving the acoustical comfort, because other tire properties, i.e. the mechanical comfort, will be affected adversely. Another possibility of improving the damping is given by the mechanical properties of the materials. The loss factor of rubber compounds can be changed considerably, and it is likely that a high loss factor increases the energy dissipation of tire vibrations.

The tread is the largest element of the tire, and thus changes of the tread compound have a considerable influence on the vibrational behavior. Figure 5 shows that the influence of the tread compound can indeed be measured by vehicle measurements. Tire A has a tread compound with higher material damping than tire B, which results in a reduction of the vertical belt vibrations at about 90 Hz.

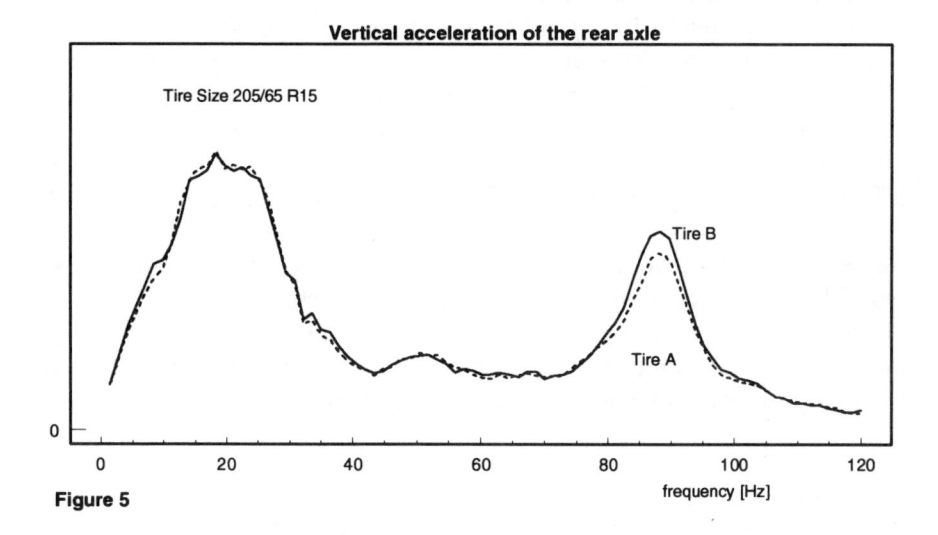

Figure 5

Though an increase in tread compound damping is a more realistic approach than increasing the tire stiffness, it has the disadvantage that the rolling resistance is substantially increased. The tread compound accounts for the largest part of the rolling resistance, and the different compounds used for tires A and B give a 25% difference in rolling resistance.

These results make it clear that the optimization of the tire is a complex process, with a multitude of conflicting tire properties which need to be considered. Vehicle testing is able to establish a relation between the subjective rating of ride comfort and physical measurements, and it can give the tire engineer an idea of the relevant tire modes. But indoor testing is to be preferred for the quantification of tire properties, because indoor testing achieves the accuracy which is needed to determine the influence of small tire design changes.

INDOOR TESTING

When the test tire is transferred from the vehicle to a test stand, the boundary conditions are changed. For a proper interpretation of the results, it is important to analyze the differences between the modes of the tire-vehicle system in comparison to the tire-test stand system.

In the vertical as well as in the fore-aft direction, the flexibility of the car's axle adds one degree of freedom which is missing when testing is performed on a test stand. It is obvious that the influence of the tire on these modes - which are important concerning the mechanical ride comfort - cannot be tested by a test stand with a rigid axle. But also those modes which exist in similar form for the car and the test stand are affected by the different boundary conditions [3]. Differences in frequency and damping must be expected when vehicle test results are compared to the results of a test stand, but also measurements performed on different test stands may show differences due to the dynamic behavior of the test stand itself.

Figure 8 shows spectra of the vertical force, measured with the same tire on two different test stands which have a rigid axle. A cleat of 4 mm height and 25 mm width was fixed on the drum, and the forces were measured by a 6 component piezoceramic measuring hub. By averaging 30 revolutions of the drum, the influence of tire non-uniformities was reduced.

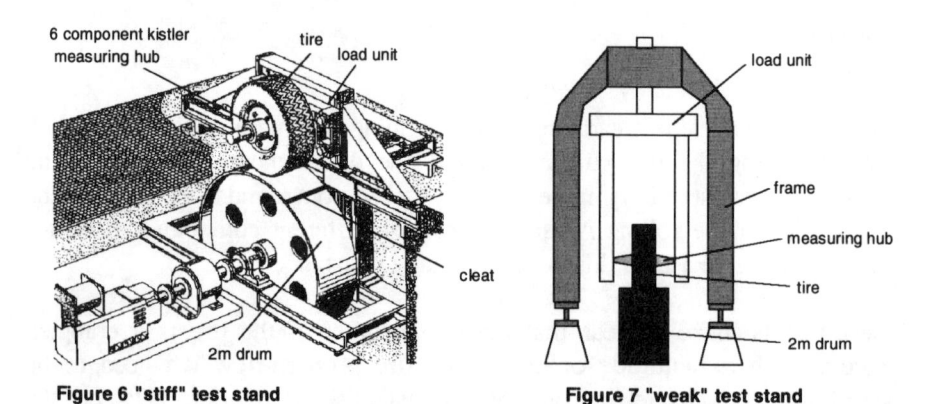

Figure 6 "stiff" test stand

Figure 7 "weak" test stand

The test stand referred to as "stiff" (figure 6) was especially designed for high speed uniformity measurements. With the help of FEM-calculations, a mechanical design was found which gives a first natural frequency of 319 Hz. Essential for the high natural frequency of the test stand is the hydraulic driven load unit which adjusts the wheel load and is then fixed to the frame. The natural frequency of the measuring hub itself is above 400 Hz.

The other test stand, referred to as "weak" (figure 7), was not optimized in respect of stiffness, because it was not designed for dynamic testing. Vibrations of the tire excite the test rig, and the interaction between tire and test rig results in a considerable decrease of the frequency of the vertical belt mode.

The conclusion is that results gained on a test stand can only be used for a relative comparison of tires. Frequency and damping of a certain tire cannot be compared directly to the results of vehicle testing, because the different boundary conditions result in differences of frequency and damping. Also results gained on different test stands cannot be compared directly. In order to reduce the influence of the test rig, the stiffness should be as high as possible.

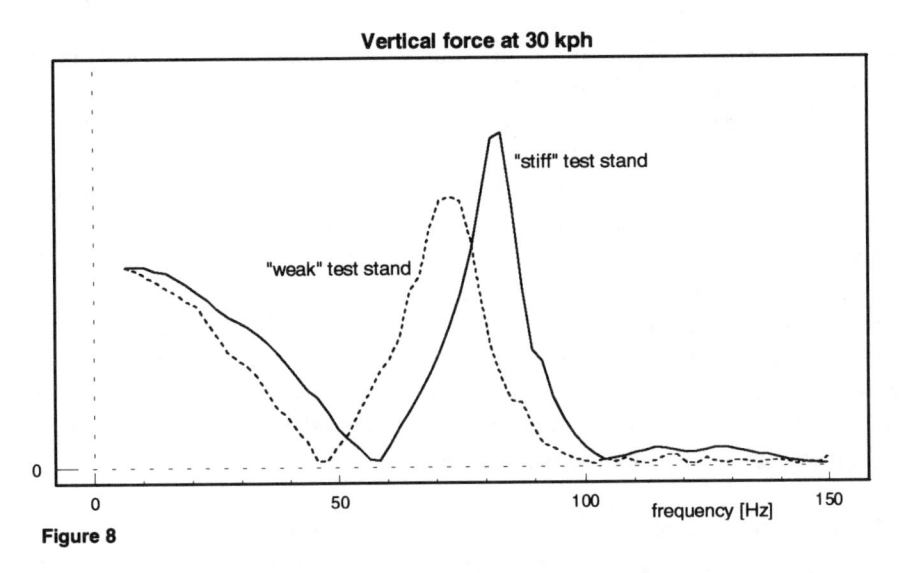

Vertical force at 30 kph

Figure 8

QUANTIFICATION OF THE VERTICAL TRANSFER BEHAVIOR

Cleat impact testing under different conditions, such as speed, load, inflation pressure and cleat geometry, gives a complex description of the dynamic behavior of the tire. A reduction of this complexity is possible, when the process is separated into two parts: the enveloping of the obstacle and the dynamic transfer behavior of the tire [4,5]. The enveloping process can be measured separately, when the tire passes the obstacle at low speed (2 kph). At this speed, the cleat generates only low-frequency forces which do not excite the tire modes. When the speed is increased, the spectrum of the forces is shifted towards higher frequencies, and the response of the tire becomes speed-dependent.

This model relies on the assumption, that the enveloping process and the transfer behavior of the tire are the same for all speeds. If this is true, it should be possible to calculate a transfer function which describes the transfer behavior of the tire. For a verification, tire forces were measured at 2 kph and at 70 kph. The sampling frequency of the measurement was adjusted to the speed, so that the number of samples per rolling distances was the same for both speeds (100 Hz at 2 kph, 3500 Hz at 70 kph). This makes it possible to calculate directly the transfer function

$$H(f) = \frac{\text{Force at Spindle}}{\text{Force at Contact Patch}} = \frac{\text{Force at 70 kph}}{\text{Force at 2 kph}}$$

Figure 9 shows a comparison of the transfer functions of the vertical forces for two different tires. The tires differ in the damping of the tread compound and are the same as shown in figure 5.

At low frequencies, the gain is 1 and the phase lag is zero, which means that forces measured at the axle are the same as those applied at the contact patch. At about 80 Hz, the gain reaches a maximum, which is due to the vertical belt mode, and the phase lag drops 180 degrees. The different tread compounds of tires A and B result in a different peak height, with the higher damping giving the lower peak. Above 80 Hz, some smaller peaks can be found, and the phase lag increases. This is probably due to higher belt bending modes. The magnitude of these peaks is much lower, because the resulting forces acting on the axle are smaller for the bending modes.

The measurement of transfer functions is not only a tool for the tire development, but also for the verification of tire models. This test procedure has the advantage that the transfer function of the rolling tire is measured, which is a more realistic approach than measurements done with standing tires. The mechanical properties of rolling and standing tires are known to differ significantly [6].

Concerning the longitudinal direction, the coherence of the transfer functions is not satisfying. This is due to the fact that at high speeds, forces are acting on the tire which cannot be found for quasi-static rolling: When the tire passes the obstacle, the effective rolling radius is changed [5]. This change in rolling radius forces the wheel to change its angular velocity. In the case of quasi-static rolling, the torque created by the small changes in angular velocity may be neglected, while at high speeds, the moments of inertia of the tire and the rim result in considerable variations of torque and longitudinal force.

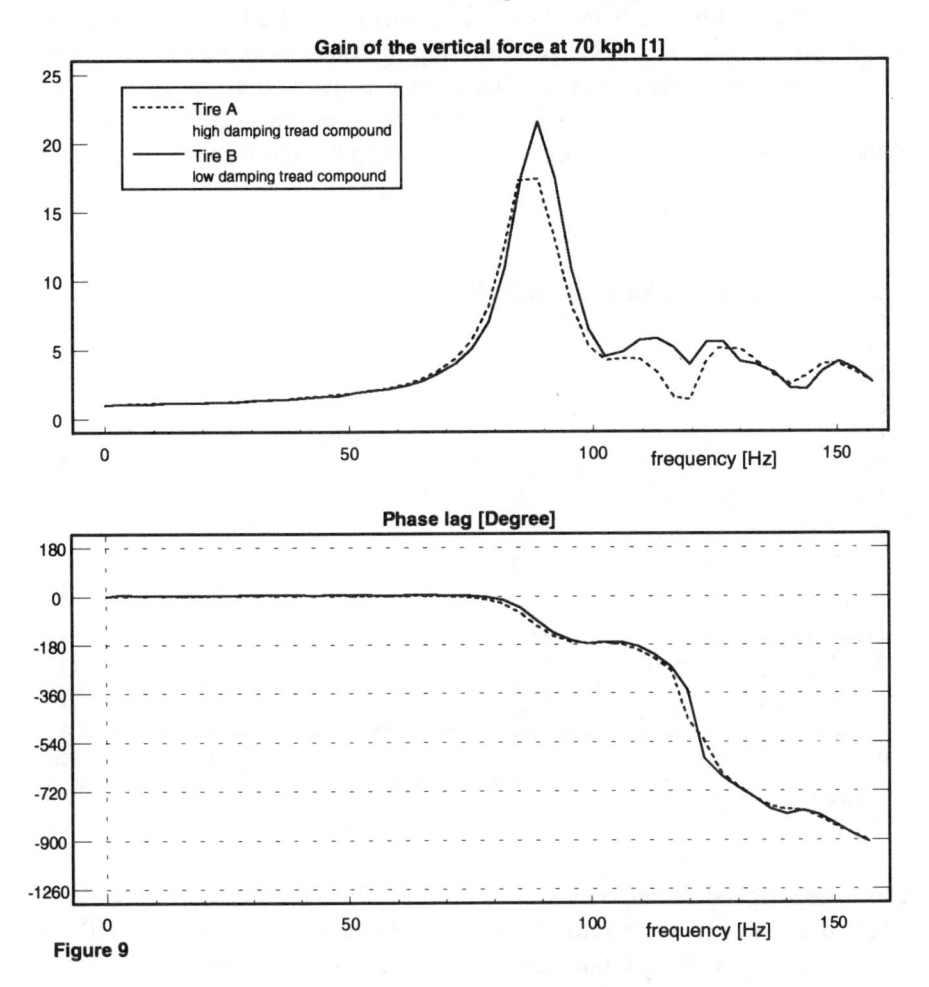

Figure 9

QUANTIFICATION OF FREQUENCY AND DAMPING

As shown in Figure 9, the influence of different tire designs can be described with the help of transfer functions. This approach has the disadvantage that it is limited to the vertical forces, as discussed above. For use as a tool in the tire development process, the quantification of frequencies and damping ratios of the different tire modes is needed.

The measurement of the decay of the tire vibrations is not easy, since the cleat excites different frequencies which are difficult to separate. A solution can be found, when the spectra of the forces at different speeds are compared (Figure 10): at 10 kph, the spectrum shows a maximum at about 25 Hz, which is created by the specific shape of the enveloping force. With increasing speed, this maximum is shifted towards higher frequencies, and at 30 kph matches the natural frequency of the vertical belt mode. This causes a strong reaction of the vertical forces, which decreases again when the speed is increased to 40 kph.

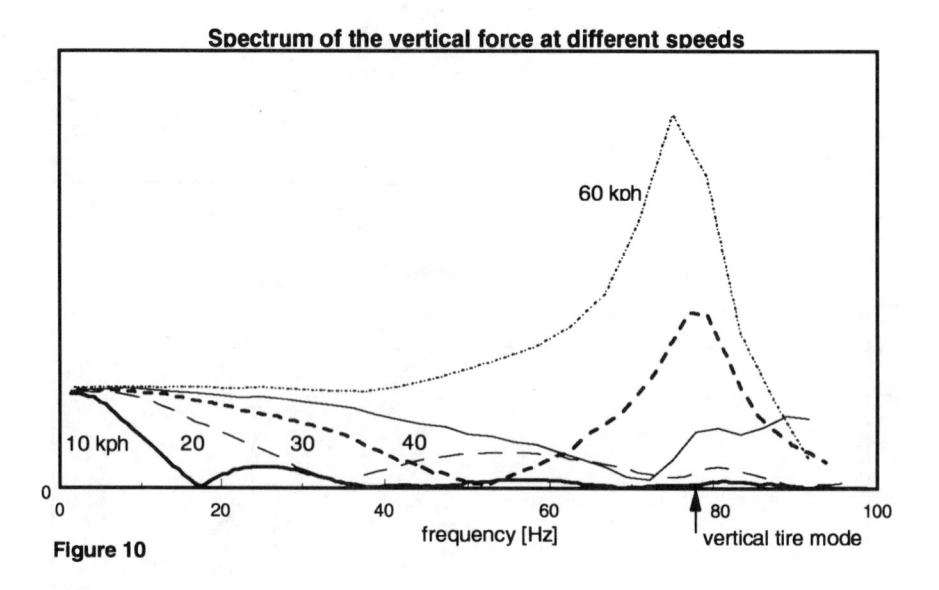

Spectrum of the vertical force at different speeds

Figure 10

This speed-dependent behavior can be used for the measurement of the decay function of the tire vibrations. When the relations shown in figure 10 are known for a specific combination of tire size, load, inflation pressure and cleat geometry, a test speed can be chosen which gives a maximum excitation of a certain tire mode, but minimum excitation at other frequencies. Figure 11 shows vertical forces, measured at 30 kph, and longitudinal forces, measured at 20 kph. In both cases, only one tire mode is excited, and the decay of the forces can easily be approximated by an exponential decay function. With the help of this method, the frequency and the decay time of the first vertical and longitudinal belt mode can be evaluated with high accuracy. The decay time T_d describes the time needed for a reduction of the amplitude of 63%.

Estimation of Decay Time Td

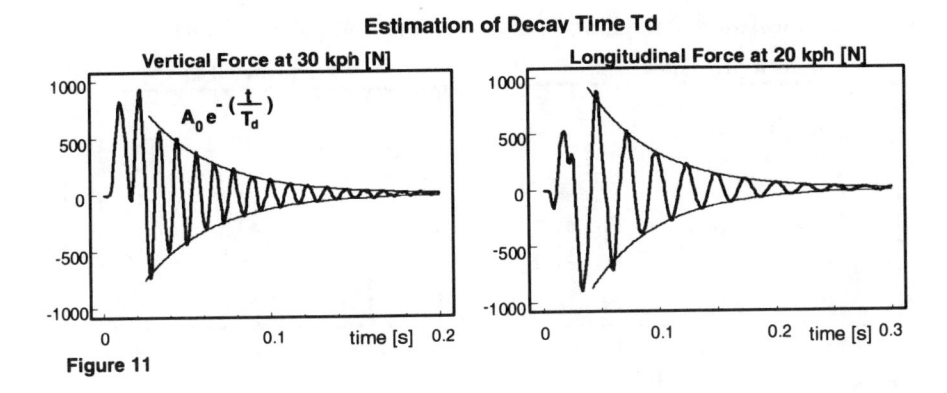

$$A_0 e^{-\left(\frac{t}{T_d}\right)}$$

Figure 11

As an example, figure 12 shows the comparison of the tires A and B which have already been discussed. The influence of the different tread compounds can be quantified as 10% concerning the decay time of the vertical belt mode, and 14% concerning the decay time of the fore-aft mode.

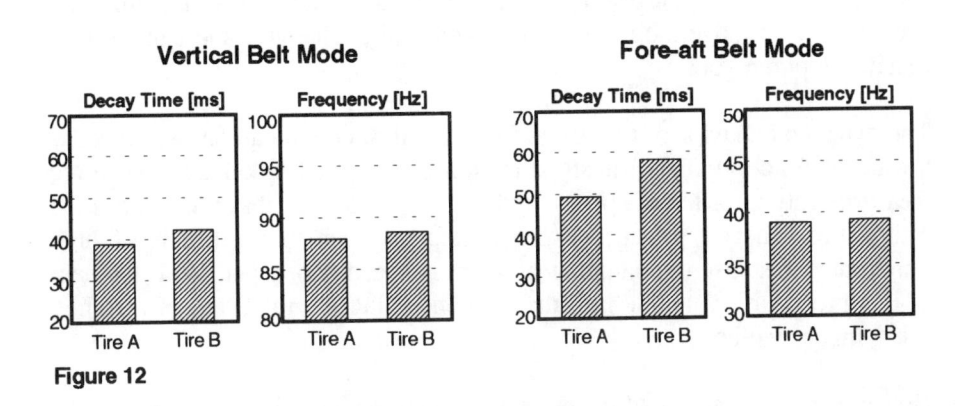

Figure 12

Figure 13 shows the influence of the inflation pressure. The frequency is increased with rising inflation pressure, while the damping is decreased. This is due to the fact that a high inflation pressure increases the stiffness of the tire, which reduces the deflections of the material, and gives lower energy losses.

Influence of Inflation Pressure on the Vertical Belt Mode

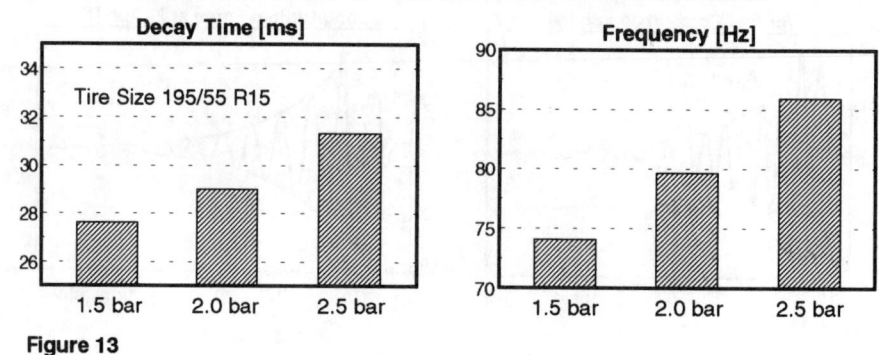

Figure 13

CONCLUSIONS

A test procedure has been developed which allows the quantification of tire belt vibrations by indoor testing. Because of its accuracy, this test enables the tire engineer to judge the influence of small design changes, and can be used as a development tool.

The dynamic behavior of the tire in the vertical direction can be described by a transfer function. This transfer function can easily be derived from force measurements at different speeds, and has the advantage that it describes the rolling tire. The difference between rolling and non-rolling tire is of high importance, because the damping of a tire is speed-dependent. With the help of this procedure, valuable information concerning the modeling of a rolling tire is made available.

However, the limitations of the test stand have been discussed. Because of the different boundary conditions, not all modes of the tire-vehicle system can be found using the test stand. The influence of the tire on the first vertical and fore-aft mode of the tire-vehicle system can only be analyzed by vehicle testing, because the interaction between the flexible axle and the tire is essential. Concerning the higher tire modes, the interpretation of indoor results must take into account that frequency and damping are affected by the different boundary conditions.

REFERENCES

1. Hazelaar, M., and Laermann, F. J., "Abrollkomfort - Schwingempfindung und Geräusch", VDI-Bericht Nr. 974, VDI-Verlag Düsseldorf 1992
2. Scavuzzo, R. W., Richards, T. R., and Charek, L. T., "Tire Vibration Modes and Effects on Vehicle Ride Quality", Tire Science and Technology, TSTCA, Vol. 21, No. 1, January-March, 1993, pp. 23-29
3. Gong, S., Savkoor, A. R., and Pacejka, H. B., "The Influence of Boundary Conditions on the Vibration Transmission Properties of Tires", Delft University of Technology, SAE-Paper 931280
4. Bandel, P., and Monguzzi, C., "Simulation Model of the Dynamic Behavior of a Tire Running Over an Obstacle," Tire Science and Technology, TSTCA, Vol. 16, No. 2, April-June, 1988, pp. 62-77
5. Zegelaar, P.W.A., and Pacejka, H.B., "The In-Plane Dynamics of Tyres on Uneven Roads", Proceedings of 14th IAVSD Symposium held in Ann Arbor; Michigan, USA, August 21-25, 1995
6. Rasmussen, R. E., and Cortese, A. D., "Dynamic Spring Rate Performance of Rolling Tires", SAE-Paper 680408
7. Kao, B. G., Kuo, E. Y., Adelberg, M. L., Sundaram, S. V., Richards, T. R., and Charek, L. T., "A new Tire Model for Vehicle NVH Analysis", International Congress and Exposition, Detroit, Michigan, 1987, SAE-Paper 870424

The Relaxation Length Concept at Large Wheel Slip and Camber

A. HIGUCHI and H. B. PACEJKA

ABSTRACT

The objective of this study is to understand the transient force and moment characteristics of tyres involving large wheel slip and camber, and to develop the tyre model which is capable of describing those characteristics. Various kinds of experiments are performed including large slip angles, camber, and also combined cases of side slip and camber. Some particular characteristics are found such as *non-lagging part* in side force response and an apparent peak in aligning torque response to a stepwise camber change. It is confirmed that the *stretched string* tyre model is able to reflect major fundamental characteristics of tyres with sufficient accuracy under conditions of relatively small slip quantities. Some aspects on the analogy of turn slip to camber are also discussed. The tyre model combines a dynamic part based on the physical aspect using the *relaxation length* concept of stretched string theory, and a steady state part which is often described empirically such as the Magic Formula. In combined situations of side slip and camber, an estimation method to determine the transient slip quantities is introduced.

1. INTRODUCTION

The study of transient tyre characteristics was initiated to investigate the *shimmy* phenomenon[6, 8]. Subsequently, the influence of transient tyre characteristics on vehicle behaviour[2] or vehicle performance on uneven roads[10] became of interest. With the progress in control technology[3], transient tyre characteristics play more important role specially in large wheel slip range. The study of camber[9] was done for a motor cycle tyre, but only few studies have been carried out for transient cases as far as the authors know. It is also important in passenger cars, for example, to study the *wandering* phenomenon[4]. The most widely used test for measuring transient tyre characteristics will be applying a sinusoidal input in slip angle or steer angle to a wheel, and the result as a frequency response function[1]. However, a stepwise input with a low travel speed is useful for the camber study since side force response contains *non-lagging part*, and the influence of the heat generation by the sliding of the tyre contact patch can be avoided particularly in large wheel slip cases.

2. STRETCHED STRING TYRE MODEL

The tyre model[5] consists of a set of endless strings representing the circumferential tyre belt which are kept under a certain pretension and elastically supported with respect to the wheel centre plane. The strings are thought to be provided with a large number of tread elements which, for simplification reasons, are assumed to be flexible in circumferential direction only. From this assumption, longitudinal deformations which arise at both sides of the wheel centre plane when the wheel is swivelled, are supposed to be taken up by the tread elements only. General expressions of the sliding velocities V_{gx} and V_{gy} of a tyre contact point are given as follows[6]:

$$\begin{cases} V_{gx} = V_{sx} - y\omega_z - V_r \dfrac{\partial u}{\partial x} + \dfrac{\partial u}{\partial t} \\ V_{gy} = V_{sy} + x\omega_z - V_r \dfrac{\partial v}{\partial x} + \dfrac{\partial v}{\partial t} \end{cases} \quad (1)$$

with

$$\begin{cases} V_{sx} = V_x - V_r \\ V_{sy} = V_y \end{cases} \quad (2)$$

where u and v denote the horizontal displacements of a contact point with respect to its horizontally undeformed position in the x and y directions respectively. They are functions of horizontally undeformed coordinates and of the time: $u = u(x, y, t)$ and $v = v(x, y, t)$. V_{sx} and V_{sy} denote the longitudinal and lateral slip velocities of the wheel respectively, V_x and V_y the longitudinal and lateral travelling velocities of the wheel, and V_r the rolling velocity.

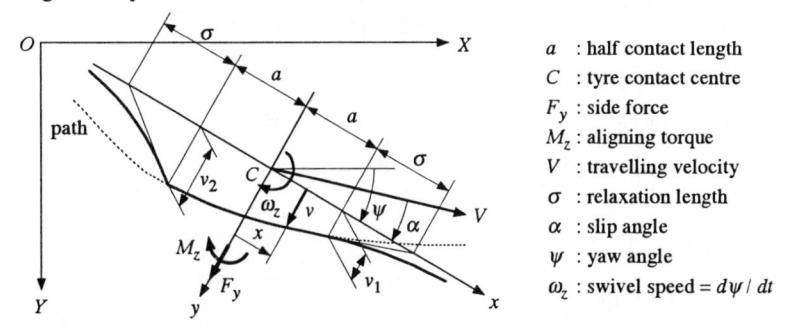

Fig. 1. Top view of string; for simplification, only lateral displacement v is depicted.

a : half contact length
C : tyre contact centre
F_y : side force
M_z : aligning torque
V : travelling velocity
σ : relaxation length
α : slip angle
ψ : yaw angle
ω_z : swivel speed = $d\psi / dt$

For the case of complete adhesion in whole contact area, zero longitudinal slip speed (free rolling) and small slip angle, the fundamental equations governing the behaviour of the horizontal displacements of the contact point become[6]

$$\begin{cases} U = \dfrac{1}{p} y \left[1 - e^{p(x-a)}\right] \Phi \\ V = \dfrac{1}{p} \left[\dfrac{-A + (\sigma + a + 1/p)\Phi}{1 + \sigma p} e^{p(x-a)} + A - \left(x + \dfrac{1}{p}\right)\Phi \right] \end{cases} \quad (3)$$

where p is the Laplace variable. Force and moment generated by these displacements are calculated through integration[5]. For the side force F_y and the aligning torque M_z' due to lateral string deflection, we obtain

$$F_y = c_c \int_{-a}^{a} v \, dx + S(v_1 + v_2)/\sigma \quad (4)$$

$$M_z' = c_c \int_{-a}^{a} v x \, dx + S(a + \sigma)(v_1 - v_2)/\sigma \quad (5)$$

where c_c denotes the lateral carcass stiffness per unit length, S effective tension force of the string. The aligning torque M_z^* due to longitudinal deformations of the tread elements is derived from the deflection u distributed over the contact area. With c'_{px} denoting the longitudinal stiffness of the elements per unit area and b half contact width,

$$M_z^* = -c'_{px} \int_{-a}^{a} \int_{-b}^{b} u y \, dx \, dy \quad (6)$$

3. CAMBER

Camber angle γ between the wheel centre plane and the normal to the road is also an important variable to understand tyre characteristics. If we observe the movement of a point on the circumferential line of a tyre in top view when the wheel is unloaded and rolling along a straight line with a camber angle, we will see it moving along the projection of the undeformed peripheral line (Fig. 2). On the other hand, in loaded situation, the contact point tends to move straight because of the existence of friction and the tyre moving along the straight line. The peripheral line in contact zone in top view therefore becomes more or less a straight line from the initial circular shape which can be described by an arc of a circle if the camber angle is small. The lateral deformation of the contact line involved produces a side force.

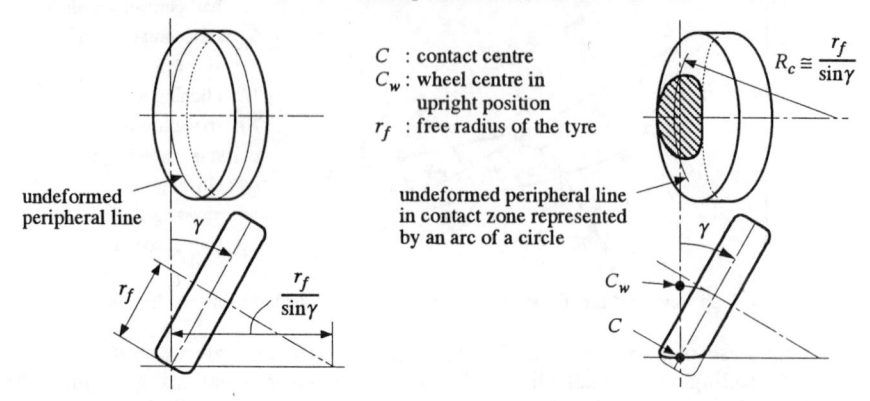

Fig. 2. Cambered tyre in unloaded (left), and in the final situation.

The development of the lateral deformation will differ if the process toward the final situation is different. The three different processes are:

1. initially loaded at zero camber angle; the wheel cambered around the intersection line of the wheel centre plane and the road surface plane; then start rolling.
2. initially unloaded; the wheel moved parallel to the wheel plane against the road surface at a constant camber angle; then start rolling.
3. initially unloaded; the wheel moved along the line normal to the road surface at a constant camber angle; then start rolling.

In all cases, the peripheral line in contact zone in top view just before rolling will already show some lateral deformation because of flattening the belt by loading and also lateral bending stiffness of the belt.

Consider an elastic body showing lateral bending purely due to vertical loading on a surface without friction (Fig. 3). Another coordinate system (C, ξ, η, z) is introduced of which the axes ξ and η lie in the (X, O, Y) plane forming the polar coordinates and z points downwards. With a radius R_c, the axis ξ simulates the vertical projection of the undeformed peripheral line of the tyre with a camber angle in top view. The horizontal displacements of a contact point with respect to its position in the horizontally undeformed situation with the coordinates (ξ, η) are indicated by u and v in the ξ and η directions respectively. They are functions of ξ, η and the independent variable: travelled distance s or the time t. The position with respect to the coordinates (ξ, η) is transformed into the one with respect to the coordinates (x, y) in the following manner:

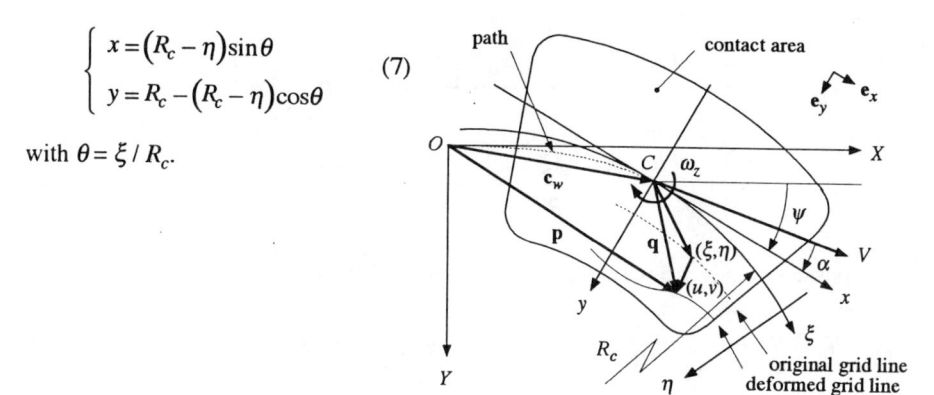

$$\begin{cases} x = (R_c - \eta)\sin\theta \\ y = R_c - (R_c - \eta)\cos\theta \end{cases} \quad (7)$$

with $\theta = \xi / R_c$.

Fig. 3. Top view of cambered tyre.

The position in space of a material point of a rolling and slipping body in contact with the road is indicated by the vector

$$\mathbf{p} = \mathbf{c}_w + \mathbf{q} \quad (8)$$

where $\mathbf{c}_w$ indicates the position of the wheel centre C_w in *upright* position in space and $\mathbf{q}$ the position of the material point with respect to the moving system (x, C, y),

$$\begin{aligned}\mathbf{q} &= (\xi + u, \ \eta + v) \\ &= (R_c - \eta - v)\sin\theta \cdot \mathbf{e}_x + \left[R_c - (R_c - \eta - v)\cos\theta\right]\mathbf{e}_y \end{aligned} \quad (9)$$

with $\mathbf{e}_x$ and $\mathbf{e}_y$ representing the unit vectors. The vector of the sliding velocity of the material point relative to the road becomes

$$\begin{aligned}\mathbf{V}_g = \dot{\mathbf{p}} &= \dot{\mathbf{c}}_w + \dot{\mathbf{q}} = \mathbf{V} + \dot{\mathbf{q}} \\ &= \mathbf{V} + \left[-(\dot{\eta} + \dot{v})\sin\theta + (R_c - \eta - v)\cos\theta \cdot \dot{\theta}\right]\mathbf{e}_x \\ &\quad + \left[(\dot{\eta} + \dot{v})\cos\theta + (R_c - \eta - v)\sin\theta \cdot \dot{\theta}\right]\mathbf{e}_y \\ &\quad + \dot{\psi}\left\{(R_c - \eta - v)\sin\theta \cdot \mathbf{e}_y - \left[R_c - (R_c - \eta - v)\cos\theta\right]\mathbf{e}_x\right\} \end{aligned} \quad (10)$$

in which $\mathbf{V}$ denotes the vector of the travelling velocity of the wheel centre C_w in upright position. We consider small ξ and u with respect to R_c, then

$$\sin\theta \cong \theta = \frac{\xi + u}{R_c}, \qquad \cos\theta \cong 1 \quad (11)$$

We introduce furthermore

$$\mathbf{V}_r = -(\dot{\xi} \cdot \mathbf{e}_x + \dot{\eta} \cdot \mathbf{e}_y) \quad (12)$$

$$\mathbf{V}_s = \mathbf{V} - \mathbf{V}_r \quad (13)$$

where $\mathbf{V}_r$ denotes the vector of the rolling velocity and $\mathbf{V}_s$ the vector of the slip velocity of the wheel with respect to the road. From Eqs. (10) to (13),

$$\mathbf{V}_g = \mathbf{V}_s + \left[\dot{u} - \frac{(\eta + \dot{v})(\xi + u) + (\eta + v)(\dot{\xi} + \dot{u})}{R_c} - (\eta + v)\dot{\psi}\right]\mathbf{e}_x$$
$$+ \left[\dot{v} + \frac{(R_c - \eta - v)(\xi + u)(\dot{\xi} + \dot{u})}{R_c^2} + \frac{(R_c - \eta - v)(\xi + u)}{R_c}\dot{\psi}\right]\mathbf{e}_y \qquad (14)$$

We consider $\xi \gg u$, $\eta \gg v$ and $R_c \gg \eta$. Equation (14) then read

$$\mathbf{V}_g = \mathbf{V}_s + \left(\dot{u} - \frac{\dot{\eta}\xi + \eta\dot{\xi}}{R_c} - \eta\dot{\psi}\right)\mathbf{e}_x + \left(\dot{v} + \frac{\xi\dot{\xi}}{R_c} + \xi\dot{\psi}\right)\mathbf{e}_y \qquad (15)$$

We realize furthermore that, for a certain material point, the deflections are functions of its horizontally undeformed coordinates and of the time: $u = u(\xi, \eta, t)$ and $v = v(\xi, \eta, t)$. Hence we have

$$\begin{cases} \dot{u} = \dfrac{du}{dt} = \dfrac{\partial u}{\partial \xi}\dot{\xi} + \dfrac{\partial u}{\partial \eta}\dot{\eta} + \dfrac{\partial u}{\partial t} \\ \dot{v} = \dfrac{dv}{dt} = \dfrac{\partial v}{\partial \xi}\dot{\xi} + \dfrac{\partial v}{\partial \eta}\dot{\eta} + \dfrac{\partial v}{\partial t} \end{cases} \qquad (16)$$

The components in the x and y directions of the sliding velocity $\mathbf{V}_g$ of a point of a rolling body in the contact area with respect to the road are in general

$$\begin{cases} V_{gx} = V_{sx} - \left(V_{rx}\dfrac{\partial u}{\partial \xi} + V_{ry}\dfrac{\partial u}{\partial \eta} - \dfrac{\partial u}{\partial t} - \dfrac{V_{rx}\eta + V_{ry}\xi}{R_c} + \eta\dot{\psi}\right) \\ V_{gy} = V_{sy} - \left(V_{rx}\dfrac{\partial v}{\partial \xi} + V_{ry}\dfrac{\partial v}{\partial \eta} - \dfrac{\partial v}{\partial t} + V_{rx}\dfrac{\xi}{R_c} - \xi\dot{\psi}\right) \end{cases} \qquad (17)$$

where (V_{sx}, V_{sy}) denotes the vector of the slip velocity of the wheel. (V_{rx}, V_{ry}) is the vector of the rolling velocity with which the wheel centre moves relative to the contact centre C.

We will restrict ourselves to small values of side slip and assume $|\alpha| \ll 1$ ($V_x \cong V$). For tyres, where only rolling in the x direction occurs, the relation holds $V_{ry} = 0$. We may write henceforth $V_{rx} = V_r$. When the longitudinal slip velocity is taken equal to zero (free rolling), which is the case when no driving or braking forces are applied, we obtain for the slip and rolling velocities (slip angle α considered small)

$$V_{sx} = 0, \quad V_{sy} = -V \cdot \alpha \quad \text{and} \quad V_r = V_x \cong V \qquad (18)$$

With relations $s = V \cdot t$ and $\phi = d\psi / ds$, and Eqs. (18), we finally obtain the following expressions for the sliding velocities of a point with the coordinates (ξ, η) for a freely rolling tyre with $|\alpha| \ll 1$:

$$\begin{cases} V_{gx} = V\left[-\eta\left(\phi - \dfrac{1}{R_c}\right) - \dfrac{\partial u}{\partial \xi} + \dfrac{\partial u}{\partial s}\right] \\ V_{gy} = V\left[-\alpha + \xi\left(\phi - \dfrac{1}{R_c}\right) - \dfrac{\partial v}{\partial \xi} + \dfrac{\partial v}{\partial s}\right] \end{cases} \qquad (19)$$

Note that the curvature $1/R_c$ of the laterally bending belt due to camber has the same effect as the curvature $-\phi$ in turn slip. Therefore camber can be represented by the turn slip with an equivalent curvature $-1/R_c \cong -\sin\gamma/r_f$ even in transient situations. In turn slip case, the path is curved while the original peripheral line is straight in top view. In camber case, the situation is just opposite. However in reality, some lateral deformation will arise before rolling as mentioned before. Therefore it is expected that some portion of the ultimate side force associated with the initial deformation responds without lag and that the side force produced by camber is different from that by the equivalent turn slip because of the bending stiffness of the belt.

4. MEASUREMENT OF TRANSIENT TYRE BEHAVIOUR

Various kinds of measurements are carried out at the laboratory of Delft University of Technology. The machine used is capable of applying a slip angle, camber angle, lateral and vertical motion to a freely rolling tyre. The *Flat Plank Tyre Tester* illustrated in Fig. 4 is composed of a flat steel track, measuring hub, turn table, axle height and lateral position adjust mechanism, camber mechanism and so on. The flat track moves at a constant speed of 47.5 mm/s. The tyre attached on the measuring hub can be rotated around the vertical axis on the turn table, therefore with a travelling speed, a slip angle arises. The camber mechanism allows the flat track to rotate around the axis on its surface, therefore no lateral movement is involved by camber.

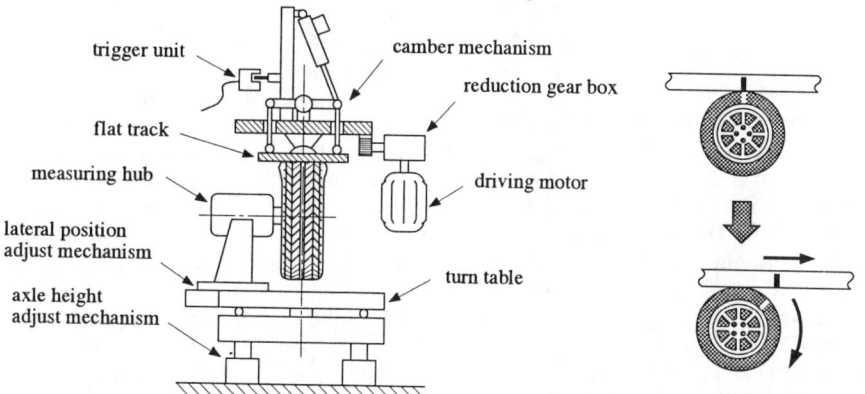

Fig. 4. Flat Plank Tyre Tester (left) and positioning of the tyre on the track.

Signals from strain gauges in the measuring hub are sent to the strain amplifier and sampled after A/D conversion by the data acquisition program. The cutoff frequency of the low-pass filter is set to 100 Hz. *Moving average* is introduced for aligning torque and consequently pneumatic trail also, because they are very noisy specially at large wheel slip. Sampling frequency is 50 Hz. The sampling starts by a triggering signal.

Tyres usually produce non-zero side force and aligning torque at zero slip and zero camber angle because of well-known effects: *ply-steer* and *conicity*. This initial force and moment vary with travelling due to non-uniformities in the tyre structure. Reference measurements are carried out to minimize unfavorable effects which arise from those phenomena in the data processing. The corrected data are obtained by subtracting the reference data from raw data. To correctly eliminate the force and moment variations associated with the angular position of the tyre and the longitudinal position of the track, the markers on the tyre and the track should coincide with each other each time before start rolling (Fig. 4).

Force and moment usually respond to the input with delay. In the case of the Flat Plank Tyre Tester, it is represented with respect to the travelled distance s. We introduce a simple fitting function to express the delay,

$$f(s) = c\left(1 - e^{-s/\tau_s}\right) \tag{20}$$

where c denotes the steady state value and τ_s space constant which represents the distance needed to reach a certain percent of the steady state value. If the response takes an exponential form with respect to the travelled distance, this function will give a good fit.

In the case of camber, a certain portion of the side force immediately responds to the camber change, which is henceforth called *non-lagging part*. To cope with such a case, the data is vertically shifted by subtracting the first value from all the values before the fitting so that it starts with zero. The part defined as the difference between the steady state value and the initial non-lagging value is called *lagging part*. The ratios of these values to the steady state value are called *non-lagging part ratio* and *lagging part ratio*. Note that non-lagging part ratio sometimes becomes negative.

Due to the limited track length, it is not always possible to reach the ultimate state value. There are two ways to find the steady state value. If the response does not present a peak, we use the steady state value c out of the fitting. Otherwise we pick the subsequent data after a fixed travelled distance and define it as their average.

5. LINEAR ANALYSIS

This chapter analyzes the theoretical predictions by the stretched string tyre model using Eqs. (3) to (6) and experimental results. The maneuvers performed are step changes in slip angle, turn slip and camber angle. We restrict ourselves to small input quantities to remain linear. A 205/60R15 tyre for a passenger car is used. The vertical load is set to 4 kN and the inflation pressure 2.2 bar. For turn slip case, due to the structural limit of the Flat Plank Tyre Tester, we actually apply an impulse change in turn slip. By integrating the result, we get the *calculated* response to a step change in turn slip. Parameters for the theoretical calculations are listed in Table 1.

Figures 5 and 6 depict side force and aligning torque responses to a step change in slip angle of 1 degree respectively, and Figs 7 and 8 response to a step change in turn slip. In both cases, theoretical and experimental results meet good agreement with each other in terms of side force, while theoretical results are somewhat larger in aligning torque. This tendency is also mentioned for the *bare string model* in earlier study[7].

Figures 9 and 10 depict responses to a step change in camber angle. Theoretical results are larger than experimental ones in both side force and aligning torque. Note that the development fashions of side force and aligning torque resemble those of turn slip case, and the non-lagging part in side force response which is also found in a motor cycle tyre case in earlier study[9]. From these results, the theoretical study of the analogy of camber to turn slip (cf. Chapter 3) is partially supported.

Figure 11 shows the space constant τ_s of side force change fitting out of experiments according to Eq. (20). The process to the final situation is *camber after loading*. The space constant has a strong relation to vertical load, but a weaker relation to camber angle. Figures 12 to 14 depict non-lagging part ratio in the three different processes (cf. Chapter 3). The space constant of side force is not shown for each process because the difference between processes is relatively small. The non-lagging part ratio has a strong relation to vertical load and the process to the final situation, but a weaker relation to camber angle. The response appears to be more or less linear in camber case within 15 degrees, therefore there is no strong dependence on camber angle.

Fig. 5. Side force; $\alpha = 1°$.

Fig. 6. Aligning torque; $\alpha = 1°$.

Fig. 7. Side force; $\phi = 8.73 \times 10^{-3}$ m^{-1}.

Fig. 8. Aligning torque; $\phi = 8.73 \times 10^{-3}$ m^{-1}.

Fig. 9. Side force; $\gamma = -2°$.

Fig. 10. Aligning torque; $\gamma = -2°$.

Fig. 11. τ_s of side force change fitting; *camber after loading.*

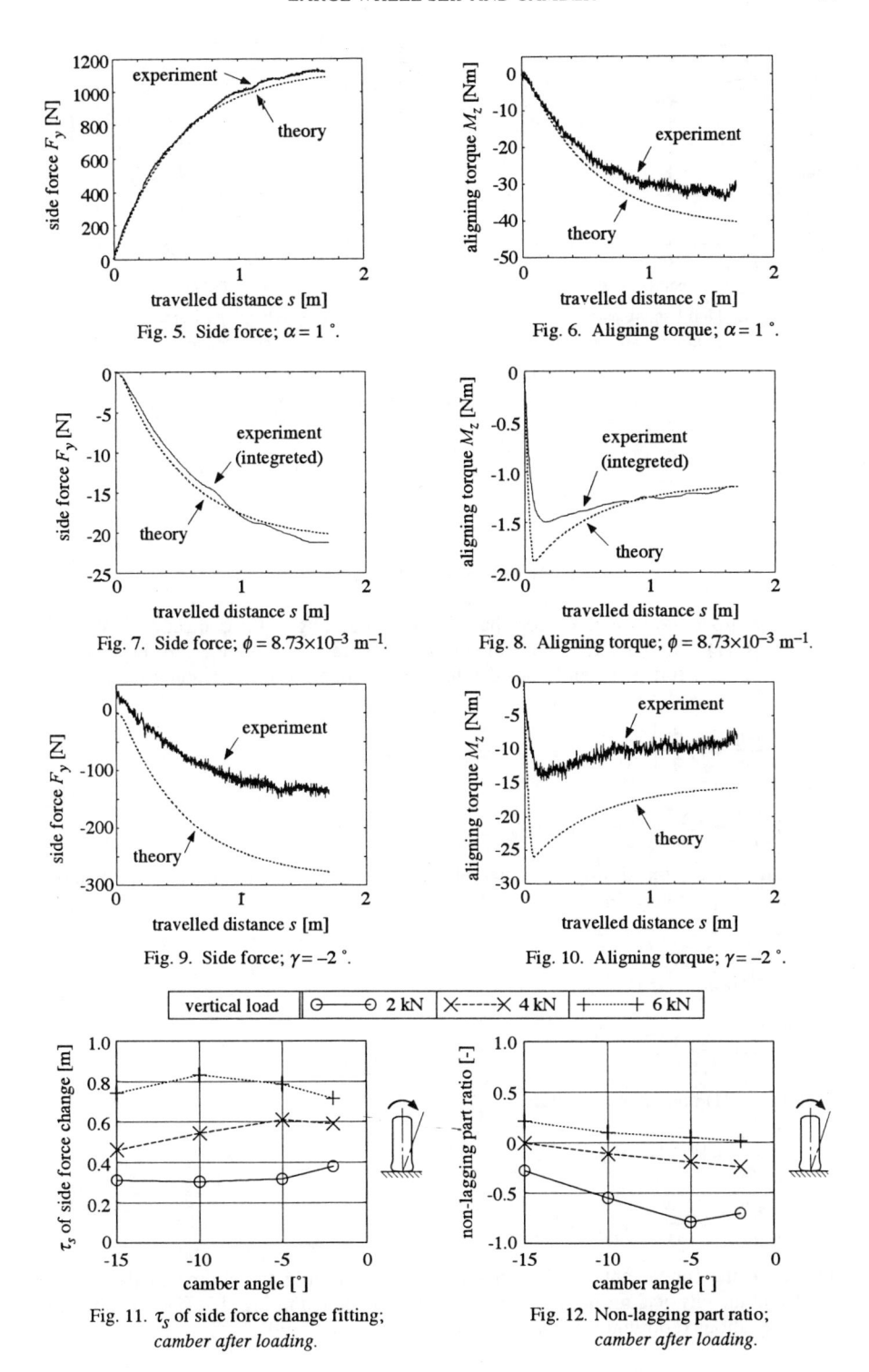

Fig. 12. Non-lagging part ratio; *camber after loading.*

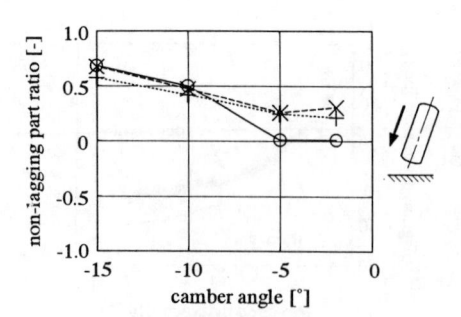

Fig. 13. Non-lagging part ratio; loading along cambered wheel.

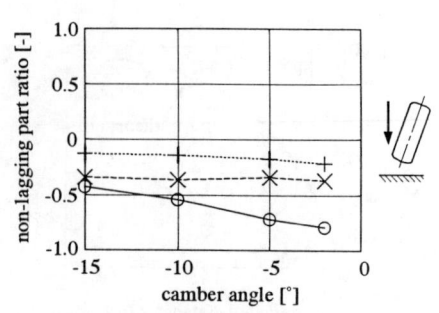

Fig. 14. Non-lagging part ratio; loading along the normal of road.

Table 1. Parameters used in the theoretical calculation.

a	0.04	m	b	0.1	m
c_c	1×10^5	N/m²	c'_{px}	6×10^7	N/m³
σ	0.53	m			

6. NONLINEAR ANALYSIS

In a large slip angle case, the theory has to consider sliding in the contact area. We first recall Eqs. (1). The equations contain sliding velocities: V_{gx} and V_{gy}. We now neglect the spin ω_z (no turn slip) and consider the equations only at the leading edge of the contact area ($x = a$, $u = u_1$, $v = v_1$).

$$\begin{cases} \dfrac{du_1}{dt} - V_r \dfrac{\partial u}{\partial x}\bigg|_{x=a} - V_{gx1} = -V_{sx} \\ \dfrac{dv_1}{dt} - V_r \dfrac{\partial v}{\partial x}\bigg|_{x=a} - V_{gy1} = -V_{sy} \end{cases} \quad (21)$$

V_{gx1} and V_{gy1} are the sliding velocities of the string at the leading edge in the x and y directions respectively. From the stretched string theory, we have

$$\begin{cases} \dfrac{\partial u}{\partial x}\bigg|_{x=a} = -\dfrac{u_1}{\sigma_\kappa} \\ \dfrac{\partial v}{\partial x}\bigg|_{x=a} = -\dfrac{v_1}{\sigma_\alpha} \end{cases} \quad (22)$$

With Eqs. (21) and (22), we obtain

$$\begin{cases} \dfrac{du_1}{dt} + V_r \dfrac{u_1}{\sigma_\kappa} - V_{gx1} = -V_{sx} \\ \dfrac{dv_1}{dt} + V_r \dfrac{v_1}{\sigma_\alpha} - V_{gy1} = -V_{sy} \end{cases} \quad (23)$$

These are the fundamental differential equations involving the sliding in the contact patch and the physical meaning of the equations is depicted in Fig. 15. The second terms of the equations represent the transient slip velocities $-V_{sx}'$ and $-V_{sy}'$.

$$\begin{cases} V_r \dfrac{u_1}{\sigma_\kappa} - V_{gx1} = -V_{sx} - \dfrac{du_1}{dt} = -V_{sx}' = V_r \zeta_x' = V_r \dfrac{\kappa'}{1+\kappa'} \\ V_r \dfrac{v_1}{\sigma_\alpha} - V_{gy1} = -V_{sy} - \dfrac{dv_1}{dt} = -V_{sy}' = V_r \zeta_y' = V_r \dfrac{\tan\alpha'}{1+\kappa'} \end{cases} \quad (24)$$

where ζ_x' and ζ_y' denote the transient *theoretical* slip quantities. Finally, we obtain the differential equations in terms of the transient slip quantities κ' and α'

$$\begin{cases} \dfrac{du_1}{dt} + V_r \dfrac{\kappa'}{1+\kappa'} = -V_{sx} \\ \dfrac{dv_1}{dt} + V_r \dfrac{\tan\alpha'}{1+\kappa'} = -V_{sy} \end{cases} \quad (25)$$

With the help of physical relationship[11], the transient slip quantities κ' and α' are found by solving the following equations:

$$\begin{cases} u_1 = \sigma_\kappa \dfrac{F_x(\kappa',\alpha')}{C_{F\kappa}} \\ v_1 = \sigma_\alpha \dfrac{F_y(\kappa',\alpha')}{C_{F\alpha}} \end{cases} \quad (26)$$

These differential equations (25) have to be solved in cooperation with the algebraic equations (26). Figure 16 depicts the calculation diagram of this model.

Equations (26) are used to find the transient slip quantities κ' and α' from given string deflections u_1 and v_1. Note the inverse usage of the steady state tyre model which is usually nonlinear. To investigate the transient behaviour of this model, we linearize the equations by considering small disturbances from steady state situation.

For convenience, when we consider the longitudinal behaviour of the string, we regard longitudinal slip velocity V_{sx} and/or vertical load F_z as varying, and other inputs (slip angle, camber angle, etc.) as constants. The lateral behaviour of the string is dealt with in the same way. Moreover, we disregard the *transient* interaction between the longitudinal and lateral motions of the string.

$$\begin{cases} u_1 = \bar{u}_1 + \tilde{u}_1 \\ \kappa' = \bar{\kappa}' + \tilde{\kappa}' \\ V_{sx} = \bar{V}_{sx} + \tilde{V}_{sx} \\ F_z = \bar{F}_z + \tilde{F}_z \end{cases} \qquad \begin{cases} v_1 = \bar{v}_1 + \tilde{v}_1 \\ \alpha' = \bar{\alpha}' + \tilde{\alpha}' \\ V_{sy} = \bar{V}_{sy} + \tilde{V}_{sy} \\ F_z = \bar{F}_z + \tilde{F}_z \end{cases} \quad (27)$$

where $^-$ above symbols denotes the steady state value, and $^\sim$ the disturbance. The steady state terms cancel out. As a result, we obtain

$$\begin{cases} \tau_\kappa \dfrac{d\tilde{\kappa}'}{dt} + \tilde{\kappa}' = \dfrac{(1+\bar{\kappa}')^2}{V_r}\left(-\tilde{V}_{sx} - \left.\dfrac{\partial u_1}{\partial F_z}\right|_{\substack{\kappa=\bar{\kappa}' \\ F_z=\bar{F}_z}} \dfrac{d\tilde{F}_z}{dt}\right) \\ \tau_\alpha \dfrac{d\tilde{\alpha}'}{dt} + \tilde{\alpha}' = \dfrac{(1+\kappa')\cos^2\bar{\alpha}'}{V_r}\left(-\tilde{V}_{sy} - \left.\dfrac{\partial v_1}{\partial F_z}\right|_{\substack{\alpha=\bar{\alpha}' \\ F_z=\bar{F}_z}} \dfrac{d\tilde{F}_z}{dt}\right) \end{cases} \quad (28)$$

with

$$\begin{cases} \tau_\kappa = \dfrac{\sigma_\kappa(\overline{F}_z)(1+\overline{\kappa}')^2}{C_{F\kappa}(\overline{F}_z)V_r} \dfrac{\partial F_x}{\partial \kappa}\bigg|_{\substack{\kappa=\overline{\kappa}'\\F_z=\overline{F}_z}} \\ \tau_\alpha = \dfrac{\sigma_\alpha(\overline{F}_z)(1+\kappa')\cos^2\overline{\alpha}'}{C_{F\alpha}(\overline{F}_z)V_r} \dfrac{\partial F_y}{\partial \alpha}\bigg|_{\substack{\alpha=\overline{\alpha}'\\F_z=\overline{F}_z}} \end{cases} \quad (29)$$

and

$$\begin{cases} \dfrac{\partial u_1}{\partial F_z} = \dfrac{F_x}{C_{F\kappa}} \dfrac{\partial \sigma_\kappa}{\partial F_z} + \dfrac{\sigma_\kappa}{C_{F\kappa}} \dfrac{\partial F_x}{\partial F_z} - \dfrac{\sigma_\kappa F_x}{C_{F\kappa}^{\,2}} \dfrac{\partial C_{F\kappa}}{\partial F_z} \\ \dfrac{\partial v_1}{\partial F_z} = \dfrac{F_y}{C_{F\alpha}} \dfrac{\partial \sigma_\alpha}{\partial F_z} + \dfrac{\sigma_\alpha}{C_{F\alpha}} \dfrac{\partial F_y}{\partial F_z} - \dfrac{\sigma_\alpha F_y}{C_{F\alpha}^{\,2}} \dfrac{\partial C_{F\alpha}}{\partial F_z} \end{cases} \quad (30)$$

The time constants τ_κ and τ_α govern the speed of the transient responses. Note that τ_κ and τ_α are proportional to the local derivatives of forces: $\partial F_x / \partial \kappa$ and $\partial F_y / \partial \alpha$.

When the slip velocities V_{sx} and V_{sy} are relatively small and the vertical load is constant, Eqs. (28) become

$$\begin{cases} \dfrac{\sigma_\kappa}{C_{F\kappa}} \dfrac{\partial F_x}{\partial \kappa}\bigg|_{\kappa=\overline{\kappa}'} \dfrac{d\widetilde{\kappa}'}{dt} + V_r \widetilde{\kappa}' = -\widetilde{V}_{sx} \\ \dfrac{\sigma_\alpha}{C_{F\alpha}} \dfrac{\partial F_y}{\partial \alpha}\bigg|_{\alpha=\overline{\alpha}'} \dfrac{d\widetilde{\alpha}'}{dt} + V_r \widetilde{\alpha}' = -\widetilde{V}_{sy} \end{cases} \quad (31)$$

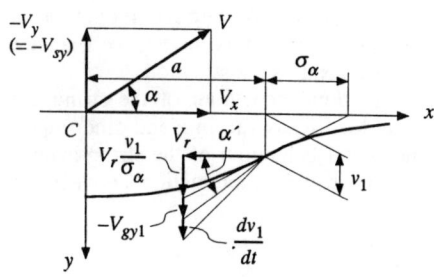

Fig. 15. Lateral velocity of the string.

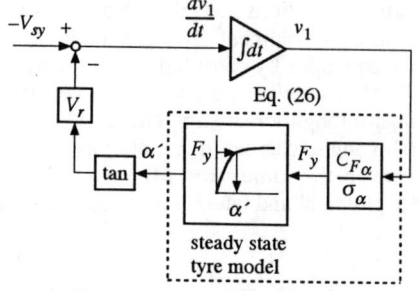

Fig. 16. Calculation diagram (for v_1).

In experiments, the linearity in the response to a large slip angle does not hold, thus the fitting by Eq. (20) is not suitable (an example shown in Fig. 17). To analyze such cases, we restrict ourselves to a small variation so that the response remains linear. The measurement procedure is as follows:

(a) Apply a rotational angle of the turn table which is 0.5 ° smaller than the final angle
(b) Press the tyre against the track until the desired vertical load is reached
(c) Move the track
(d) After steady state is reached, stop the track
(e) Increase the rotational angle by 0.5 ° so that the final angle is reached
(f) Move the track again

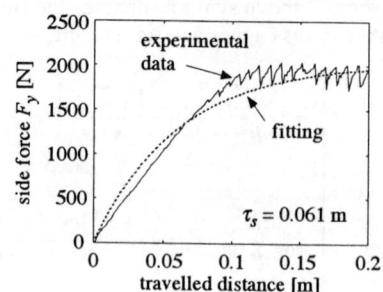

Fig. 17. Side force fitting; $F_z = 4$ kN, $\alpha = 10$ °.

The reference measurement procedure is the same but skipping (*e*). The data is sampled during *f*). An additional small torsional angle between rim and tyre belt is involved in (*e*) whose consequences are a little longer space constant of side force fitting in the case of small slip angles, and non-zero initial aligning torque.

Figures 18 and 19 depict side force response to step changes (from 0 °) and to *small* step increases (0.5 °) in slip angle respectively. Through the fitting of the result shown in Fig. 19, we get the local derivative and the space constant of side force which are summarized in Figs. 20 and 21 respectively. It is confirmed that the space constant of side force has a proportional relationship to the local derivative which is also theoretically found in Eqs. (29). The space constant in side slip case is almost the same as the one in camber case within small a slip angle range (Figs. 11 and 21).

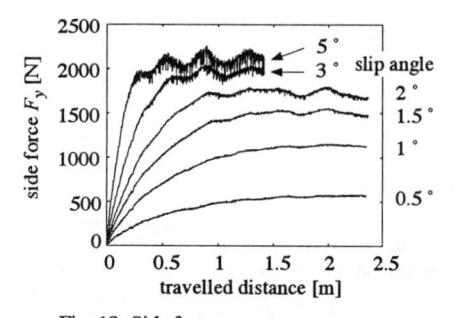

Fig. 18. Side force response to step change in slip angle; $F_z = 4$ kN.

Fig. 19. Side force response to *small* step increase in slip angle; $F_z = 4$ kN.

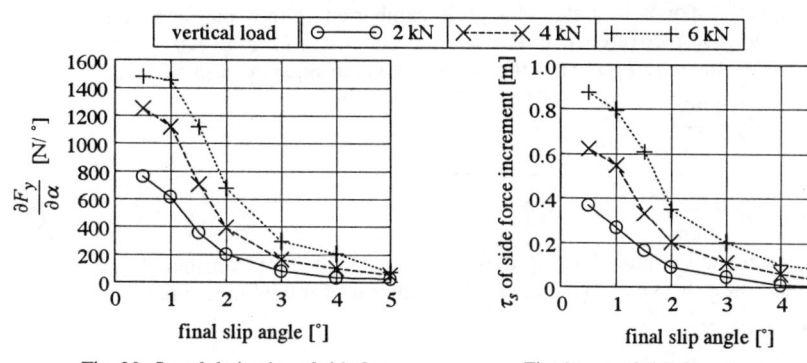

Fig. 20. Local derivative of side force.

Fig. 21. τ_s of side force increment fitting.

7. TYRE MODEL FOR COMPUTER SIMULATION

Equations (29) indicate that the speed of the response in Eqs. (25) becomes infinitely fast at the peak of F_x vs. κ or F_y vs. α curve, and due to the inevitable error in the numerical calculation, Eqs. (26) may lose real solutions.

This problem is attributed to the fact that Eqs. (25) do not consider the decreasing string deflections (due to decreasing force characteristics) with increasing slip quantities, although they do consider the sliding in the contact patch. To avoid the above problem, we apply corrections to the relations represented by Eqs. (26). Since we can not alter the steady state tyre characteristics, we actually correct the relations between forces (F_x or F_y) and string deflections (u_1 or v_1). The corrections are made in such a way that the string deflections keep rising with slip quantities.

Two correction methods are introduced. For the part where the slope $\partial v_1 / \partial \alpha'$ of the original curve is smaller than the minimum value ε, we use the correction line or curve v_1^*. Figure 22a illustrates the first method by a straight line v_1^* given by Eq. (32) where (α_c, v_{1c}) denotes the connection point and F_{yc} the side force at α_c. The problem of this method is the computational complexity in finding the connection point. Fig. 22 b depicts a simpler method given by Eq. (33), whose computation is much simpler. The meaning of this correction is that we simply add the product of α' and the difference between ε and the slope $\partial v_1 / \partial \alpha'$ of the original curve. The disadvantage of this method is the discontinuity in the slope at the connection point.

$$v_1^* = \varepsilon(\alpha' - \alpha_c) + \frac{\sigma_\alpha}{C_{F\alpha}} F_{yc} \qquad (32)$$

$$v_1^* = \frac{\sigma_\alpha}{C_{F\alpha}} F_y + \left(\varepsilon - \frac{\sigma_\alpha}{C_{F\alpha}} \frac{\partial F_y}{\partial \alpha'}\right)\alpha' \qquad (33)$$

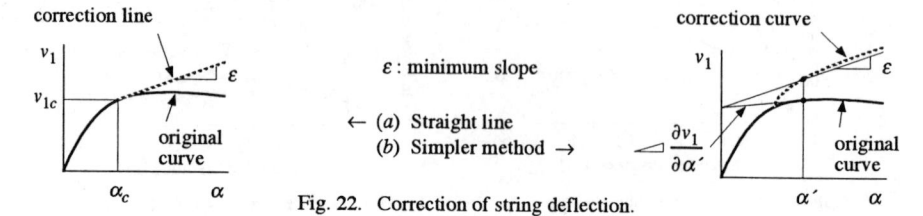

Fig. 22. Correction of string deflection.

(a) Straight line
(b) Simpler method
ε: minimum slope

Side force responds to a step change in camber angle in a similar way to the side slip case. The major difference is the presence of a *non-lagging part*. We apply the same method to the *lagging part* of the side force as the side slip case.

$$\frac{dv_{\gamma 1}}{dt} + V_r \frac{\tan \alpha'_\gamma}{1 + \kappa'} = -V_{\gamma sy} \qquad (34)$$

$$v_{\gamma 1} = \sigma_\gamma \frac{F_y(\gamma'_{lag})}{C_{F\alpha}} \qquad (35)$$

where $v_{\gamma 1}$ denotes the lateral string deflection, α'_γ equivalent transient slip angle, and $V_{\gamma sy}$ effective side slip velocity which are the quantities for the lagging part.

$$\begin{cases} \alpha'_\gamma = \varepsilon_{\gamma\alpha} \gamma'_{lag} \\ V_{\gamma sy} = -|V_x| \tan\left[(1 - \varepsilon_{Nlag})\varepsilon_{\gamma\alpha}\gamma\right] \end{cases} \qquad (36)$$

where $\varepsilon_{\gamma\alpha} = C_{F\gamma} / C_{F\alpha}$ denoting the ratio of camber stiffness to the cornering stiffness which is introduced to estimate the equivalent slip angle due to camber, and ε_{Nlag} non-lagging part ratio. Transient camber angle γ' is

$$\gamma' = \varepsilon_{Nlag} \gamma + \gamma'_{lag} \qquad (37)$$

In practice, it is convenient to determine transient slip quantities by estimation. Figure 23 shows the estimation method. In this situation, camber angle is not zero, therefore F_y vs. α curve has an offset value α_{off} on the horizontal axis. If the total side force F_y is given, we estimate $F_{y\alpha}$ which is the contribution by the slip angle α, by the following manner:

$$\frac{F_{y\alpha}}{F_y} = \frac{\alpha}{\alpha - \alpha_{off}} \qquad (38)$$

From the above relation, Eq. (26) becomes

$$v_1 = \sigma_\alpha \frac{F_y(\kappa', \alpha', \gamma')}{C_{F\alpha}} \frac{\alpha'}{\alpha' - \alpha'_{off}} \tag{39}$$

In the numerical calculation, the current transient slip angle $\alpha'_{(n)}$ is determined by using the current lateral string deflection $v_{1(n)}$ due to slip angle, and the previous state variables designated by $(n-1)$. Figure 24 shows the calculation diagram of the estimation.

$$v_{1(n)} = \sigma_{\alpha(n-1)} r_{estm(n-1)} \alpha'_{(n)} \tag{40}$$

$$v_{\gamma 1(n)} = \sigma_{\gamma(n-1)} r_{estm(n-1)} \alpha'_{\gamma(n)} \tag{41}$$

$$r_{estm(n-1)} = \frac{F_y(\kappa', \alpha', \gamma')_{(n-1)}}{C_{F\alpha(n-1)}\left(\alpha'_{(n-1)} - \alpha'_{off(n-1)}\right)} \tag{42}$$

where $r_{estm(n-1)}$ represents the reduction ratio of the relaxation lengths. We apply the same approach to the camber case in Eq. (41). Because, from the combined experiments of side slip and camber, the space constant τ_s for camber mainly depends on the slip angle.

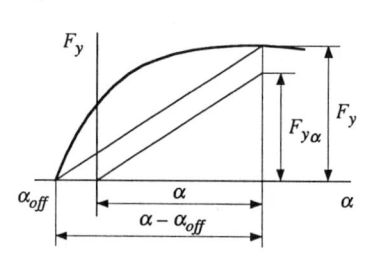

Fig. 23. Estimation method.

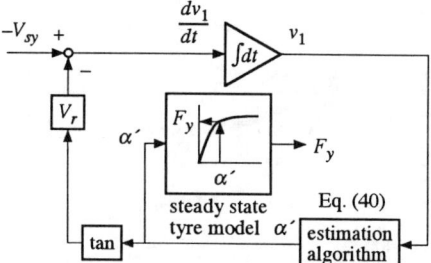

Fig. 24. Calculation diagram (estimation of α').

In the case of turn slip, according to the stretched string theory, there are two contributions to the aligning torque generation: asymmetric lateral deformation of the string and opposite (in left and right) longitudinal deformations of the tread elements. The former exists only in transient states and the steady state value is only due to the latter. With the analogy of camber to turn slip, we may assume that the aligning torque response consists of two similar mechanisms in camber case, too. Thus, the model partly contains an assumption. The first contribution M_{zs} corresponds to the string deformation. For the detection of the transient state, we use the time derivative of $v_{\gamma 1}$,

$$\frac{dM_{zs}}{dt} = \frac{V_r}{\tau_{Mzs}} \left(C_{Mzs} \frac{dv_{\gamma 1}}{dt} - M_{zs} \right) \tag{43}$$

where τ_{Mzs} and C_{Mzs} are tyre parameters. The second one M_{zt} corresponds to the tread element deformations. Since non-lagging part does not exist in the aligning torque response, we use $\gamma'_{lag} / (1 - \varepsilon_{Nlag})$ as the transient camber angle for the aligning torque.

$$M_{zt} = M_z\left(\kappa', \alpha', \frac{\gamma'_{lag}}{1 - \varepsilon_{Nlag}}\right) \quad (\varepsilon_{Nlag} \neq 1) \tag{44}$$

Total aligning torque reads

$$M_z = M_{zs} + M_{zt} \tag{45}$$

Figures 25 and 26 show the comparison between the model and experiments on side force and aligning torque responding to step changes in slip angle. The Magic Formula is used to express the steady state characteristics of experimental results. The model well describes the transient behaviour in side force, but in aligning torque, peak values of the model are larger at large slip angles. In camber case, the model well agrees with the experimental results in both side force and aligning torque (Figs. 27 and 28).

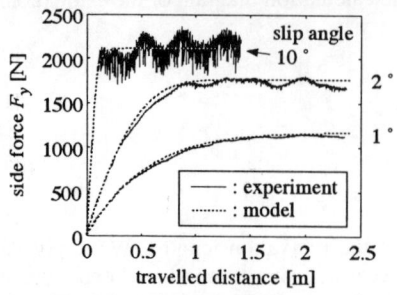

Fig. 25. Side force response to step change in slip angle; $F_z = 4$ kN.

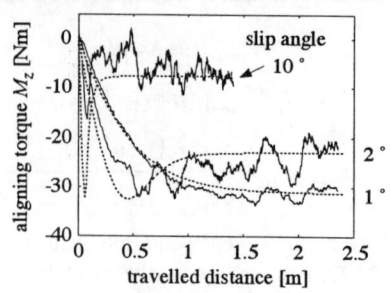

Fig. 26. Aligning torque response to step change in slip angle; $F_z = 4$ kN.

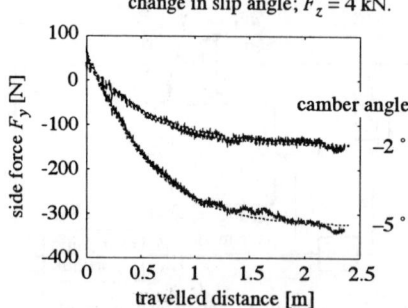

Fig. 27. Side force response to step change in camber angle; $F_z = 4$ kN.

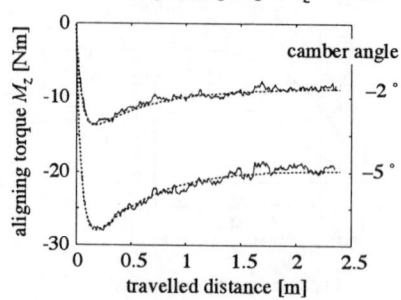

Fig. 28. Aligning torque response to step change in camber angle; $F_z = 4$ kN.

REFERENCES

1. Barson, C. W. and Osborne, D. J., *Dynamic properties of tyres*, C277/83 © I Mech. E., 1983.
2. Heydinger, G. J., Garrott, W. R. and Chrstos, J. P., *The importance of tire lag on simulated transient vehicle response*, SAE Paper 910235, 1991.
3. Koibuchi, K., Yamamoto, M., Fukada, Y. and Inagaki, S., *Vehicle stability control in limit cornering by active brake*, SAE Paper 960487, 1996.
4. Nagai, M. and Koike, K., *Theoretical study of vehicle wandering phenomenon induced by dented road cross profile*, Int. J. of Vehicle Design, Vol. 1, No. 2, 1994, pp. 182-194.
5. Pacejka, H. B., *Analysis of the dynamic response of a rolling string-type tire model to lateral wheel-plane vibrations*, Vehicle System Dynamics 1, 1972, pp. 37-66.
6. Pacejka, H. B., *The wheel shimmy phenomenon*, Doctoral thesis, Delft University of Technology, Delft, The Netherlands, 1966.
7. Pacejka, H. B., *Yaw and camber analysis*, Sec. 9.5. of "Mechanics of Pneumatic Tires" (editor S. K. Clark), Washington D. C., U.S., 1981, pp. 785-871.
8. Schilippe, B. von and Dietrich, R., *Das Flattern eines bepneuten Rades*, Bericht 140 der Lilienthal-Gesellschaft (1941), English translation: NACA TM 1365, 1954, pp. 125-147.
9. Segel, L. and Wilson, R., *Requirements for describing the mechanics of tire used on sigle-track vehicles*, Proceedings of IUTAM Symposium on the Dynamics of Vehicles, 1975, pp. 173-186.
10. Takahashi, T. and Pacejka, H. B., *Cornering on uneven roads*, Proceedings of 10th IAVSD Symposium, Prague, 1987, pp. 469-480.
11. Takahashi, T. and Hoshino M., *The tyre cornering model on uneven roads for vehicle dynamics studies*, Proceedings of AVEC '96, 1996, pp. 941-953.

Dynamic Tyre Responses to Brake Torque Variations

P.W.A. ZEGELAAR[*] and H.B. PACEJKA[*]

ABSTRACT

An in-plane tyre model to study the tyre responses to brake torque variations is presented. This model conceives a rigid ring (representing the tyre tread band) on an elastic foundation (representing the tyre sidewalls). The contact of the ring with the road is governed by a vertical residual stiffness representing the large deformation of the tyre in the contact patch and a slip model that generates the horizontal forces in the contact patch. This slip model, which is based on the relaxation length principle, is formulated such that it can handle zero velocity conditions. Experiments were carried out on a test drum. The tyre was excited by applying variations of brake torque. The measured frequency response functions of the longitudinal force show clearly two modes of vibration of the tyre at 33 and 77 Hz. The frequency response functions are used to estimate the tyre parameters. Large variations of brake torque are used to validate the tyre model. Two responses are discussed: successive step increases in brake pressure and braking with wheel lock.

1. INTRODUCTION

The dynamics of pneumatic tyres is important for the dynamic behaviour of road vehicles. There are three major sources for tyre in-plane vibrations: brake torque fluctuations, road unevenesses, and horizontal and vertical oscillations of the axle. This paper presents the tyre vibrations due to intermittent braking only. A previous study presented the tyre responses on uneven roads [7].

The tyre model described in Section 2 is able to generate the typical tyre vibrations in the frequency range 0-100 Hz. In this frequency range the tyre tread-band behaves as a rigid body with respect to the rim. Consequently, the model is called 'rigid ring model'. Above 100 Hz the tread-band shows large deformations during vibration. These modes, which are neglected in this paper, are the flexible mode shapes of the tyre.

The rigid ring model is developed to be used in vehicle simulations; therefore, the model should be compact, accurate and robust. The model adds the dynamics of the rigid ring to measured stationary slip characteristics; although, for the sake of simplicity, the theoretical characteristics of a brush type model are used instead.

Experiments have been carried out on a test drum. The parameters of the model are estimated from the measured frequency response functions. The non-linear behaviour of the model is validated with stepwise increases of the brake pressure. A brake manoeuvre with wheel lock is used to show the robustness of the model.

---

[*] Vehicle Research Laboratory, Delft University of Technology, Delft, The Netherlands

2. THE RIGID RING TYRE MODEL

The rigid tyre ring model, shown in Figure 1, represents a pneumatic tyre-wheel system. It is modelled as three components: the tyre tread-band, the rim, and the sidewalls. The tyre tread-band is modelled as a rigid circular ring with three degrees of freedom: the displacement in longitudinal direction x_b; the displacement in vertical direction z_b; and the rotation about the wheel axis (y-axis) θ_b. The rim is modelled as a rigid body that rotates with an average speed Ω. Small deviations of the rim angle are denoted with θ_a. The tyre belt and the rim are connected by a third component: the sidewalls and pressurised air. This component is modelled as horizontal, vertical, and rotational springs and dampers.

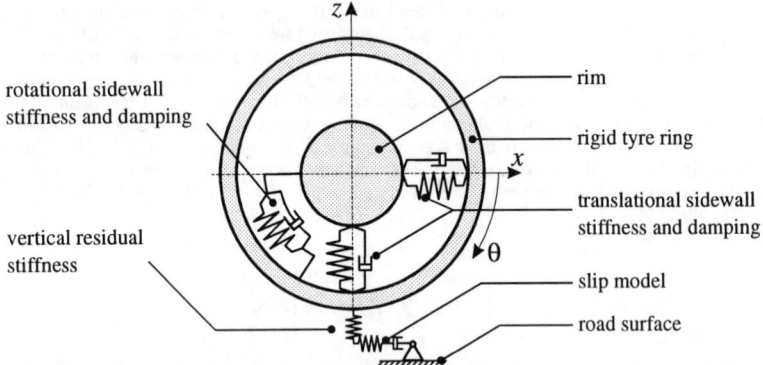

Fig. 1. The basic tyre model: the tyre is modelled as circular rigid ring, supported on an elastic foundation.

The equations of motions of the tyre-wheel system read:

$$m_b \ddot{x}_b = F_{cx} - F_{bx} \tag{1a}$$

$$m_b \ddot{z}_b = F_{cz} - F_{bz} \tag{1b}$$

$$I_{ay} \ddot{\theta}_a = M_{by} - M_{ay} \tag{1c}$$

$$I_{by} \ddot{\theta}_b = -r_e F_{cx} - M_{by} \tag{1d}$$

where m_b is the mass of the tyre belt, and I_{by} and I_{ay} are the moments of inertia about the y-axis of the tyre belt and the rim, respectively. The applied brake torque is denoted with M_{ay}. The situation at the rim or axle is indicated with the index "a"; the displacement of the tyre belt with "b"; and the situation in the contact patch with "c". The horizontal force F_{bx}, the vertical force F_{bz}, and the moment M_{by} in the tyre sidewalls depend on the tyre sidewall deformation:

$$F_{bx} = k_b(\dot{x}_b - \dot{x}_a) + c_b(x_b - x_a) - k_b(\Omega + \dot{\theta}_a)(z_b - z_a) \tag{2a}$$

$$F_{bz} = k_b(\dot{z}_b - \dot{z}_a) + c_b(z_b - z_a) + k_b(\Omega + \dot{\theta}_a)(x_b - x_a) \tag{2b}$$

$$M_{by} = k_{b\theta}(\dot{\theta}_b - \dot{\theta}_a) + c_{b\theta}(\theta_b - \theta_a) \tag{2c}$$

where c_b denotes the horizontal and vertical stiffnesses; and k_b denotes the horizontal and vertical damping coefficients. The symbols $c_{b\theta}$ and $k_{b\theta}$ denote the rotational stiffness and damping coefficient. During the experiments the axle motions are constrained, so: $\dot{x}_a = 0$; $x_a = 0$; $\dot{z}_a = 0$. Only the axle height z_a is prescribed to load the tyre on the drum. The terms with the rotational velocity times the tire sidewall deflection in Equations (2a,b) are caused by the rotating damper.

The total vertical stiffness of the tyre is about six times smaller than the stiffness of the first mode of vibration c_b. To match the vertical stiffness of the model with the real tyre the vertical residual stiffness c_{cz} is introduced. The vertical force in the contact patch F_{cz} depends on the deformation of the residual stiffness:

$$F_{cz} = c_{cz}(z_r - z_b) \tag{3}$$

The road height z_r is set equal to zero, because the tyre rolls over an even road surface. Rather than the linear equation, a second order polynomial is used, because the vertical force increases slightly more than proportional with the tyre deflection.

A brush type model is used to represent the stationary slip characteristics. This approach gives realistic slip characteristics using three parameters only: half the contact length a, the tread element stiffness per unit of length c_{px} and the friction coefficient μ. The longitudinal force F_{cx} in the contact patch reads as function of the slip ζ_{cx} in the contact patch:

$$\begin{aligned} F_{cx} &= \mu F_{cz}\left\{3|\theta_x \zeta_{cx}| - 3|\theta_x \zeta_{cx}|^2 + |\theta_x \zeta_{cx}|^3\right\}\text{sgn}(\zeta_{cx}) & for\ |\zeta_{cx}| \le 1/\theta_x \\ F_{cx} &= \mu F_{cz} \cdot \text{sgn}(\zeta_{cx}) & for\ |\zeta_{cx}| > 1/\theta_x \end{aligned} \tag{4}$$

Introduced is the tyre parameter θ_x:

$$\theta_x = \tfrac{2}{3}\frac{c_{px}a^2}{\mu F_{cz}} \tag{5}$$

The slip velocity of an elastic tyre may be defined as the absolute speed of an imaginary point S, which is attached to the wheel at distance r_e below the wheel centre, see Figure 2. The effective rolling radius r_e is defined such that the slip velocity is zero for free rolling. At braking the point S has a forward speed. This slip velocity $V_{c,sx}$ in the contact patch equals the difference between the forward velocity V_{cx} and the rolling velocity V_{cr} in the contact patch:

$$V_{c,sx} = V_{cx} - V_{cr} = (V_x + \dot{x}_b) - r_e(\Omega + \dot{\theta}_b) \tag{6}$$

where V_x is the average forward velocity of the tyre, or in our case the forward velocity of the drum (see Figure 6); and Ω the average rotational velocity of the wheel.

The elastic deformations of the tyre induce the tyre transient responses. The transient response of the tyre can be modelled as the tyre sidewall springs and the tread stiffness in series with a damper, see Figure 3. The elastic deformation of the tyre sidewalls causes the difference in the slip velocity V_{sx} based on rim velocities and the slip velocity in the contact patch $V_{c,sx}$. The elastic deformation of the tread elements in series with the damper are modelled as a first order slip model:

$$\sigma_c \dot{\zeta}_{cx} + V_{cr} \zeta_{cx} = -V_{c,sx} \qquad (7)$$

The stationary solution of the first order slip definition equals the normal slip definition $(\zeta_{cx} = -V_{c,sx}/V_{cr})$. The relaxation length of the contact patch σ_c equals the slip stiffness divided by the longitudinal tread element stiffness c_{cx} ($c_{cx} = 2ac_{px}$):

$$\sigma_c = \frac{C_\kappa}{c_{cx}} \qquad (8)$$

In case of full adhesion the relaxation length of the contact patch σ_c equals half the contact length a. The relaxation length decreases with increasing slip. The minimum value of the relaxation length is limited to insure numerical stability. The slip stiffness C_κ is defined as the local derivative of the slip characteristics:

$$C_\kappa = \frac{\partial F_x}{\partial \zeta_{cx}}, \quad C_\kappa|_{(\zeta_{cx}=0)} = 2c_{px} a^2 \qquad (9)$$

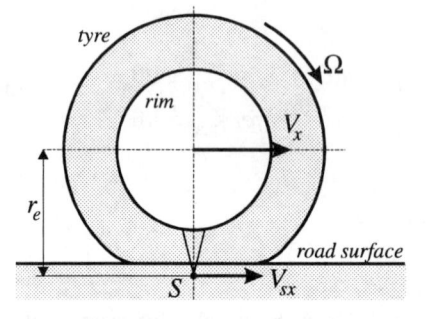

Fig. 2. The kinematics of rolling.

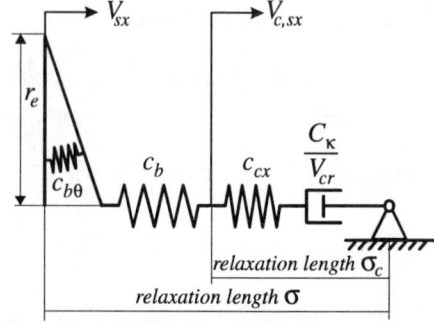

Fig. 3. The transient slip model.

The overall relaxation length of the tyre σ is not used directly in the model. It results from the sidewall and tread element deformations. This relaxation length equals the slip stiffness times all longitudinal compliances in series (see Figure 3):

$$\sigma = C_\kappa \left(\frac{1}{c_{cx}} + \frac{1}{c_b} + \frac{r_e^2}{c_{b\theta}} \right) \qquad (10)$$

The overall relaxation length σ is three to seven times larger than the relaxation length of the contact patch, depending on the vertical load.

The disadvantage of this formulation of the transient tyre response is that horizontal force in the contact patch reacts directly on vertical load variations (*cf.*

Equation 4) without using the first order slip model of Equation (7). The transient response to vertical load variations is only due to the dynamics deformations of the tyre sidewalls. This results in a relaxation length for vertical load variations that is 15%-30% smaller than the relaxation length for slip variations.

An additional advantage of the formulation of Equation (7) is that the first order slip model is able to handle a zero velocity condition. In that case the derivative of the slip is proportional to the slip velocity. Or, if we integrate this equation, the slip (and thus the force) in the contact patch is proportional to the deformation in the contact patch:

$$\sigma_c \dot{\zeta}_c = -V_{c,sx} \quad , \quad \zeta_c = -\int V_{c,sx}/\sigma_c \, dt \qquad (11)$$

In conclusion, if the rotational velocity of the tyre equals zero, the tyre will behave as a horizontal spring. Naturally, the generated horizontal force is limited by the friction coefficient.

As mentioned previously, the imaginary point S is used to define the slip velocity (cf. Equation 6). Simultaneously, this point should be used to transform the longitudinal force F_{cx} to a torque acting on the tyre (cf. Equation 1d). This idea, introduced in a previous study [6], is substantiated by measurements. Figure 4 shows the three tyre radii as function of the vertical load: the loaded tyre radius, which is defined as the axle height above the road surface; the effective rolling radius, which is obtained from the forward and rotational velocities at free rolling; and finally, the point of application of the longitudinal force, which is defined as brake torque divided by longitudinal force.

An alternative way of formulating the torque about the wheel axle is by taking into account the horizontal force acting at the loaded radius and vertical pressure distribution acting in the contact patch. This approach requires a detailed tyre model that describes the geometry of the tyre accurately.

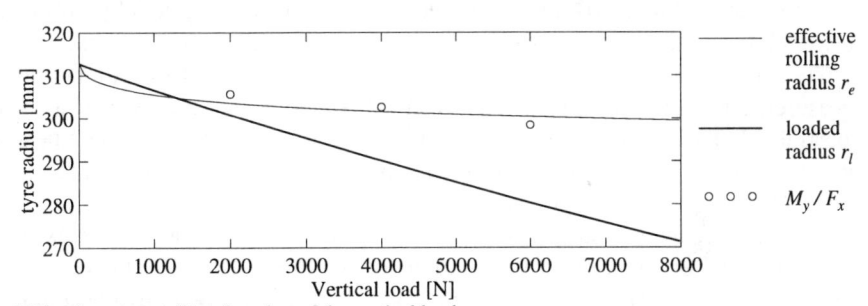

Fig. 4. The three tyre radii as function of the vertical load.

Figure 5 shows the modes of vibration of the rigid ring tyre model. The natural frequencies result from the parameters given in Table 1. The vertical mode ($f = 79$ Hz) is excited mainly by road unevenesses. The two longitudinal modes arise from brake torque variations. We may distinguish between an in-phase rotation of the rim and the tyre ring and an out-of-phase rotation. The three modes of the model correspond to the modes obtained experimentally from modal analysis [8].

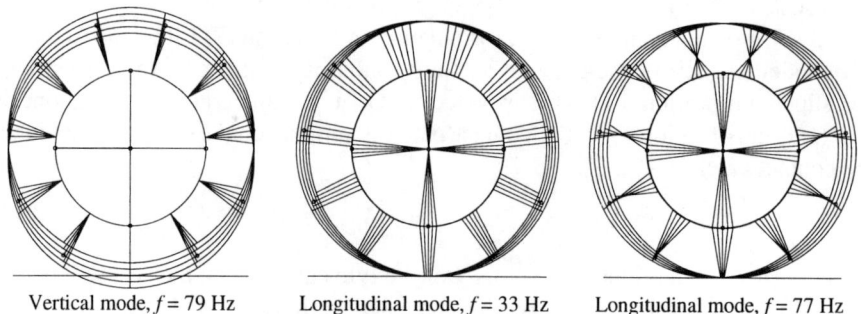

Vertical mode, $f = 79$ Hz

Longitudinal mode, $f = 33$ Hz
(in-phase rotation rim and tyre)

Longitudinal mode, $f = 77$ Hz
(out-of-phase rotation rim and tyre)

Fig. 5. The modes of vibration of the rigid ring model.

3. EXPERIMENTAL SET-UP

The experiments were carried out on the drum test stand in the Vehicle Research Laboratory. This test stand, shown in Figure 6, consists of a tyre mounted on a rotating drum representing the road surface. The drum test stand is used to examine the dynamic behaviour of the tyre in the frequency range 0-100 Hz. Obviously, the natural frequencies of the test stand are higher than the frequency range of interest.

The drum has a steel surface. Even though a more realistic road surface (*i.e.* safety walk paper) generates better stationary slip characteristics, the tyre properties will deteriorate fast due to excessive tread wear. Conversely, the stationary slip characteristics on a steel surface differ substantially from road measurements, but the dynamic tyre properties will remain more constant.

The vertical axle height can be adjusted to load the tyre on the drum. During the measurements, the horizontal and vertical motions of the wheel axle are constrained. The reaction forces of the tyre are measured at both wheel bearings with piezo electric elements. These elements measured the variations in the forces only and not the static components.

The disc brake is mounted in a separate structure. The reaction forces of the brake are counterbalanced by the bearings of the brake axle. The brake axle and the wheel axle are connected with an intermediate shaft and two flexible couplings. This arrangement results in an application of a pure brake torque to the wheel without residual forces due to an offset in the alignment of the two axles. The applied brake torque is measured with strain gauges in the intermediate shaft.

To estimate the wheel slip, the wheel and drum velocities are measured. The wheel velocity is measured with a dynamometer. This sensor has a good dynamic response. The drum velocity is measured with an incremental angle encoder that generates 2000 pulses each revolution. These pulses are converted into an analog signal by a frequency converter.

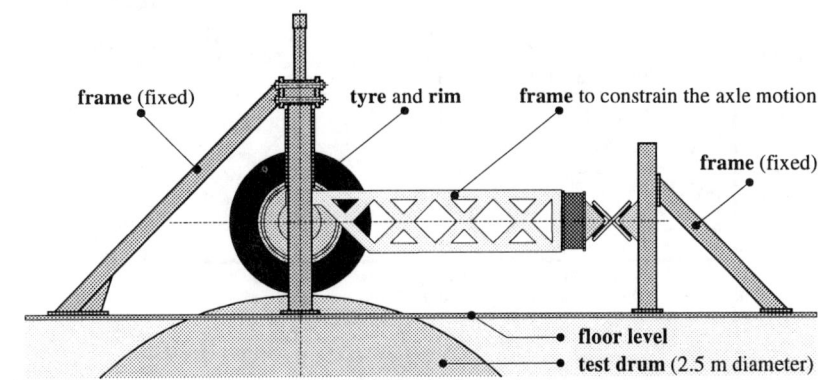

Fig. 6a. Schematic view of the drum test stand (side view).

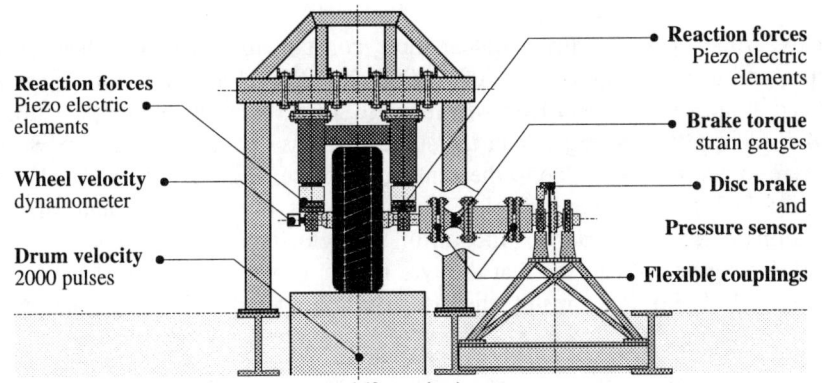

Fig. 6b. Schematic view of the drum test stand (front view).

A hydraulic servo system is used for controlling the brake pressure. The servo valve is controlled by a feedback of the difference between the measured and desired brake pressure. The desired brake pressure signals are generated by a computer. This set-up allows us to be flexible in the kind of brake torque excitations used: sinusoidal, block pulses, successive steps, random variations, or sinusoidal sweeps.

Figure 7 shows the hydraulic scheme. We may discern three pressures: the high pressure generated by a pump; the low pressure of the reservoir; and the controlled pressure in the brake cylinder. The accumulator and pressure relieve valve are used to maintain a constant high pressure.

Fig. 7. The hydraulic scheme.

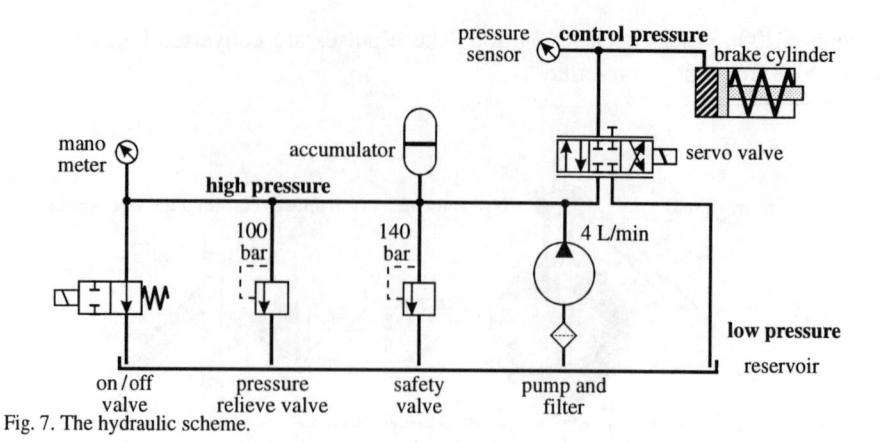

4. FREE ROLLING EXPERIMENTS

All experiments were carried out at three constant axle heights. These heights correspond to a vertical load of 2000, 4000 and 6000 N for a non-rotating tyre. If the tyre rotates the vertical load will increase. First, the dynamic vertical stiffness of the rotating tyre is larger than the static stiffness of the standing tyre. Second, the tyre radius increases due to the centrifugal force. Both effects were taken into account in the rigid ring model. Figure 8 shows the measured and simulated variations of vertical load as function of the velocity for a constant axle height. From this figure we can see that the tyre radius grows with 2 mm if the velocity range 0-150 km/h. As a result, the effective rolling radius increases also with 2 mm.

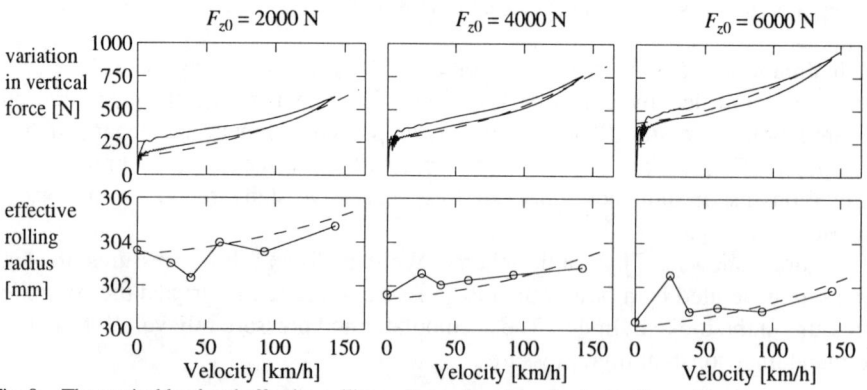

Fig. 8. The vertical load and effective rolling radius as function of velocity (free rolling).
—⊙— measurement − − − simulation

5. FREQUENCY RESPONSE FUNCTIONS

To estimate the Frequency Response Functions (FRFs) the tyre was excited by small variations of brake torque. Two kinds of FRFs were estimated: the response of the longitudinal force to brake torque variations and the response of the longitudinal force to wheel slip variations.

Three axle heights, corresponding to 2000, 4000 and 6000 N vertical load, were used; and five velocities: 25, 39, 59, 92, 142 km/h. The standard deviation of the brake pressure variation was 1 bar, and the average brake pressure was varied between 5 and 60 bar. The total number of conditions investigated was 35. For each condition 10 measurements were averaged to be able to estimate the coherence functions as well. The duration of each measurement was 16 seconds at 256 Hz sampling rate.

Figure 9a shows one of the measured and simulated FRFs of longitudinal force to brake torque variations. These FRFs, show clearly two modes of vibration of the tyre. The first mode is the in-phase rotational mode; and the second one is the out-of-phase rotational mode. The estimated coherence functions show that the measured FRFs are valid until 80 Hz. The measured FRFs are similar to the ones presented by Kobiki [1].

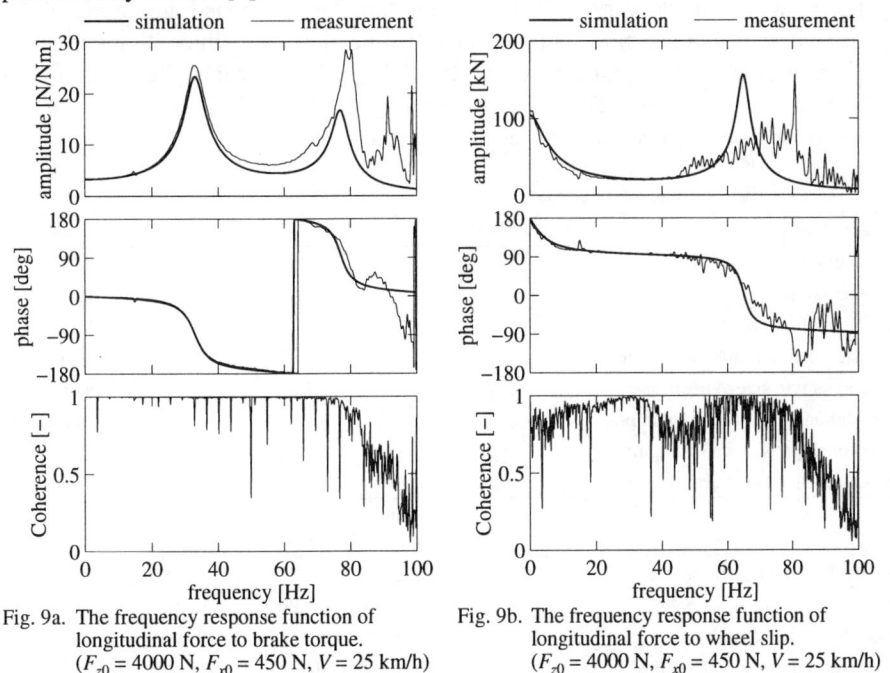

Fig. 9a. The frequency response function of longitudinal force to brake torque.
(F_{z0} = 4000 N, F_{x0} = 450 N, V = 25 km/h)

Fig. 9b. The frequency response function of longitudinal force to wheel slip.
(F_{z0} = 4000 N, F_{x0} = 450 N, V = 25 km/h)

The FRFs of the longitudinal force to slip variations ($\zeta_x = -V_{sx}/V_r$) are shown in Figure 9b. From this figure two important tyre properties can be found: the slip stiffness from the amplitude of the FRF at frequency zero; and the relaxation length of the tyre which can be estimated from the phase shift of 90 degrees in the frequency range 0-40 Hz. The coherence functions show that these measured FRFs could not be estimated very accurately.

The rigid ring tyre model is linearised to simulate the FRFs. Some parameters of the model were measured directly: the mass and moment of inertia of the tyre, the contact length; the effective rolling radius; and the overall tyre vertical stiffness. Three other parameters were used to fit the dynamics of the tyre: the translational sidewall stiffness c_b, the rotational sidewall stiffness $c_{b\theta}$, and the element stiffness c_{px}. Varying only these three parameters six properties of the tyre were fit:

- frequency of the vertical mode (from cleat experiments [7])
- frequency of the in-phase rotational mode (from FRF F_x/M_y)
- frequency of the out-of-phase rotational mode (from FRF F_x/M_y)
- relative damping of the in-phase rotational mode (from FRF F_x/M_y)
- slip stiffness (from FRF F_x/ζ_x)
- relaxation length of the overall tyre σ (from FRF F_x/ζ_x)

The simulated and measured frequency response functions are compared in Figure 9 as well. The parameters used are given in Table 1. The model represents the in-phase rotational mode very well (see Figure 9a). The simulated amplitude of the out-of-phase rotational mode is lower than the measured amplitude. Somehow, the excitation of the out-of-phase mode of the real tyre is larger than the excitation of this mode in the model. The relative damping of the mode is the same for the measurements and the simulation, according to the phase shift

Table 1. The parameters of the tyre model.

Description	symbol	value	unit
mass of tyre ring	m_b	7.18	kg
moment of inertia tyre ring	I_{by}	0.636	kg m^2
moment of inertia rim	I_{ay}	0.627	kg m^2
translational tyre sidewall stiffness	c_b	1.57 10^6	N/m
rotational tyre sidewall stiffness	$c_{b\theta}$	7.7 10^4	Nm/rad
translational tyre sidewall damping	k_b	320	Ns/m
rotational tyre sidewall damping	$k_{b\theta}$	50	Nms/rad
longitudinal tread stiffness	c_{px}	17 10^6	N/m^2
		value at F_z = 2000 N value at F_z = 4000 N value at F_z = 6000 N	
vertical residual stiffness	c_{cz}	201000 225000 271000	N/m
effective rolling radius	r_e	0.3044 0.3029 0.3021	m
half the contact length	a	0.0355 0.0534 0.0685	m
load dependency rolling radius	η	-1.38 10^{-6} -0.62 10^{-6} -0.29 10^{-6}	m/N

The model represents the measured FRF of force to slip variations well in the frequency range 0-50 Hz, see Figure 9b. This figure shows the phase shift of 90 degrees due to the modelling of a relaxation length system, and a mode of vibration corresponding to the natural vibration of the belt of the loaded tyre in case the rim is fixed in rotation. The simulations shows this mode clearly at 63 Hz. The measured amplitude hardly shows this mode, but from the phase we can identify this mode at 65 Hz.

Figure 10a shows the tyre properties as function of the vertical load; Figure 10b shows the tyre properties as function of the average brake torque; and Figure 10c shows the tyre properties as function of the velocity. These figures show that the slip stiffness increases when the vertical load increases; and decreases when the average brake torque increases. The relaxation length σ varies proportionally with the slip stiffness (cf. Equation 10). The relative damping of the in-phase rotational mode is proportional with the velocity and inversely proportional with the slip stiffness. In all these aspects, the model represents the measurements rather well.

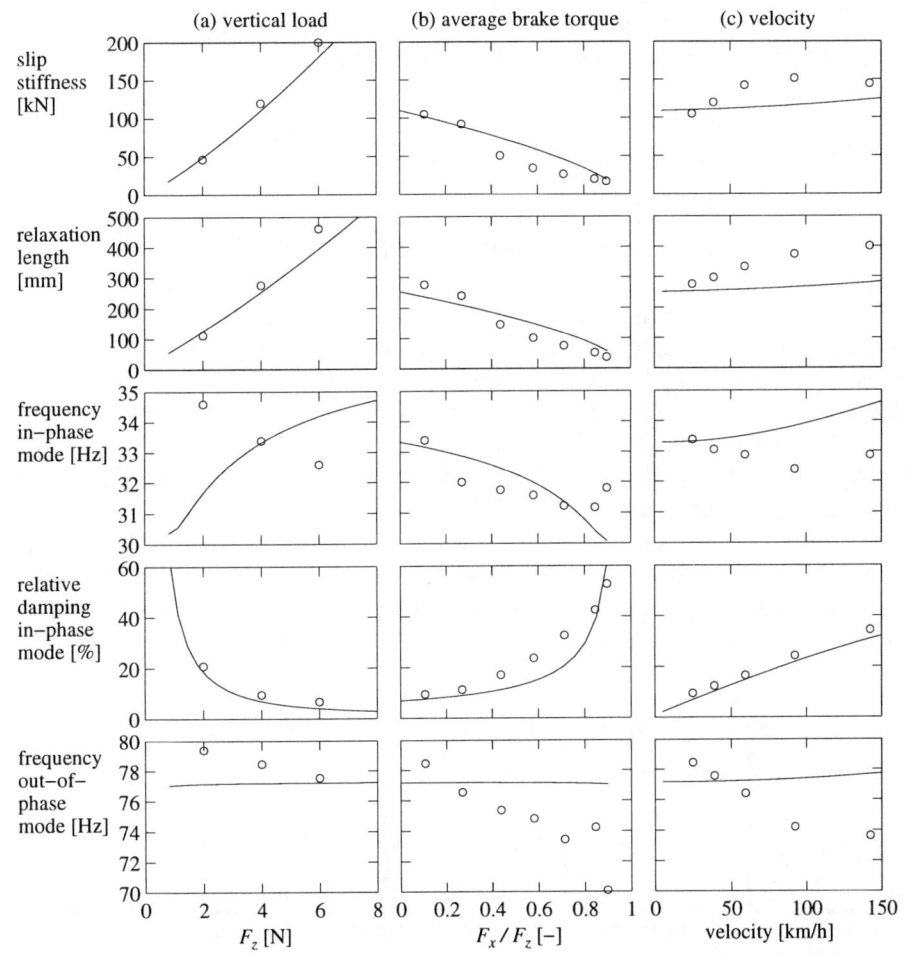

Fig. 10. The tyre properties as function of (a) vertical load; (b) average brake force; and (c) velocity. Standard conditions: $F_{z0} = 4000$ N, $F_{x0} = 450$ N, $V = 25$ km/h. —— model ○ ○ ○ measurement

There are also some deviations between the model and the measurements. The slip stiffness, for instance, increases with velocity. This effect is probably caused by an increase of tyre temperature at high velocity. As a result, the relaxation length increases with the velocity as well.

Another effect is the decrease of the natural frequencies with velocity and vertical load. Other studies report the same dependency of the tyre natural frequencies with velocity [1,2,3,4]. Kobiki *et al.* [1] found that the frequency of the out-of-phase rotation dropped from 89 to 84 Hz when the velocity increases from 0 to 40 km/h. Mills *et al.* [2] excited the wheel axle longitudinally: the longitudinal frequencies decreased by 15% in the velocity range 0-60 km/h. Ushijima *et al.* [3] excited the tyre by a hammer and a road obstacle. He found that the frequency of the vertical mode of vibration decreased by 15% in the velocity range 0-80 km/h. Finally, Vinesse [4] measured the reaction forces at the hub. He identified the same three modes of vibration as presented in Figure 5. He concluded that the natural frequencies of the rotating tyre are significantly lower than the frequencies of the standing tyre. Furthermore, the frequency of the out-of-phase rotational mode decreases with the velocity in the investigated velocity range 25-55 km/h.

6. NON-LINEAR TYRE RESPONSES

The responses of the tyre were validated for non-linear conditions: large variations of brake torque and large variations of wheel velocity. In that situation, the non-linear equations should be used; and the tyre responses are investigated in the time domain rather than the frequency domain. The model is simulated with a fourth order Runga Kutta Integration method.

The non-linear simulation model is based on the equations given in Section 2. The major non-linearities, shown schematically in Figure 11, are: (1) the non-linear slip characteristics of the brush type model; (2) the vertical load dependency of the loaded tyre radius r_l, and the effective rolling radius r_e; (3) the slight increase of the tyre radius due to the centrifugal force; (4) the horizontal shift of the contact point duo to a horizontal force, influencing the vertical force; and (5) the dry friction in the disc brake.

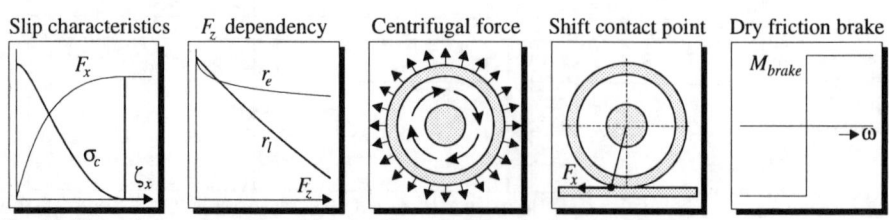

Fig. 11. The major non-linearities of the simulation model of the tyre-wheel system.

Figures 12 and 13 show the measured and simulated responses of the tyre due to successive increases in brake pressure. The duration of each measurement was 4 seconds at 1024 Hz sampling rate. The filter frequency was 400 Hz. The measurements were averaged 10 times to reduce the influence of noise. The measured brake torque was used as inputs for the simulation model.

From t is 0 to 2 seconds the brake pressure is stepwise increased. Each step in brake pressure causes oscillations of the tyre. The dominant oscillation is the in-phase rotational mode at 33 Hz. These oscillations are shown clearly in the dynamic force-slip characteristics of Figures 12 and 13. After a few oscillations a new steady state value of the slip characteristics is found.

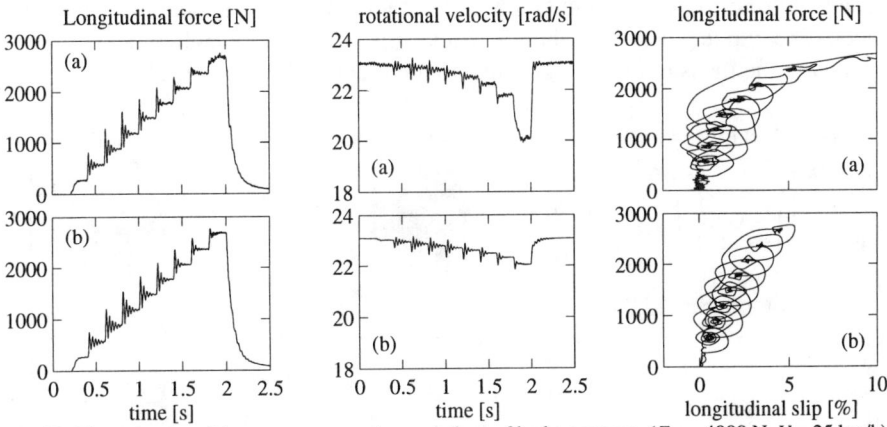

Fig. 12. The response of the tyre on successive variations of brake pressure. (F_{z0} = 4000 N, V = 25 km/h) (a) measurements, (b) simulation.

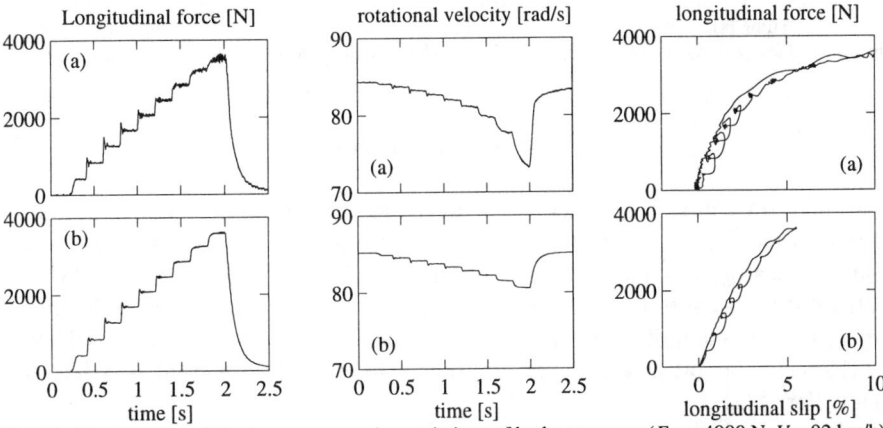

Fig. 13. The response of the tyre on successive variations of brake pressure. (F_{z0} = 4000 N, V = 92 km/h) (a) measurements, (b) simulation.

The model represents the measurements rather well. Only at very large levels of slip the model deviates from the measurements. This is because the used slip characteristics (brush type model) do not represent the stationary slip characteristics accurately enough. The performance of the model will increase if a better representation of the slip characteristics (*e.g.* Magic Formula) is used. The results achieved by the rigid ring model are comparable to the results achieved with tyre models showing distributed deflections over the contact length [5].

Figure 14 shows the response of the tyre during wheel lock. The input of the simulation was the measured brake pressure. From $t = 0.5$ to 1.5 seconds the brake pressure was high; causing the wheel lock-up within 0.25 seconds. The model represents the measurements rather well. We see wheel lock-up within 0.25 seconds. Furthermore, the simulations show same vibrations at spinning up of the wheel. The model does not show a stick slip behaviour which is shown very severely in the measurements, again due to the µ constant brush model.

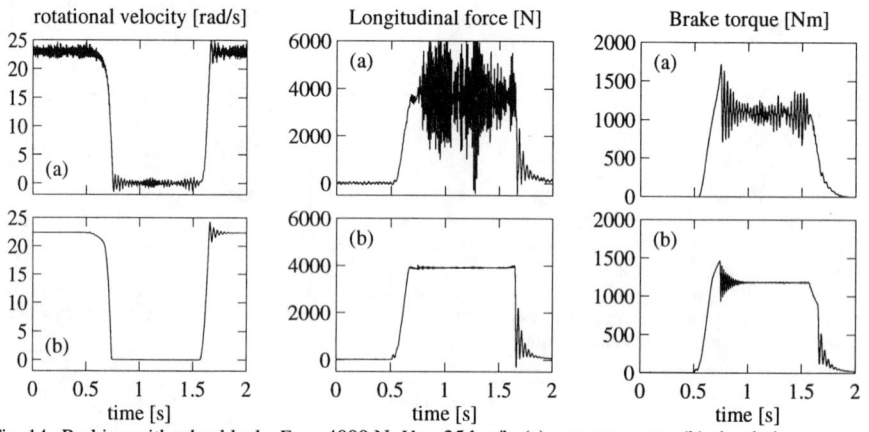

Fig. 14. Braking with wheel lock, $F_{z0} = 4000$ N, $V_0 = 25$ km/h, (a) measurements, (b) simulation.

7. CONCLUSIONS

The rigid ring model represents the dynamic tyre responses to brake torque variations rather well. The parameters of the model were successfully estimated from the measured frequency response functions. The tyre parameters should vary with velocity, because the measured natural frequencies decrease with velocity. A further improvement of the stationary performance of the model could be achieved by changing the brush type characteristics into a better representation like the Magic Formula.

8. NOTATION

a	half the contact length	η	load dependency of the effective rolling radius	indices:	
c	stiffness			0	normal conditions
F	Force			a	wheel axle or rim
I	moment of inertia	μ	friction coefficient	b	tyre belt
k	damping	θ	angle about y-axis	c	contact patch
m	mass	σ	relaxation length	e	effective
r	radius	Ω	rotational velocity	l	loaded
M	moment	ζ	longitudinal slip	r	rolling
V	velocity			s	slip
x	longitudinal displacement			x	in longitudinal direction
z	vertical displacement			y	about the y-axis
				z	in vertical direction

9. REFERENCES

1. Y. Kobiki, A. Kinoshita, and H. Yamada, "Analysis of interior booming noise caused by tyre and powertrain-suspension system vibration", Int. J. of Vehicle Design, Vol. 11, no. 3, pp. 303-313, 1990.
2. B. Mills and J.W. Dunn, "The mechanical mobility of rolling tyres", Paper no C104/71, Proceedings of Vibration and Noise in Motor Vehicles, Institution of Mechanical Engineers, London, July 6-7, 1971.
3. T. Ushijima and M. Takayama, "Modal analysis of tire and system simulation", SAE paper 880585, 1988.
4. E. Vinesse, "Tyre vibration testing from modal analysis to dispersion relations", ISATA 88, Proceeding, Vol. 1, Paper 88048, 1988.
5. A. van Zanten, W.D. Ruf, and A. Lutz, "Measurement and Simulation of Transient Tire Forces", SAE Paper 890640, 1989.
6. P.W.A. Zegelaar, S. Gong, H.B. Pacejka "Tyre model for the study of in-plane dynamics", 13th IAVSD Symposium of Vehicles on Roads and Tracks, Chengdu, P.R. of China., August 23-27, 1993, Vehicle System Dynamics, Vol. 23 supplement.
7. P.W.A. Zegelaar, H.B. Pacejka, "The In-Plane Dynamics of Tyres on Uneven Roads", 14th IAVSD Symposium of Vehicles on Roads and Tracks, Ann Arbor, U.S.A., August 21-25, 1995, Vehicle System Dynamics, Vol. 25 supplement.
8. P.W.A. Zegelaar, "Modal Analysis of Tire In-Plane Vibrations", SAE International Congress and Exposition, Detroit, U.S.A., February 24-27, 1997.

Modelling and Simulation of Non-Steady State Cornering Properties and Identification of Structure Parameters of Tyres

K. GUO and Q. LIU

ABSTRACT

A theoretical model of nonsteady state tyre cornering properties (NSSTCP) with small lateral inputs and its experimental validation are presented. The flexibility of carcass composed of translating, bending and twisting parts is considered. Tyre structure parameters in the model can be simplified as four nondimensional factors which associate with stiffness of tread and carcass, tyre width, length of contact patch respectively. The tests of NSSTCP including pure yaw motion and pure lateral motion are designed and realized by step lateral inputs. Then the structure parameters are identified according to the expressions of the analytical model and frequency response data resulting from the test data in spatial domain. The derived model is validated by experiment results.

1. INTRODUCTION

In the analysis and application of vehicle dynamics and tyre dynamics, it is necessary to establish accurate models of tyre cornering properties [1-13]. A tyre can be assumed as a control system, in which lateral displacement and yaw angle, or slip angle and path curvature [2][6], are considered as the inputs, and lateral force and aligning moment as the outputs, illustrated in Fig.1. If the inputs are constant, tyre properties appears the steady state cornering properties. Conversely, when the inputs are varied with time, tyre will show nonsteady state cornering properties [2][7]. In practice, when a vehicle is running over road surface, tyre must appear nonsteady state properties. The NSSTCP affect vehicle properties, especially for the controllability and shimmy of vehicles.

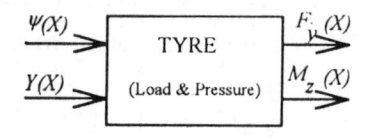

Fig. 1 A tyre control system

NSSTCP are associated closely with rolling distance and path frequency (reduced or spatial frequency) when they are discussed in spatial domain [2-10]. In this sense, they have not direct relation to travel velocity of tyres.

2. MODEL DESCRIPTION

The coordinate systems of modelling is shown in Fig. 2. $P_c(X_c, Y_c)$ is the point of carcass; $P_t(X_t, Y_t)$ is the corresponding point of tread to point P_c. Y_c can be expressed as the function of X and x, i.e. $Y_c = Y_c(X, x)$; similarly $Y_t = Y_t(X, x)$. When $x = 0$, point P_t should be in the same position as point P_c. On the other hand, there is no slip in the contact patch with assumption of small tyre motion. Hence the relations between Y_t and Y_c are obtained as [3]:

$$\begin{cases} Y_t(X, 0) = Y_c(X, 0) \\ Y_t(X, x) = Y_t(X - x, 0) \end{cases} \quad (1)$$

The lateral deformation of tread related to carcass, $\Delta y(X, x)$, can be calculated based on Fig. 2 as follows:

$$\begin{cases} \Delta y(X, x) \approx Y_t(X, x) - Y_c(X, x) \\ Y_c(X, x) \approx Y(X) + (a - x) \cdot \psi(X) + y_c(X, x) \\ y_c(X, x) = y_{c0}(X) + y_{cb}(X, x) + y_\theta(X, x) \end{cases} \quad (2)$$

where $y_c(X, x)$ denotes the total deflection of carcass, which includes the translating part $y_{c0}(X)$, the bending part $y_{cb}(X, x)$ [1] and the twisting part $y_\theta(X, x)$. They can be defined as:

$$\begin{cases} y_{c0}(X) = F_y(X)/K_{c0} \\ y_{cb}(X, x) = \xi(x/a) \cdot F_y(X)/K_{cb} \\ y_\theta(X, x) \approx (a - x) \cdot \theta(X) \\ \theta(X) = \Sigma M_z(X)/N_\theta \end{cases} \quad (3)$$

The lateral force $F_y(X)$ and the aligning moment $M_z(X)$ are given as [1][3]:

$$\begin{cases} F_y(X) = k_{ty} \cdot \int_0^{2a} \Delta y(X, x) dx \\ M_z(X) = k_{ty} \cdot \int_0^{2a} (a - x) \cdot \Delta y(X, x) dx \end{cases} \quad (4)$$

Combining Eq.(2) with Eq.(1) and Eq.(3), $\Delta y(X, x)$ is rewritten as:

$$\Delta y(X, x) = Y(X - x) - Y(X) + y_{c0}(X - x) - y_{c0}(x)$$
$$+ a \cdot [\psi(X - x) - \psi(X)] + a \cdot [\theta(X - x) - \theta(X)] \quad (5)$$
$$+ x \cdot [\psi(X) + \theta(X)] - y_{cb}(X, x)$$

Considering the effect of tyre width (see Fig. 3) and from Eq. (1) we can derive:

$$\begin{cases} X_t(X, 0, y) = X_c(X, 0, y) \\ X_t(X, x, y) = X_t(X - x, 0, y) \end{cases} \quad (6)$$

The longitudinal deformation of tread related to carcass, $\Delta x(X, x, y)$, is determined by:

$$\begin{cases} \Delta x(X,x,y) \approx -[X_t(X,x,y) - X_c(X,x,y)] \\ X_c(X,x,y) \approx X + a - x - y \cdot \psi(X) \end{cases} \quad (7)$$

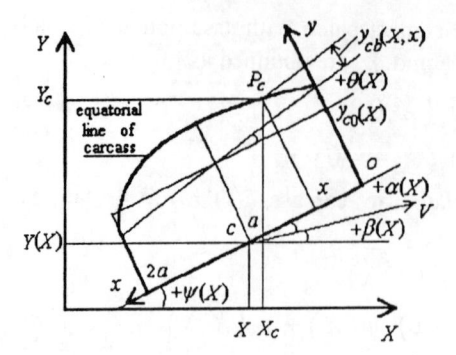

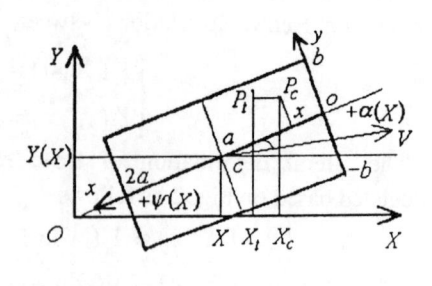

Fig. 2   The coordinate systems of Modelling

Fig. 3.   The coordinate systems when the tyre width is considered

The additional aligning moment, $\Delta M_z(X)$, is expressed as [4]:

$$\begin{cases} \Delta M_z(X) = \dfrac{k_{tx}}{2b} \cdot \int_0^{2a} \int_{-b}^{b} y \cdot \Delta x(X,x,y) dy dx \\ \Sigma M_z(X) = M_z(X) + \Delta M_z(X) \end{cases} \quad (8)$$

According to Eq.(6) and Eq.(7), $\Delta x(X,x,y)$ is rewritten as:

$$\Delta x(X,x,y) = \dot{y} \cdot [\psi(X-x) - \psi(X)] \quad (9)$$

The initial conditions of modelling are assumed as follows [4-5]:

$$\begin{cases} \psi(X)|_{X<0} = 0 & Y(X)|_{X<0} = 0 \\ F_y(X)|_{X<0} = 0 & \Sigma M_z(X)|_{X<0} = 0 \end{cases} \quad (10)$$

The *Laplace* transform of a general function $F(X)$, which is indicated in the form of $F(s)$, is defined through:

$$F(s) = \int_0^\infty F(X) \cdot e^{-sX} dX \quad (11)$$

where s denotes the factor of *Laplace* transformation in spatial domain.

The *FUNCTIONS E* of s are defined as:

$$\begin{cases} E(s) = \dfrac{1}{2a} \cdot \int_0^{2a} (1 - e^{-sx}) dx \\ E_1(s) = \dfrac{1}{2a^2} \cdot \int_0^{2a} x \cdot (1 - e^{-sx}) dx \\ \Delta E_t(s) = 3 \cdot [E_1(s) - E(s)] \end{cases} \quad (12)$$

The $4-\varepsilon$ parameters, i.e. the nondimensional structure constants, are defined as follows:

$$\begin{cases} \varepsilon_0 = K_{ty}/K_{c0} = 2ak_{ty}/K_{c0} \\ \varepsilon_b = K_{ty}/K_{cb} = 2ak_{ty}/K_{cb} \\ \varepsilon_\theta = N_{ta}/N_\theta = \frac{2}{3}a^3 k_{ty}/N_\theta \\ \varepsilon_w = 2b^2 k_{tx}/C_{ta} = (2b^2 k_{tx})/(2a^2 k_{ty}) \end{cases} \quad (13)$$

Then $\Delta M_z(s)$ can be derived from Eq.(8) to Eq.(13) as follows:

$$\Delta M_z(s) = -N_{ta} \cdot \varepsilon_w \cdot E(s) \cdot \psi(s) \quad (14)$$

the Eq.(14) is the same as that presented by Pacejka [2].

The *FUNCTIONS D* of u (where $u = x/a$ denotes the nondimensional length unit of contact patch) are defined as [1]:

$$\begin{cases} D(u) = \frac{1}{2} \cdot \int_0^u \xi(u) du \\ D_1(u) = \frac{1}{2} \cdot \int_0^u u \cdot \xi(u) du \\ \rho_t = 3 \cdot [D_1(2) - 1] \end{cases} \quad (15)$$

in which $\xi(u)$ should satisfy the conditions $D(2) = 1$ and $\xi(0) = 0$. Since $\xi(u)$ is a symmetric function about the center of contact patch [1], $\rho_t = 0$ can be derived from Eq.(15).

By calculating, substituting, *Laplace* transforming and synthesizing with Eq.(4), Eq.(5), and Eq.(8)~Eq.(15), the final expressions of lateral force and aligning moment are obtained as below:

$$\begin{cases} F_y(s) = C_{ta} \cdot \left\{ [1 - E(s)] \cdot \psi(s) - \frac{E(s)}{a} \cdot Y(s) \right\} \\ \qquad - [\varepsilon_0 \cdot E(s) + \varepsilon_b] \cdot F_y(s) + \varepsilon_\theta \cdot [1 - E(s)] \cdot \frac{\Sigma M_z(s)}{a/3} \\ -\frac{\Sigma M_z(s)}{a/3} = C_{ta} \cdot \left\{ [1 - \Delta E_t(s) + \varepsilon_w \cdot E(s)] \cdot \psi(s) - \frac{\Delta E_t(s)}{a} \cdot Y(s) \right\} \\ \qquad - \varepsilon_0 \cdot \Delta E_t(s) \cdot F_y(s) + \varepsilon_\theta \cdot [1 - \Delta E_t(s)] \cdot \frac{\Sigma M_z(s)}{a/3} \end{cases} \quad (16)$$

The corresponding transfer function diagram to Eq.(16) is shown in Fig. 4. Then the transfer functions can be derived directly from Eq.(16) or Fig. 4. The pure yaw motion ($Y(s) \equiv 0$) yields:

$$\begin{cases} G_{f\psi}(s) = \dfrac{F_y(s)}{\psi(s)} = C_{t\alpha} \cdot [1-E(s)] \cdot \dfrac{1-\varepsilon_w \cdot \varepsilon_\theta \cdot E(s)}{B(s)+\varepsilon_\theta \cdot N(s)} \\ G_{\Sigma m\psi}(s) = \dfrac{\Sigma M_z(s)}{\psi(s)} = -N_{t\alpha} \cdot \dfrac{N(s)+\varepsilon_w \cdot E(s) \cdot B(s)}{B(s)+\varepsilon_\theta \cdot N(s)} \end{cases} \quad (17)$$

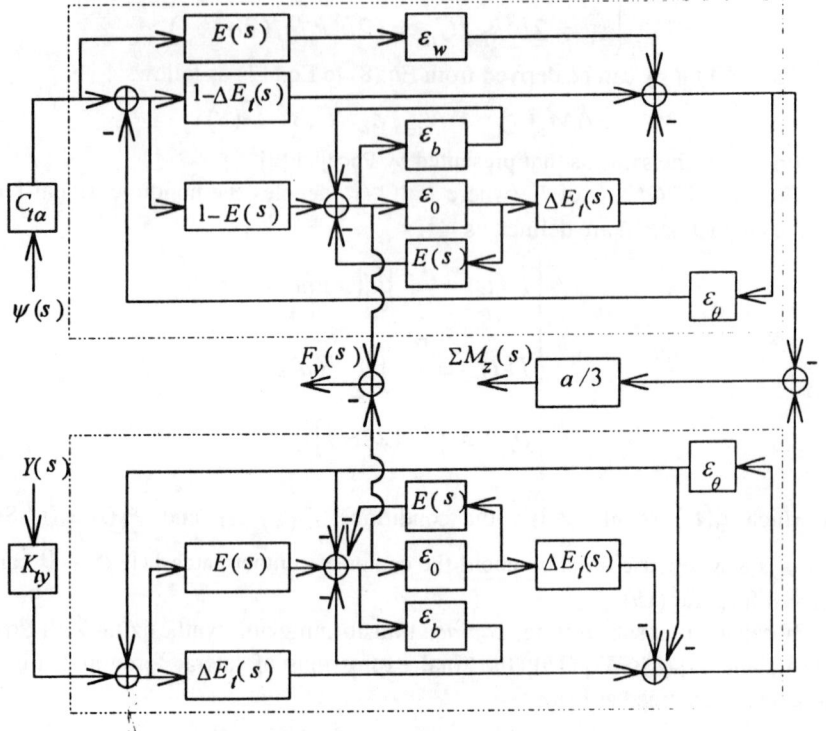

Fig. 4 The transfer function diagram of the derived model

and the pure lateral motion ($\psi(s) \equiv 0$) yields:

$$\begin{cases} G_{fy}(s) = \dfrac{F_y(s)}{Y(s)} = -C_{t\alpha} \cdot \dfrac{(1+\varepsilon_\theta) \cdot E(s) - \varepsilon_\theta \cdot \Delta E_t(s)}{a \cdot [B(s)+\varepsilon_\theta \cdot N(s)]} \\ G_{\Sigma my}(s) = \dfrac{\Sigma M_z(s)}{Y(s)} = N_{t\alpha} \cdot \dfrac{(1+\varepsilon_b) \cdot \Delta E_t(s)}{a \cdot [B(s)+\varepsilon_\theta \cdot N(s)]} \end{cases} \quad (18)$$

where

$$\begin{cases} B(s) = 1+\varepsilon_b + \varepsilon_0 \cdot E(s) \\ N(s) = B(s) \cdot [1-\Delta E_t(s)] - \varepsilon_0 \cdot \Delta E_t(s) \cdot [1-E(s)] \end{cases} \quad (19)$$

$\beta(X)$ and $\varphi(X)$ are defined as path angle and path curvature (turn slip) respectively [2]. From Fig. 2, we have:

$$\begin{cases} \beta(X) = dY(X)/dX \\ \varphi(X) = d\psi(X)/dX \\ \alpha(X) = \psi(X) - \beta(X) \end{cases} \quad (20)$$

Then the relationships represented by transfer functions with different inputs can be derived as [2][4-5]:

$$\begin{cases} G_{f\alpha}(s) = -\dfrac{1}{s} \cdot G_{fy}(s) & G_{f\varphi}(s) = \dfrac{1}{s} \cdot \left[G_{f\psi}(s) - G_{f\alpha}(s) \right] \\ G_{\Sigma m\alpha}(s) = -\dfrac{1}{s} \cdot G_{\Sigma my}(s) & G_{\Sigma m\varphi}(s) = \dfrac{1}{s} \cdot \left[G_{\Sigma m\psi}(s) - G_{\Sigma m\alpha}(s) \right] \end{cases} \quad (21)$$

Substituting with $s = 0$ to Eq. (17) yields the steady state properties as [2-7]:

$$\begin{cases} C_\alpha = G_{f\psi}(0) = G_{f\psi}(s) \big|_{s=0} = C_{t\alpha}/D_f \\ N_\alpha = -G_{\Sigma m\psi}(0) = -G_{\Sigma m\psi}(s) \big|_{s=0} = N_{t\alpha}/D_m \end{cases} \quad (22)$$

where $D_f = (1+\varepsilon_b)(1+\varepsilon_\theta)$ and $D_m = 1+\varepsilon_\theta$. Therefore, the nondimensional transfer functions are interpreted as $\overline{G}_{f\psi}(s) = G_{f\psi}(s)/G_{f\psi}(0)$ and $\overline{G}_{\Sigma m\psi}(s) = G_{\Sigma m\psi}(s)/G_{\Sigma m\psi}(0)$.

3. IDENTIFICATION OF STRUCTURE PARAMETERS

The $4 - \varepsilon$ parameters in Eq.(16) or Fig. 4 must be identified before applying this model. So the nonsteady state tyre cornering tests, including step (quasi-step) yaw angle test and step slip angle (also named ramp lateral displacement) test, are designed and realized on a platform-type tyre test rig (shown in Fig. 5). The tracks of the contact patch are shown in Fig. 6.

In order to obtain frequency response, the test data of step inputs should be processed specially so that they can be performed by *Fourier* transformation. The following procedures named self-delayed superposition [1] should be completed (see Fig.7):

$$\widetilde{P}(X, X_0) = \begin{cases} 0 & X = 0 \\ P(X) & 0 < X < X_0 \\ P_0 & X_0 \le X \le +\infty \end{cases} \quad (23)$$

where $\widetilde{P}(X, X_0)$ is a general continuous function featured as a step one. And $\hat{P}(X, X_0)$ is given as follows:

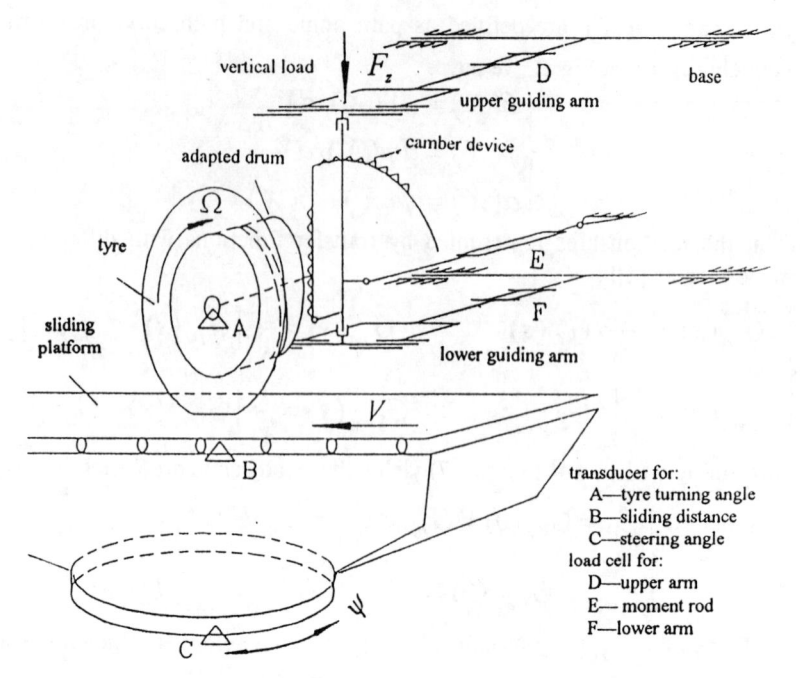

Fig. 5 A platform-type tyre test rig

$$\hat{P}(X, X_0) = \begin{cases} 0 & 0 \le X \le X_0 \\ -P(X - X_0) & X_0 < X < 2X_0 \\ -P_0 & 2X_0 \le X \le +\infty \end{cases} \quad (24)$$

$\Sigma P(X, X_0)$ is satisfied with the condition of *Fourier* transformation for zero values at two ends (absolutely integrable).

$$\Sigma P = \tilde{P} + \hat{P} = \begin{cases} 0 & X = 0 \\ P(X) & 0 < X < X_0 \\ P_0 & X = X_0 \\ P_0 - P(X - X_0) & X_0 < X < 2X_0 \\ 0 & 2X_0 \le X \le +\infty \end{cases} \quad (25)$$

ω_s is defined as path frequency. As a consequence, the processed $F_y(j\omega_s)$ takes the forms as:

$$\begin{aligned} F_y(j\omega_s) &= \int_0^\infty F_y(X) \cdot \exp(-j\omega_s X) dX \\ &= \Delta X \cdot \sum_{N=1}^{2N_0-1} \left[F_y(N) \cdot \exp(-jN\omega_s \Delta X) \right] \end{aligned} \quad (26)$$

Fig. 6 Comparison of two different step inputs

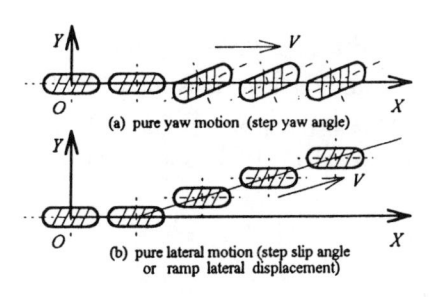

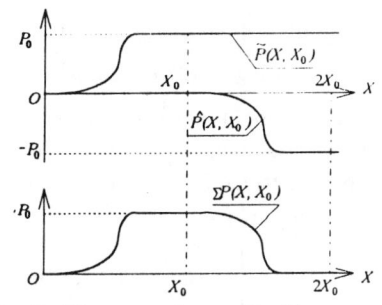

Fig. 7 The procedures of self-delayed superposition

$\psi(j\omega_s)$ and $\Sigma M_z(j\omega_s)$ can be calculated in the same way as Eq.(26). Therefore the frequency responses, $G_{f\psi}(j\omega_s) = F_y(j\omega_s)/\psi(j\omega_s)$ and $G_{\Sigma m\psi}(j\omega_s) = \Sigma M_z(j\omega_s)/\psi(j\omega_s)$ are derived.

Complex Method is utilized to identify the $4-\varepsilon$ parameters according to the theoretical model expressions and test data of frequency properties [5]. Defining a state variable $U = \begin{bmatrix} \varepsilon_0 & \varepsilon_b & \varepsilon_\theta & \varepsilon_w \end{bmatrix}^T$, the four objects are amplitudes and phases of $G_{f\psi}(j\omega_s)$ and $G_{\Sigma m\psi}(j\omega_s)$ respectively. Each specific objective function is expressed as follows:

$$J_i(U) = \left\{ \frac{1}{I} \sum_{j=0}^{I} \left[V_{\omega_{sij}}^{(m)}(U) - V_{\omega_{sij}}^{(t)} \right]^2 \right\}^{1/2} \quad (i = 0 \sim 3) \quad (27)$$

The total objective function is achieved as:

$$F(U) = \sum_{i=0}^{3} [w_i \cdot J_i(U)] \quad \left(\sum_{i=0}^{3} w_i = 1 \right) \quad (28)$$

The constrain conditions for the above function $F(U)$ are given as:

$$\begin{cases} 0 < \varepsilon_0 < 10 & \left| \varepsilon_b - (\overline{D}_x - 1) \right| / (\overline{D}_x - 1) < 15\% \\ 0 < \varepsilon_\theta < 1 & 0 < \varepsilon_w < 1 \end{cases} \quad (29)$$

where $\overline{D}_x = \dfrac{N_\alpha/C_\alpha}{a/3}$, denotes the relative pneumatic trail.

4. COMPARISON OF SIMULATION RESULTS AND TEST DATA

Substituting the identified parameters to Eq.(17), Eq.(18) or Eq.(21), with relation $s = j\omega_s$, the theoretical frequency response can be calculated, shown in Fig. 8.

Here the identified $4-\varepsilon$ parameters are: $\varepsilon_0 = 5.862$, $\varepsilon_b = 0.774$, $\varepsilon_\theta = 0.525$ and $\varepsilon_w = 0.428$ (Tyre: *Jinglun* 6.50R16; Load: 8.97kN; Pressure: 320kPa). Meanwhile, the numerical simulation of NSSTCP is also realized in spatial domain, based on the integration expressions of the analytical model in spatial domains, which are derived from Eq.(4) and Eq.(8).

Defining $\Delta f(X, x) = 2ak_{ty} \cdot \Delta y(X, x)$ yields:

$$\begin{cases} F_y(X) = \frac{1}{2} \cdot \int_0^2 \Delta f(X, au)du \\ \frac{\Sigma M_z(X)}{a/3} = \frac{3}{2} \cdot \int_0^2 (1-u) \cdot \Delta f(X, au)du \\ \quad + b^2 k_{tx} \cdot \left[\int_0^2 \psi(X-au)du - 2\psi(X) \right] \end{cases} \quad (30)$$

When $Y(X) \equiv 0$ (pure yaw motion), it can be assumed as below:

$$P(X) = \psi(X) + \varepsilon_0 \cdot \overline{F}_y(X) - \varepsilon_\theta \cdot \overline{\Sigma M}_z(X) \quad (31)$$

where $\overline{F}_y(X) = F_y(X)/C_{la}$ and $\overline{\Sigma M}_z(X) = -\Sigma M_z(X)/N_{la}$. So the simulation expressions are obtained as:

$$\begin{cases} \overline{F}_y(X) = \frac{1}{2(1+\varepsilon_0+\varepsilon_b)} \cdot \int_0^2 P(X-au)du \\ \overline{\Sigma M}_z(X) = \frac{1}{1+\varepsilon_\theta} \cdot \left\{ (1+\varepsilon_w) \cdot \psi(X) - \frac{1}{2}\varepsilon_w \cdot \int_0^2 \psi(X-au)du \right. \\ \left. \quad + \frac{3}{2} \cdot \int_0^2 u \cdot P(X-au)du - 3(1+\varepsilon_0+\varepsilon_b) \cdot \overline{F}_y(X) \right\} \end{cases} \quad (32)$$

If $\psi(X) \equiv 0$ (pure lateral motion), it can be assumed as:

$$R(X) = \overline{Y}(X) + \varepsilon_0 \cdot \overline{F}_y(X) - \varepsilon_\theta \cdot \overline{\Sigma M}_z(X) \quad (33)$$

where $\overline{Y}(X) = Y(X)/a$. Then the simulation formula is:

$$\begin{cases} \overline{F}_y(X) = \frac{1}{1+\varepsilon_0+\varepsilon_b} \cdot \left[-\overline{Y}(X) + \frac{1}{2} \cdot \int_0^2 R(X-au)du \right] \\ \overline{\Sigma M}_z(X) = \frac{1}{1+\varepsilon_\theta} \cdot \left\{ -3\overline{Y}(X) + \frac{3}{2} \cdot \int_0^2 u \cdot R(X-au)du \right. \\ \left. \quad -3(1+\varepsilon_0+\varepsilon_b) \cdot \overline{F}_y(X) \right\} \end{cases} \quad (34)$$

Eq.(32) and Eq.(34) are the basic simulation expressions, realized by an iterative integration. Exact numerical simulation can be performed through discretizating and processing these Eqs. [5]. Fig.9 to Fig.12 show several typical results in different conditions, and good agreement of simulation results with test data in spatial domain can be seen in Fig. 9. Also the simulation data can provided for the analysis and simulation of vehicle dynamics.

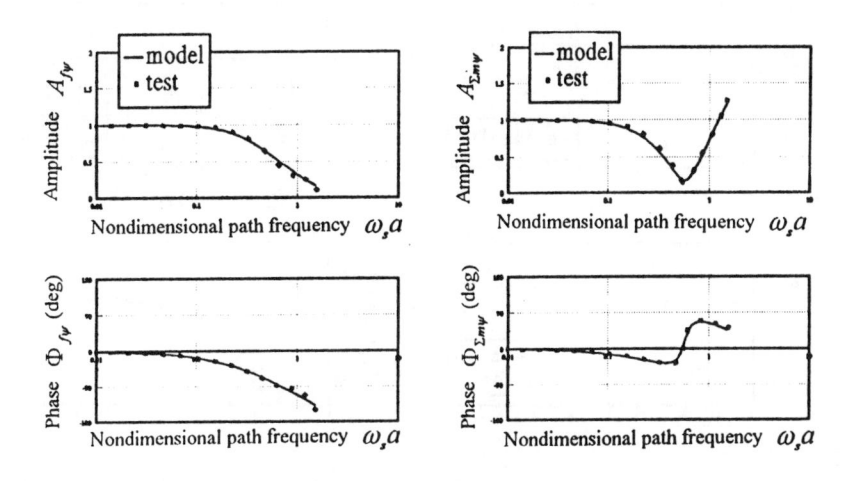

Fig. 8 Comparison of the computational results and test data in frequency domain

5. CONCLUSIONS

(1) A theoretical tyre model of nonsteady state cornering properties with small-amplitude inputs (without slipping) is obtained as Eq.(16) or Fig. 4, in which the flexibility of both tread and carcass is taken into account and the deformation of carcass consists of translating part, bending part and twisting part. Meanwhile the influence of tyre width is considered in the model. The transfer functions of the model are given as Eq.(17), Eq.(18) and Eq.(21).

(2) The basic simulation expressions of NSSTCP are derived as Eq.(32) and Eq.(34) in spatial domain. Based on them, the numerical simulation is implemented and any simulation of NSSTCP can be performed for small lateral inputs.

(3) The tests of NSSTCP are designed and realized with step lateral inputs on a platform-type tyre test rig. The inputs include step yaw angle and step slip angle (ramp lateral displacement). Then the $4 - \varepsilon$ parameters of the model, which are nondimensional structure quantities, are identified according to the model expressions and test data in frequency domain.

(4) The model are validated in frequency domain and spatial domain respectively. The theoretical model and simulation data can be applied to the analysis and simulation of vehicle dynamics and tyre dynamics.

Acknowledgments

The authors would like to thank the National Natural Science Foundation of China for the financial support. The help from the staff of Tyre Laboratory of Changchun Automobile Institute, the First Automobile Works of China is also highly appreciated.

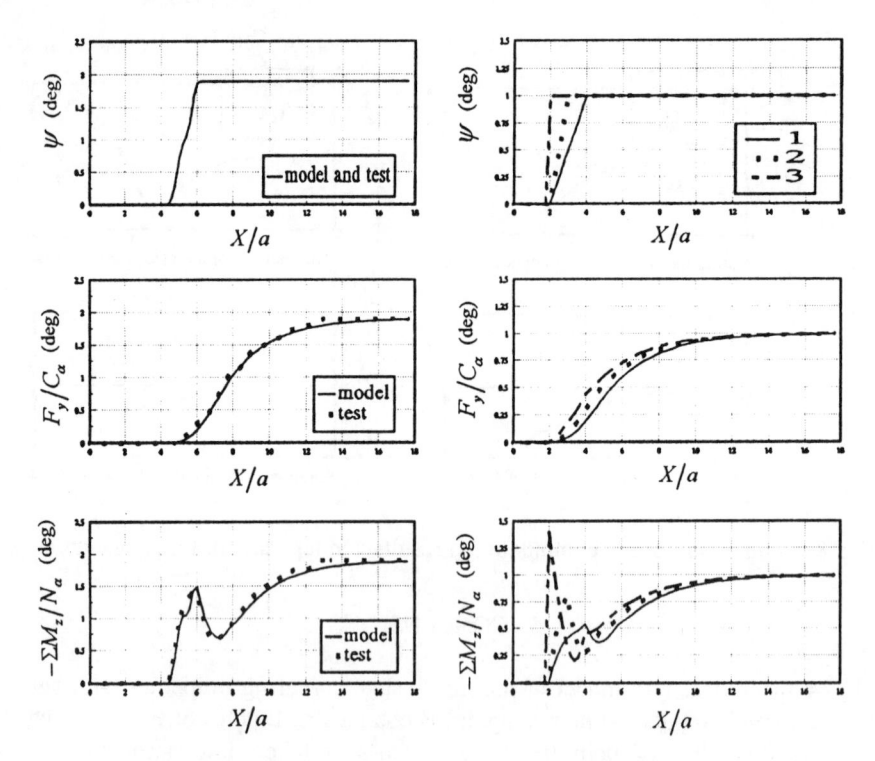

Fig. 9 Comparison between test data and model results with the same step lateral inputs

Fig. 10 The simulation results with step yaw angle as inputs under different slopes

REFERENCES

[1] Guo, Konghui, *"Handling Dynamics of Automobiles"* (in Chinese), Jilin Press of Science & Technology, 1991.

[2] Pacejka, H. B., "Analysis of Tire Properties", Chap. 9 of *"Mechanics of Pneumatic Tires"* (ed. S. K. Clark), U. S. DOT HS 805 952 (new edition), NHTSA, 1981, pp.721-870.

[3] Guo, Konghui and Liu, Qing, "Modelling of Nonsteady State Cornering Properties of Tyres" (in Chinese), *Automobile Technology*, No.2, 1996.

[4] Guo, Konghui and Liu, Qing, "Nonsteady State Cornering Properties of Tyres with Tyre Width Under Consideration" (in Chinese), *Automobile Technology*, No.5, 1996.

[5] Liu, Qing, "Analysis, Modelling and Simulation of Nonsteady-State Tire Cornering Properties" (in Chinese), Ph. D. Dissertation, Jilin University of Technology, 1996.

[6] Pacejka, H. B., "In-Plane and Out-of-Plane Dynamics of Pneumatic Tyres", *Vehicle System Dynamics*, Vol.10, Nos.4/5, 1981.

[7] Segel, L., "Force and Moment Response of Pneumatic Tires to Lateral Motion Inputs", Journal of Engineering for Industry (*Transactions of the ASME*), Vol.88B, Series B, No.1, Feb. 1966.

[8] Barson, C. W. and Osborne, D. J., "Dynamic Properties of Tyres", IMechE Conference on Automobile Wheels and Tyres, C277/83, 1983.

[9] Schuring, D. J., "Dynamic Response of Tires", *Tire Science and Technology*, TSTCA. Vol.4, No.2, May 1976.

[10] Weber, R. and Persch, H. -G., "Frequency Response of Tires—Slip Angle and Lateral Force", *SAE Trans.*, Vol.85, 760030, 1976.

[11] Furuichi, T. and Sakai, H., "Dynamic Cornering Properties of Tires", SAE Paper 780169, 1978.

[12] Bergman, W. and Clemett, H. R., "Tire Cornering Properties", *Tire Science and Technology*, TSTCA, Vol.3, No.3, Aug. 1975.

[13] Clark, S. K., Dodge, R. N. and Nybakken, G. H., "Dynamic Properties of Aircraft Tires", *Journal of Aircraft*, Vol.11, No.3, Mar. 1974.

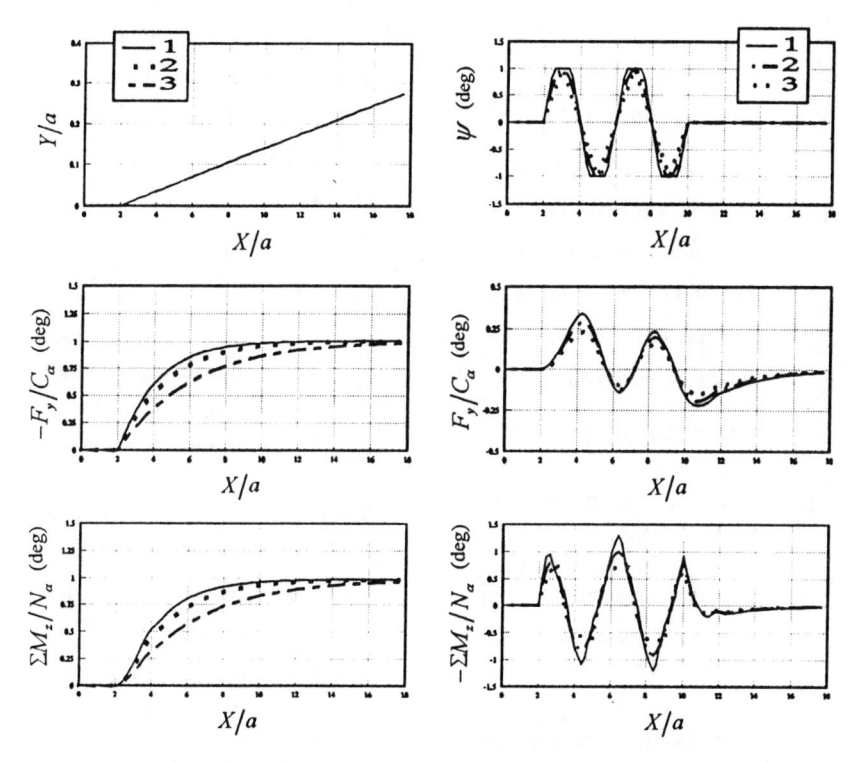

1: $\varepsilon_b = 1.18$; 2: $\varepsilon_b = 0.65$; 3: $\varepsilon_b = 0$

Fig. 11 The simulation with lateral ramp displacement inputs

1-trapzoidal; 2-sinusoidal; 3-trianglular

Fig.12 The simulation with yaw angle of different waveform as inputs

NOTATIONS

k	tread stiffness per unit length	X_t, Y_t	coordinates of point P_t in XOY
K	lateral stiffness		
C	cornering stiffness	X_c, Y_c	coordinates of point P_c in XOY (see Fig.2)
N	aligning or twisting stiffness		
a	half length of contact patch	$Y(X)$	lateral displacement
b	half width of contact patch	$\psi(X)$	yaw angle (see Fig.2)
k_{ty}	tread lateral (y) stiffness per unit length	$\alpha(X)$	slip angle (see Fig.2)
k_{tx}	tread longitudinal (x) stiffness per unit length	$\beta(X)$	path angle
		$\varphi(X)$	path curvature (turn slip)
K_{ty}	lateral stiffness of tread	$\theta(X)$	twisting angle
K_{c0}	translating stiffness of carcass	$\overline{Y}(X)$	see Eq.(33)
K_{cb}	bending stiffness of carcass	$F_y(X)$	lateral force
N_θ	twisting stiffness of carcass	$M_z(X)$	aligning moment
$C_{t\alpha}$	cornering stiffness of tread	$\Delta M_z(X)$	additional aligning moment (see Eq.(8))
$N_{t\alpha}$	aligning stiffness of tread		
C_α	cornering stiffness of tyre	$\Sigma M_z(X)$	total aligning moment
N_α	aligning stiffness of tyre	$\overline{F}_y(X)$	see Eq.(31)
ε_0	translating characteristic ratio	$\overline{\Sigma M_z}(X)$	see Eq.(31)
ε_b	bending characteristic ratio	$\Delta y(X,x)$	relative lateral deformation of tread to carcass
ε_θ	twisting characteristic ratio		
ε_w	width characteristic ratio	$y_{c0}(X)$	translating deformation of carcass
$\overline{D}_x$	nondimensional pneumatic trail (see Eq.(29))	$y_{cb}(X,x)$	bending deformation of carcass
XOY	coordinate system of contact center c with respect to system fixed in space (see Fig. 2)	$y_\theta(X,x)$	twisting deformation of carcass
xoy	coordinate system with respect to moving system (see Fig.2)	$y_c(X,x)$	total lateral deformation of carcass
P_t, P_c	point in tread and carcass respectively (see Fig.2)	$\Delta x(X,x,y)$	relative longitudinal deformation of tread to carcass
c	central point of contact patch		
V	speed of travel (see Fig.2)	u	nondimensional longitudinal coordinate (see Eq.(15))
X	rolling distance of tyre in space		
x	coordinate of point P_c in xoy	$\xi(u)$	shape function of carcass bending deformation (see Eq.(15))

ρ_t	see Eq.(15)		with respect to $Y(s)$
s	factor of *Laplace* transformation in spatial domain	$G_{f\alpha}(s)$	transfer function of $F_y(s)$ with respect to $\alpha(s)$
j	imaginary unit of a complex number	$G_{\Sigma m\alpha}(s)$	transfer function of $\Sigma M_z(s)$ with respect to $\alpha(s)$
ω_s	path, reduced or spatial frequency	$G_{f\varphi}(s)$	transfer function of $F_y(s)$ with respect to $\varphi(s)$
$E(s)$	see Eq.(12)	$G_{\Sigma m\varphi}(s)$	transfer function of $\Sigma M_z(s)$ with respect to $\varphi(s)$
$\Delta E_t(s)$	see Eq.(12)		
$G_{f\psi}(s)$	transfer function of $F_y(s)$ with respect to $\psi(s)$	ΔX	sampling interval
$G_{\Sigma m\psi}(s)$	transfer function of $\Sigma M_z(s)$ with respect to $\psi(s)$	X_0	sampling length when all sampling quantities reach to steady values
$\overline{G}_{f\psi}(s)$	nondimensional $G_{f\psi}(s)$	N_0	number of sampling points corresponding to X_0
$\overline{G}_{\Sigma m\psi}(s)$	nondimensional $G_{\Sigma m\psi}(s)$	U	state variable (see Eq.(27))
A	amplitudes of $\overline{G}_{f\psi}(j\omega_s)$ or $\overline{G}_{\Sigma m\psi}(j\omega_s)$	$V^{(m)}(U)$	model values (amplitudes or phases)
Φ	phases of $\overline{G}_{f\psi}(j\omega_s)$ or $\overline{G}_{\Sigma m\psi}(j\omega_s)$	$V^{(t)}$	test values (amplitudes or phases)
$G_{fy}(s)$	transfer function of $F_y(s)$ with respect to $Y(s)$	$J_i(U)$	objective function
		w_i	weight coefficient
$G_{\Sigma my}(s)$	transfer function of $\Sigma M_z(s)$	$F(U)$	total objective function

Dynamical Tyre Forces Response to Road Unevennesses

M. GIPSER, R. HOFER, and P. LUGNER

ABSTRACT

The contribution deals with the quantitative prediction of tyre forces and moments due to short-waved road unevennesses, as far as they influence vehicle dynamics. Safety loss frequently is supposed to be closely correlated to the RMS value of dynamical wheel load changes. But this doesn't take into account the frequency dependency of the amount of loss in lateral and longitudinal forces caused by wheel load fluctuations. In order to investigate this effects mentioned, extensive simulations with the semi physical tyre model BRIT (**B**rush and **Ri**ng **T**ire Model) have been performed and a variety of frequency responses is presented and discussed. Thereby the input consists of a sine shaped road surface with varying wave lengths. Before, some modifications of BRIT, done especially for this investigations, are described in detail and the effects of these improvements on the simulation results are presented.

1. INTRODUCTION

Since the beginning of investigations of automobile dynamics it is well known, that the tyre behaviour has an essential effect on the vehicle motion. Therefore there are a great number of approximations for different areas of application available and also a vast number of measurements.

Most of the investigations and approximations of the tyre behaviour are centered on the lateral force transfer and the transient aspects by changing side-slip angles. In reality the more essential feature seems to be the effect of wheel load fluctuations – e.g. due to random road unevenness – on the lateral force transfer. Consequently the following investigations have the intention to develop tyre characteristics approximations comprising lateral and longitudinal slip behaviour in the presence of normal load fluctuations up to about 30 Hz. This should provide the possibility to evaluate the active safety on a rough road e.g. during cornering and braking in a more precise way than the common wheel load RMS-value.

2. FEATURES OF THE TYRE MODEL BRIT

According to [4] tyre models can be divided into four main categories:
- very simple models with e.g. (nonlinear) spring-damper configurations for the vertical dynamics
- mathematical approximations of measured tyre characteristics – e.g. [8], [2]
- mechanical models based on the inclusion of the main physical properties and effects
- complex, 3D-FE-models

BRIT as a so called brush model (see also [9]) is one of the third category. The principal features and properties of this model (see [5] or [1]) are:
- rigid ring-shell (part of a toroid) for the tyre belt connected with spring-damper-elements to the wheel rim, **Fig. 1** left side
- geometrical algorithm for the calculation of the tyre contact area by road surface profile and tyre belt (ring-shell) position
- approximation of quasi-static deformation of tyre patch due to contact forces and moments, **Fig. 1** right side
- assumption of normal stress distribution in the tyre patch taking into account the rolling resistance
- detailed modelling of the tyre patch by tread stripes consisting of discrete, massless tread blocks with tangential stiffness and damping (dynamic tread model)
- calculation of lateral and longitudinal forces by tread block displacements

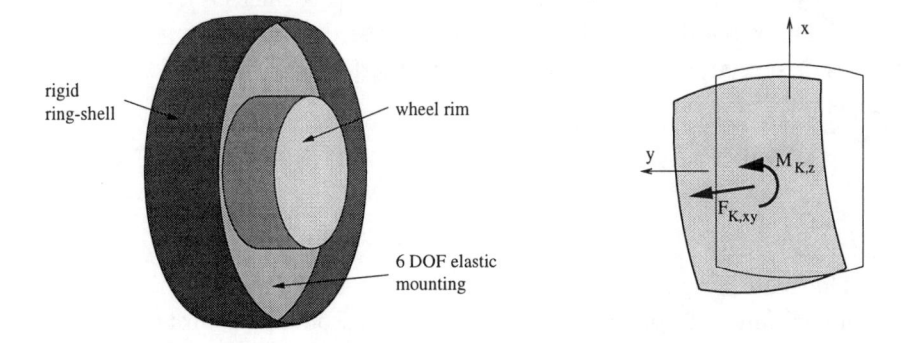

Figure 1: Tyre model of BRIT: rigid ring-shell with 6 DOF mounting (left) and additional deformation of the tyre patch (right), cf. [5]

The rigid ring-shell has a mass and is fixed to the rim by massless elements with elastic and damping properties (given by stiffness and damping matrices).

The kinematics of the shell is described by 6 generalized coordinates and the corresponding velocities whereas the dynamics is determined using Newton-Euler equations.

For the determination of the tyre contact area the intersection of the undeformed tyre belt (ring-shell) with the road surface is calculated. In the basic version of BRIT the road surface is presented by the local tangential plane that can be derived by the description $z = z(x, y)$ of the road surface. By superimposing quasi-static deformations due to the overall force transfer in the tangential plane the form and location of the patch can be found. The subdivision of the patch provides the foundations of the massless, deformable tread blocks.

The normal stress distribution over the patch is approximated based on measured data and calculations by a FE-tyre model. In the basic version of BRIT for the shape contours of these approximations, for longitudinal and lateral direction, monomial functions are used. The distribution created in this way is further weighted by a linear function to get a forward shift which represents the rolling resistance, at the same time fulfilling the condition, that the sum of the normal forces on the blocks is equal to the overall normal force of the tyre.

After the so defined local vertical stresses p_{lok} between blocks and road surface, by the tangential stiffness c_{St}, the tangential damping k_{St} and the local friction coefficient μ_{lok} the lateral and longitudinal displacement of each block can be calculated. The mass of the block is neglected but the distinction between sticking and slipping had to be taken into account. For the proper distinction a special switching algorithm was developed that proved to be numerically stable for all evaluations. The friction coefficient μ_{lok} is given as a function of relative velocity and normal stress determined by characteristic values at special conditions. By the tangential forces of the blocks the overall tangential forces of tyre can be summed up.

3. MODIFICATIONS OF BRIT FOR EXTENDED RANGE OF APPLICATION

3.1. Normal Stress Distribution

The basic description of the normal stress distribution in the patch by monomial functions did not allow to take into account the effects of the value of the tyre normal force and the local road surface curvature with respect to the character of these distribution functions. To overcome this disadvantages the distribution functions had to be modified.

For the derivation of the modified shape of these functions precalculations with a FE-tyre model (DNS-Tire, see [3]) and comparisons with measurements and other published results were used.

The essential qualitative characteristics of the stress distribution for the two mentioned influences are (in the following the term curvature κ always represents the curvature of the surface contour in the longitudinal x-direction):

- stress distribution in lateral direction: With increasing normal force and decreasing curvature κ ($\kappa < 0$ corresponds to a test rig with the test tyre on the outside of the driven drum) the stresses in the tread shoulder increase relatively to those in the longitudinal center line of the patch.
- stress distribution in longitudinal direction: For a relatively small normal force this distribution shows a maximum in the center of the patch. With increasing normal force and decreasing curvature this one maximum is replaced by two increasing maxima that shift more and more to the edges of the patch. In the middle of the patch a minimum is created.

By optimization, using the available data, shape functions of the type

$$f(x) = \sum_i a_i |x|^{b_i} \qquad (1)$$

for both, lateral and longitudinal direction, could be defined.

Notation of the specific quantities:

x	position in longitudinal direction of the patch, $-\frac{L(y)}{2} \leq x \leq \frac{L(y)}{2}$
y	position in lateral direction of the patch, $-\frac{b}{2} \leq y \leq \frac{b}{2}$
b	patch-width
$L(y)$	length of tread stripe
L_{max}	reference length of patch (= constant, determined in BRIT)
$p(x,y)$	normal pressure in the patch
$F_{K,z}$	normal contact force
F_N	nominal tyre load (characteristic quantity of type of tyre)
κ	road surface curvature in longitudinal direction
r_G	maximum geometric tyre radius.

Moreover standardized quantities are defined:

$$\bar{x} = \frac{x}{L(y)/2} \;;\quad \bar{y} = \frac{y}{b/2} \;;\quad \bar{F} = \frac{F_{K,z}}{F_N} \;;\quad \bar{\kappa} = \kappa \cdot r_G \qquad (2a)\text{-}(2d)$$

In a first step it was found that

$$p(0,y) = \max\left[0 \,,\, p(0,0) \cdot (1 + v \cdot |\bar{y}|^4) \cdot \frac{L(y)}{L_{max}} \right] \qquad (3)$$

is a fine approximation for the lateral pressure distribution. By an optimization procedure, the characteristic quantity v could be determined as a function of normal force and curvature:

$$v = -1 + (0.65 \cdot \bar{F}^2) \cdot (1 - 0.65 \cdot \bar{\kappa}) . \tag{4}$$

For the longitudinal direction ($y = y_k = const$) the corresponding shape function

$$p(x, y_k) = \max\left[0 , \; p(0, y_k) \cdot (1 - |\bar{x}|^{a(\bar{y})}) \cdot (1 + k(\bar{y}) \cdot |\bar{x}|^{a(\bar{y})})\right] \tag{5}$$

makes it possible to take into account the inversion from maximum to local minimum in the center of the patch. Once again an optimization leads to the characteristic factors:

$$k(\bar{y}) = k_0 - (k_0 - 1) \cdot \bar{y}^2 , \tag{6}$$

$$k_0 = k(\bar{y} = 0) = 1 + [0.9 - 0.645 \cdot (\bar{\kappa} + 0.381)^2] \cdot \bar{F}^2 , \tag{7}$$

$$a(\bar{y}) = a_0 - (a_0 - a_{00}) \cdot \bar{y}^2 , \tag{8}$$

$$a_0 = a(\bar{y} = 0) = 1 + 0.7 \cdot e^{(1.31 \cdot \bar{\kappa})} \cdot \bar{F} , \tag{9}$$

$$a_{00} = a_0(\bar{\kappa} = 0) = 1 + 0.7 \cdot \bar{F} . \tag{10}$$

Summing up, the complete description for the normal stress distribution in the patch can be described by

$$p(x, y) = \max\left[0 , \; p(0, 0) \cdot \left(1 + v(\bar{F}, \bar{\kappa}) \cdot |\bar{y}|^4\right) \cdot \frac{L(y)}{L_{max}} \cdot \right.$$
$$\left. \cdot \left(1 - |\bar{x}|^{a(\bar{y}, \bar{F}, \bar{\kappa})}\right) \cdot \left(1 + k(\bar{y}, \bar{F}, \bar{\kappa}) \cdot |\bar{x}|^{a(\bar{y}, \bar{F}, \bar{\kappa})}\right) \right] . \tag{11}$$

For an approximation like (11) it is essential to give the limits of the range of application.

- standardized curvature $-0.75 \leq \bar{\kappa} \leq 0.75$: that means that the local radius of the road surface contour is at least 33% larger than the tyre radius
- standardized normal force ($0 \leq \bar{F} \leq \approx 2.0$).

Since in (11) there are only standardized quantities involved, this approximation is not fixed for a special type of automobile or truck tyre, presupposing that geometric similar dimensions of a tyre imply a geometric similar pressure distribution. A qualitative comparison of different known pressure distributions validate this assumption.

As example **Fig. 2** shows the pressure distribution on a flat surface with low and very high normal force. The influence of the surface curvature can be seen by the comparison of both diagrams in **Fig. 3**. Notice especially the different lengths of the patches (range of the x-coordinates)!

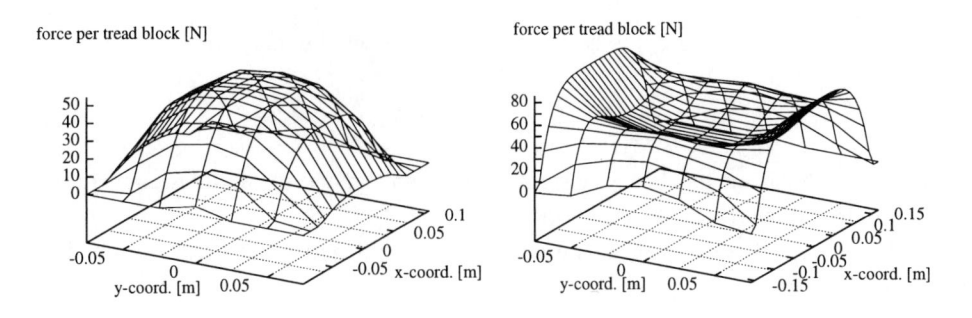

Figure 2: Calculated normal stress distribution in the patch described by contact force per tread block: flat surface ($\bar{\kappa} = 0$), $F_{K,z} = 3$ kN (left) and $F_{K,z} = 10$ kN (right)

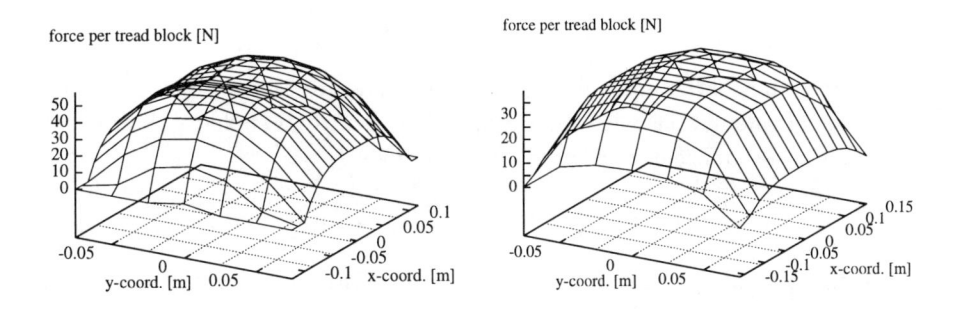

Figure 3: Like Fig. 2 but $F_{K,z} = 5$ kN with flat surface $\bar{\kappa} = 0$ (left) and concave drum $\bar{\kappa} = 0.75$ (right)

3.2. Further Modifications to BRIT

In order to apply the new normal stress distribution laws, the local curvature of the roadway must be taken into account. As a consequence, its influence onto the radial tyre spring characteristics has to be modelled, and, furthermore, the calculation of the contact patch length (foot print length) has to be generalized.

3.2.1. Roadway Height Profile Curvature

The second derivative of a function $z = f(x)$ with $z_i = f(x_i)$ and $x_{i+1} = x_i + \frac{\Delta x}{2}$ can be approximated at x_0 by the second difference operator $z'' = \frac{4}{\Delta x^2} \cdot (z_{-1} - 2z_0 + z_1)$.

This formula has been used here, where $z = f(x)$ is the height profile of the road in longitudinal direction. In order to observe the filtering properties of the footprint, a modification has been applied, using the function values z_{-2} and z_2 as well. Therefore the arithmetic mean of the two values $\frac{4}{\Delta x^2} \cdot (z_{-1} - 2z_0 + z_1)$

and $\frac{1}{\Delta x^2} \cdot (z_{-2} - 2z_0 + z_2)$ is calculated, resulting in the formula

$$z'' = \frac{1}{2 \cdot \Delta x^2} \cdot (z_{-2} + 4z_{-1} - 10z_0 + 4z_1 + z_2) \, . \tag{12}$$

The step width Δx has been chosen to 0.1 m, so the actual span $2\,\Delta x$ sums up to about the length of the footprint.

Due to small slope values the curvature is sufficiently accurately approximated by $\kappa = z''$.

3.2.2. Radial Spring Characteristics

In BRIT, the quasi-static dependency between the contact force $F_{K,z}$ and the belt deflection e_G, somewhat simplified, is expressed by the relation

$$F_{K,z} = C_1 \cdot e_G + C_2 \cdot e_G{}^2 \, . \tag{13}$$

Clearly, the contact force depends on the curvature of the roadway. Again, by means of calculations with the FE model DNS-Tire, is could be shown that this relation is sufficiently accurate described by

$$F_{K,z} = C_1 \cdot e_G + C_2(\bar\kappa) \cdot e_G{}^2 \quad \text{with} \quad C_2(\bar\kappa) = C_2(\bar\kappa = 0) \cdot e^{1.64 \cdot \bar\kappa} \, . \tag{14}$$

Using this formula, the resulting spring characteristics for a tyre of dimension 195/65R15 are shown in **Fig. 4**.

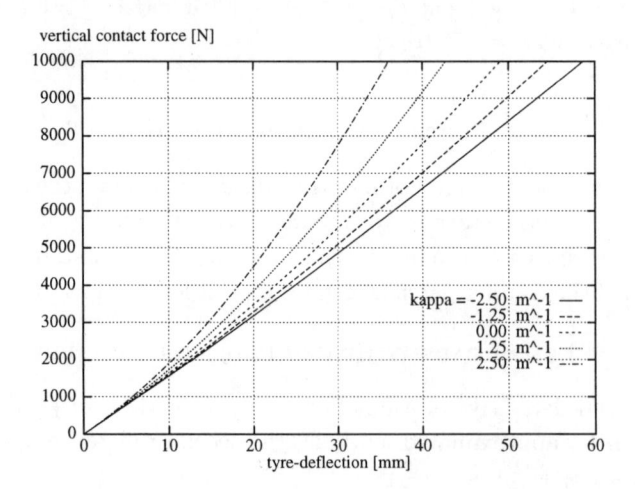

Figure 4: Radial spring characteristics of a 195/65R15 tyre for different curvatures $\kappa = -2.5, -1.25, 0, 1.25, 2.5$ m^{-1}

3.2.3. Correction of the Contact Patch Length

In order to be able to take into account the change in footprint length through roadway curvature, a new factor has been introduced in BRIT. This factor can be derived by simple geometrical considerations, see **Fig. 5**:

$$A_{LL} = \frac{L(y)}{L(y)_{even}} = \sqrt{\frac{r_G^2 - \left(\frac{\kappa \cdot \left(r_G^2 - r_G \cdot s + \frac{s^2}{2}\right) - r_G + s}{\kappa \cdot (r_G - s) - 1}\right)^2}{r_G^2 - (r_G - s)^2}} . \quad (15)$$

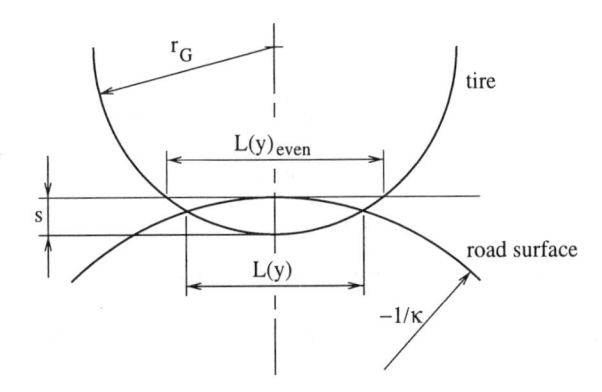

Figure 5: Geometrical relations between footprint lengths with and without curvature of the roadway

Just like in the original BRIT version, a first guess of the footprint length is calculated through the intersection of the belt ring with the local roadway tangential plane. In a second step, the resulting lengths $L(y)$ are multiplied with the curvature factor A_{LL}.

Comparisons of the resulting footprint lengths with reference values showed a fairly good correspondence.

3.2.4. Influence of Camber Angle

Early versions of BRIT showed an unsatisfactory influence of large camber angles, visible especially in steady state lateral force characteristics.

In order to improve the model the lateral deflection of the tread blocks through camber angle as a function of the footprint longitudinal coordinate has been described more precisely by introducing a new parabolic term, see **Fig. 6**.

This term is such that the deflection e in the middle of the patch is given by

$$e = k \cdot h = k \cdot s \cdot \sin \gamma \quad (16)$$

where γ denotes the camber angle, s the belt deflection, and k a constant correction factor. Lateral force characteristics with a satisfactory influence of large camber angles result for values of $k \approx 0.4$.

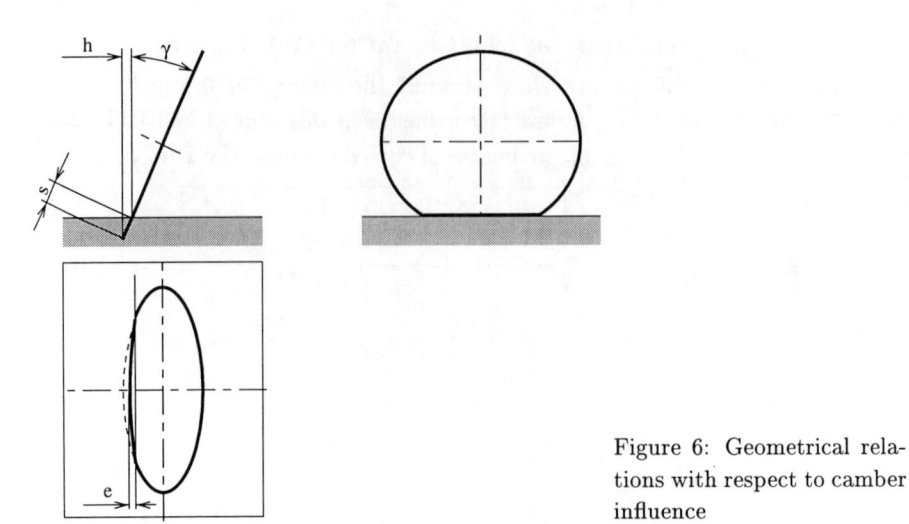

Figure 6: Geometrical relations with respect to camber influence

3.3. Effects of the Implemented Modifications

There were two main reasons to first try to improve BRIT before analyzing travel over road unevennesses:

On the one hand, the fact that the tyre contact patch was modelled to be part of the linearized tangential plane turned out to be no longer a valid assumption for short waved unevennesses. Thus, the validity range of BRIT was somewhat restricted for the investigations in mind. On the other hand, the stationary tyre characteristics should be predicted by BRIT as precise as possible, being the basic requirement for investigations of the dynamic properties to be performed.

By means of the modifications, BRIT now can be used with outstanding precision even on road unevennesses with wave lengths up to the length of the contact patch. Concerning the stationary tyre characteristics for lateral force, longitudinal force, and aligning torque, a very good agreement with measurement results is observed, whereby even the influence of the kind of test rig (inner drum, outer drum, or flat belt) is correctly reproduced.

Fig. 7, e.g. shows lateral force characteristics for different wheel loads on even roadway (left hand), and aligning torque characteristics for different values of the test drum diameter (right hand).

For both diagrams, the side slip angle variation speed deliberately had been chosen to be relatively large. By that, both characteristics show clear hysteresises, just as in 'real life' measurements. One simultaneously can recognize the stationary properties together with some aspects of dynamic behaviour.

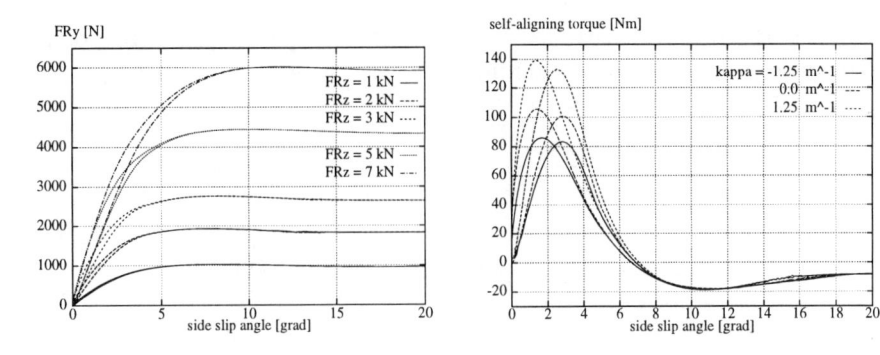

Figure 7: Lateral force vs. side slip angle for different wheel loads $F_{R,z}$ on even roadway (left hand); aligning torque vs. side slip angle for different test drum diameters, positive values of curvature κ (kappa) for inner drums (right hand)

4. NON STEADY TYRE BEHAVIOUR - RESULTS FOR SINUSOIDAL UNEVENNESSES

4.1. LATERAL DYNAMICS

These simulations are to show fundamental effects, introduced by the tyre/-road interaction, that arise in a similar manner at (quasi) steady state cornering of passenger cars on an uneven road.

Let the roadway profile height z (which, for a vertically fixed hub, coincides with the tyre deflection), as a function of the wheel travel s, being given as

$$z = \bar{z} + \hat{z} \cdot \sin(\frac{2\pi}{\lambda} \cdot s) \,. \qquad (17)$$

In the following examples, $\bar{z}$ had been chosen such that the stationary wheel load $F_{R,z\,stat}$ was about 4 kN, whereas $\hat{z} = \frac{1}{2} \cdot \bar{z}$.

Among others, the mean value $F_{R,y\,m}$ of the transferred lateral force had been calculated as a function of velocity v and (constant) side slip angle α. In the following diagrams, the mean lateral forces had been divided by the respective lateral force $F_{R,y\,stat}$ with constant tyre deflection $\bar{z}$ and plotted vs. frequency in path domain $1/\lambda = f/v$ (f ... frequency in time domain), cf. [7].

Fig. 8 clearly shows that the loss in lateral force consists of two portions: a quasi steady state portion introduced by the degressive lateral force/wheel load characteristics (this portion best can be seen for $(1/\lambda) \to 0$), and a second frequency dependent portion produced by the complicated nonlinear dynamics of lateral force increase/decrease.

It turns out that, for all speeds, this second portion is negligible up to a certain 'cutoff' frequency. Beyond this frequency the 'dynamic' portion of loss in lateral force increases strongly. Of course, due to larger time domain frequencies, the dynamic portion increases with speed.

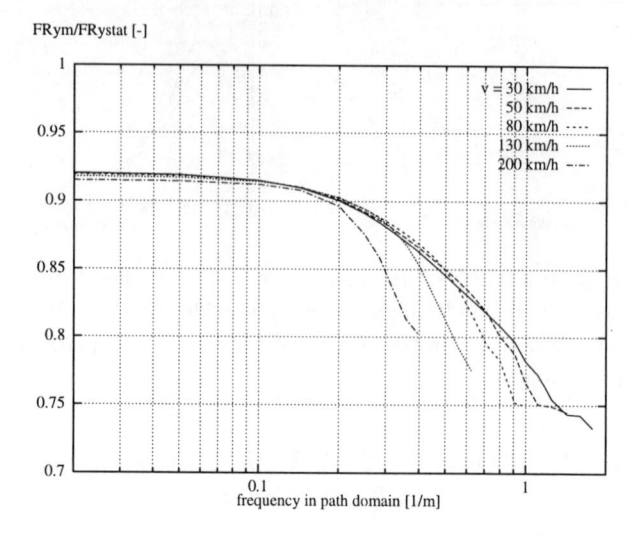

Figure 8: Relative mean lateral force at different speeds for a sinusoidal height profile ($\alpha = 4\,^\circ$, $F_{R,z\,stat} \approx 4$ kN, $\hat{z} = \frac{1}{2} \cdot \bar{z}$)

Thus, to simplify further investigations, the functions in **Fig. 8** could be approximated by two intersecting straight lines, respectively.

Fig. 9 again shows the loss in lateral force, with an additional variation of the side slip angle α.

One can see that for larger slip angles the quasi steady state portion of the loss in lateral force is less (due to the fact that the lateral force vs. wheel load characteristics is less degressive in this case). On the other hand, the dynamic portion of loss in lateral force increases considerably.

The simulation results for small slip angles ($\alpha \leq 2\,^\circ$), at very short waved unevennesses, show a slight increase in relative mean lateral force. This effect, however, is not of great importance, because for $\alpha \to 0$ the absolute values of the transferred lateral force $F_{R,y}$, together with their amounts of decrease, are very small anyway.

4.2. Longitudinal Dynamics

In analogy to the lateral dynamics, the mean value of the transferred longitudinal force $F_{R,x\,m}$, at constant longitudinal slip s and different velocities v, had been calculated. For the representation in the according diagrams, the mean longitudinal force was divided by the respective longitudinal force $F_{R,x\,stat}$ with constant tyre deflection $\bar{z}$.

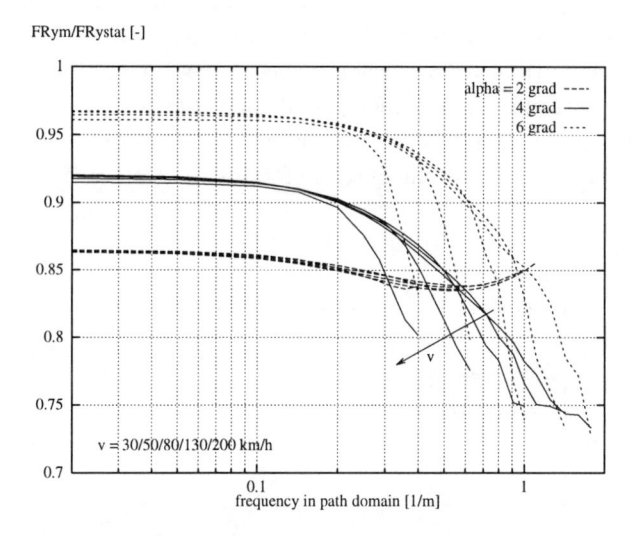

Figure 9: Relative mean lateral force at different speeds and several side slip angles for a sinusoidal height profile ($\alpha = 2/4/6\ °$, $F_{R,z\ stat} \approx 4$ kN, $\hat{z} = \frac{1}{2} \cdot \bar{z}$)

Thereby effects that also arise at braking when driving straight ahead can be observed. Because the slip s is defined by $s = \frac{r\omega - v}{\max[|r\omega|, |v|]}$, braking leads to negative slip values. The according simulation result is shown in **Fig. 10**.

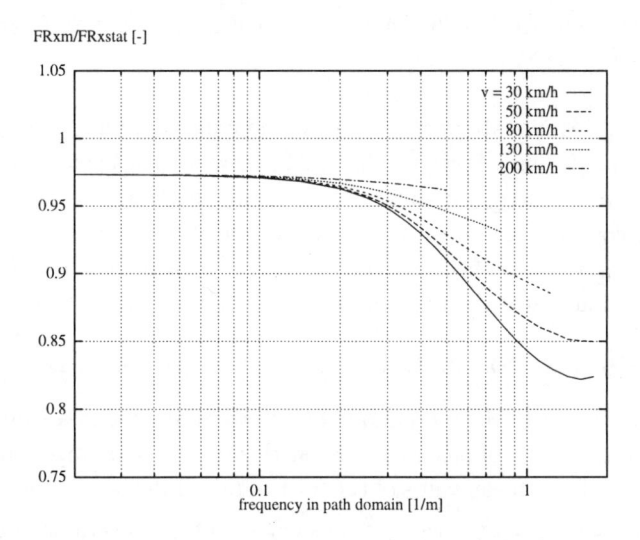

Figure 10: Relative mean longitudinal force at different speeds for a sinusoidal height profile ($s = -4\ \%$, $F_{R,z\ stat} \approx 4$ kN, $\hat{z} = \frac{1}{2} \cdot \bar{z}$)

In principle the effects occurring are similar to the ones at lateral dynamics. That leads to a quasi steady state and a frequency dependent portion of loss in longitudinal force. But in this case, the 'dynamic' loss at high frequencies becomes smaller with increasing velocity, because as a result of a higher sliding portion in the patch (due to a higher velocity and constant slip) a faster increase of transmitted longitudinal force is observed.

Fig. 11 shows the according diagram for different values of the longitudinal slip s.

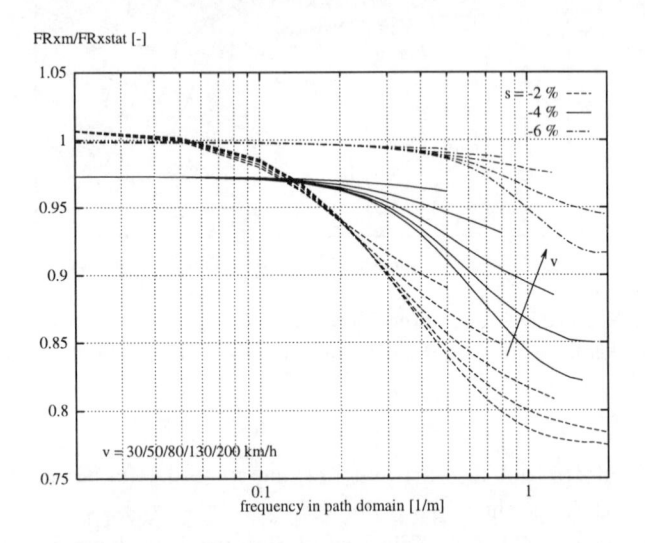

Figure 11: Relative mean longitudinal force at different speeds and several longitudinal slip values for a sinusoidal height profile ($s = -2/-4/-6$ %, $F_{R,z\,stat} \approx 4$ kN, $\hat{z} = \frac{1}{2} \cdot \bar{z}$)

The quasi steady state portion of loss in longitudinal force changes strongly for different slip values. At a slip of $s = -2$ % there is even a small increase of the mean longitudinal force, due to temporary more advantageous normal pressure and sliding conditions. Increasing the slip causes effects similar to those when increasing the velocity, because this results also in a higher sliding portion which again dominates the influence on the longitudinal dynamics behaviour.

4.3. Combination of Lateral and Longitudinal Dynamics

The following is to show fundamental effects that arise in a similar manner at driving manoeuvres of passenger cars with simultaneous cornering and braking. To this end, the mean value of the horizontal component of the transmitted force $F_{R,h\,m} = \sqrt{F_{R,x\,m}^2 + F_{R,y\,m}^2}$ was calculated for several longitudinal slip values s, constant slip angle α, and different velocities v. In the sense of comparability, this mean value then was divided by the respective horizontal force $F_{R,h\,stat}$ with constant tyre deflection $\bar{z}$.

According results for $\alpha = 4°$ and $s = 0/-2/-4/-6\,\%$ are shown in **Fig. 12**, cf. braking with increasing braking torque while cornering.

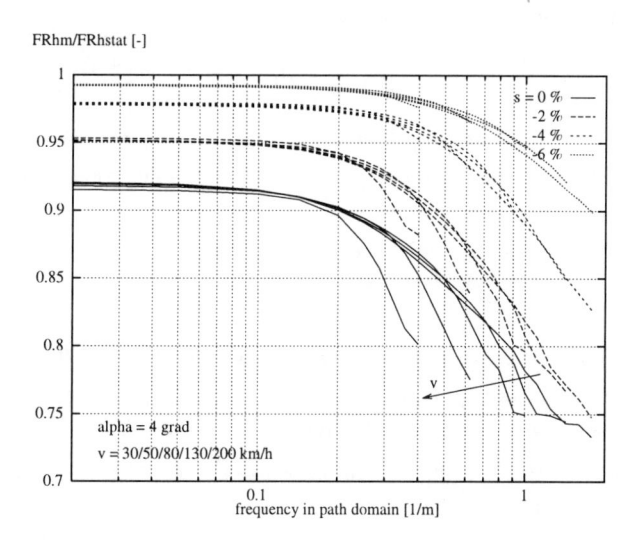

Figure 12: Relative mean horizontal force at different speeds and several longitudinal slip values for a sinusoidal height profile ($s = 0/-2/-4/-6\,\%$, $\alpha = 4°$, $F_{R,z\,stat} \approx 4$ kN, $\hat{z} = \frac{1}{2} \cdot \bar{z}$)

With increasing braking-slip, in this case at $\alpha = 4°$, both the (relative) quasi steady state portion as well as the 'dynamic' portion of loss in horizontal force decrease (again due to a higher sliding portion in the patch).

It has to be mentioned, that at simultaneous occurrence of lateral and longitudinal forces both components are decreased by different amounts due to wheel load fluctuations, whereby the direction of the resulting horizontal force is changed. For illustrating the dimension of the change in direction the example $\alpha = 4°$ and $s = -4\,\%$ is given, where the lateral force is reduced more than the longitudinal force resulting in an angular change between 3 and 6°, depending on the excitation frequency.

5. CONCLUSIONS

The influence of wheel load fluctuations on the force transmission behaviour of a tyre in the contact area is highly depending on the excitation frequency, but in any case the mean value of the transmitted force in the contact plane is decreased. Of course, this has an immediate influence on the active driving safety, which is closely correlated to the transmittable forces.

As a consequence, for a more accurate evaluation of the driving safety (see e.g. [6]), frequency dependent criteria, rather than relatively simple, frequency independent characteristic values (e.g. the RMS value of the wheel load fluctuation) should be used.

REFERENCES

[1] Ammon, D., Gipser, M., Rauh, J., and Wimmer J., *"Effiziente Simulation der Gesamtsystemdynamik Reifen-Achse-Fahrbahn"*, Reifen, Fahrwerk, Fahrbahn, VDI-Berichte Nr. 1224, 1995

[2] Bakker, E., Nyborg, L., and Pacejka H. B., *"Tyre modelling for use in vehicle dynamics studies"*, SAE Paper No. 870421, 1987

[3] Gipser, M., *"DNS-Tire - ein dynamisches, räumliches, nichtlineares Reifenmodell"*, Reifen, Fahrwerk, Fahrbahn, VDI-Berichte 650, 1987

[4] Gipser, M., *"Zur Modellierung des Reifens in CASCaDE"*, Berechnung im Automobilbau, VDI-Berichte Nr. 816, 1990

[5] Gipser, M., *"Dokumentation zum Reifenmodell BRIT"*, Interner Bericht der Daimler Benz AG, Esslingen, 1994

[6] Hofer, R., *"Untersuchung zur Fahrsicherheit von Personenkraftwagen bei Radlastschwankungen mit besonderer Berücksichtigung des Reifenverhaltens"*, Dissertation, TU-Wien, 1996

[7] Laermann, F.-J., *"Seitenführungsverhalten von Kraftfahrzeugreifen bei schnellen Radlaständerungen"*, Dissertation, TU-Braunschweig, 1986

[8] Lugner, P., *"Numerische Erfassung von Reifenkennfeldern zur Berechnung von Fahrzeugbewegungen"*, ATZ 74, pp. 17-23, 1972

[9] Popp K., and Schiehlen W., *"Fahrzeugdynamik"*, B.G.Teubner-Verlag, Stuttgart, 1993

A Ride Comfort Tyre Model for Vibration Analysis in full Vehicle Simulations

M. EICHLER

ABSTRACT

A computer-aided simulation of the tyre, with given accuracy specifications, can be realised by accessing a family of physical and empirical tyre models. This model hierarchy offers tyre models of varying complexity to calculate the forces and moments of rolling contact. These models can be coupled adaptively to any vehicle model by way of a standardised mechanical interface, e.g. to the ADAMS MBS. The change of model on the individual wheels is initiated automatically at the time of simulation by a road surface scanner. The comparison of measurement and calculation of horizontal and vertical acceleration of two different vehicle rear axles shows that short-wave irregularity of the ground such as cleats can also be calculated with physical tyre models.

REQUIREMENTS OF TYRE MODELS WITH REGARD TO GROUND TOPOLOGY

In the field of physical tyre models for short-wave rolling contact, a distinction can be made between models with discretised contact patches and models with discretised belts. The characteristic features of the first model group are that the shape of the contact area and the pressure distribution cannot be calculated from the degrees of freedom of the models. The discretised contact patch relates to a known contact plane, which means that unambiguous determination of the ideal centrepoint of tyre contact and of the contact area must be specified. The calculable ground topology is thus limited to solid road surfaces with long-wave irregularity in relation to the contact length.

The representable shear deformation in the contact plane is dependent on the contact patch discretisation. With simple differential equations, Figure 1 is applied to calculate only in the area of stationary tyre behaviour; a characteristic K_{Bh} can be defined, which describes the ratio of the contact length $2h$ to the permissible horizontal wave excitation λ_{Bh}:

$$K_{Bh} = \frac{2h}{\lambda_{Bh}} < 5\%$$

Short-wave rolling states in the road surface plane can be calculated if the wavelength of the shear deformation in the contact patch is taken into account by means of adequate discretisation.

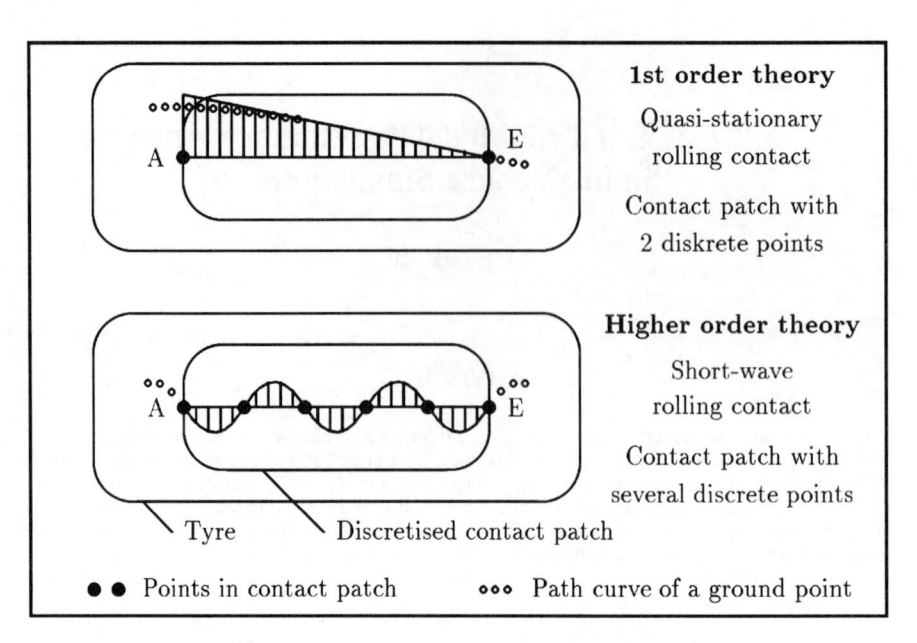

Figure 1: Representable shear deformation in the contact area

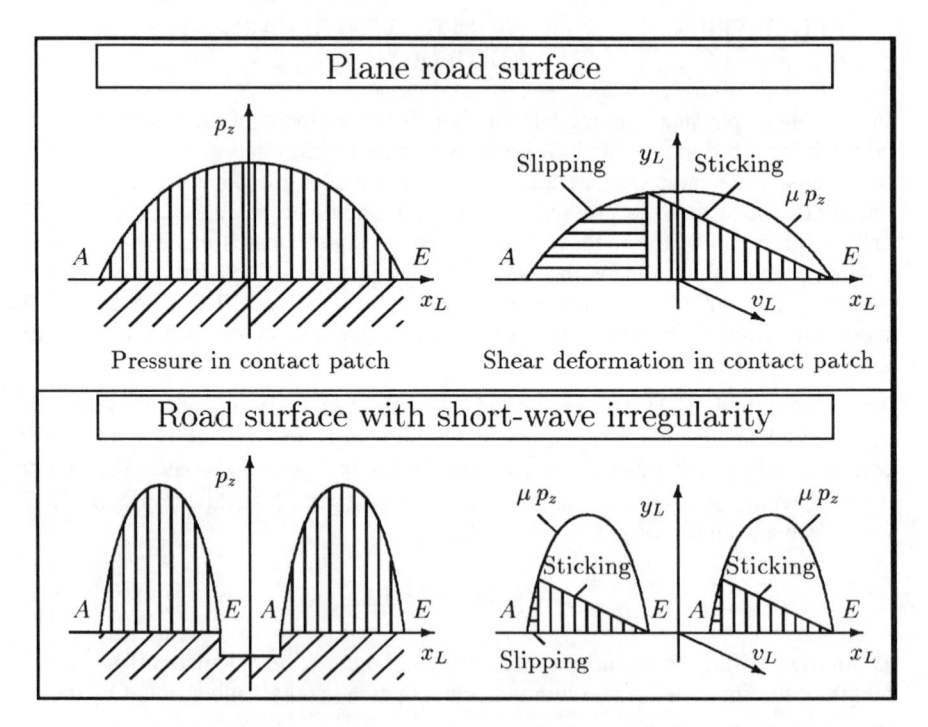

Figure 2: Representable pressure distribution in the contact area

For the characteristic of horizontal wave excitation in this case the following applies:

$$K_{Bh} = \frac{2h}{\lambda_{Bh}} > 100\%$$

Rolling on road surfaces with short-wave irregularity such as cleats or potholes can only be calculated with tyre models from the second group, in which parts of the tyre structure are flexibly modelled. The discretisation describes the tread of the tyre as a single- or multi-track model and, for complex models, also describes the sidewall. The shape of the contact area and the pressure distribution are calculated autonomously in this case in the integration of degrees of model freedom of the tyre model, as shown in Figure 2. According to horizontal excitation a characteristic K_{Bv} for short-wave vertical excitation can be defined:

$$K_{Bv} = \frac{2h}{\lambda_{Bv}} > 100\%$$

Multiple contacts, occurring when the individual obstacles cited are run over, can also be simulated. The number of degrees of model freedom must be adapted to the wavelength of the ground irregularities.

DRIVING DYNAMICS TYRE MODEL WITH DISCRETISED CONTACT SURFACE

The tyre model for calculation of the short-wave rolling contact on long-wave road surfaces as per Figure 3 includes a rigidly modelled belt with 6 degrees of freedom, which is elastically bedded against the rim. The model thus takes into account the basic vibration behaviour of the tyre up to a frequency range of about 100 Hz, according to the modes of the rigid ring. The stiffness of the bedding c_u, c_s, c_γ and c_φ can be calculated by means of the measured free natural frequencies of the belt:

$$c_u = 4\pi^2 f_u^2 m_G = c_v$$
$$c_s = 4\pi^2 f_s^2 m_G$$
$$c_\gamma = 4\pi^2 f_\gamma^2 J_{G\gamma} = c_\psi$$
$$c_\varphi = 4\pi^2 f_\varphi^2 J_{G\varphi}$$

The tyre force $\vec{F}_F$ and tyre moment $\vec{M}_F$ in relation to the rim are calculated from the elastic restoring force of the belt and the restoring moment:

$$\underline{F}_F^{(G)} = \underline{C}_F^{(G)} \underline{r}_{FG}^{(G)} + \underline{D}_F^{(G)} \underline{\dot{r}}_{FG}^{(G)}$$
$$\underline{M}_F^{(G)} = \underline{C}_M^{(G)} \underline{\varphi}_{FG} + \underline{D}_M^{(G)} \underline{\omega}_{FG}^{(G)}$$

To construct the equations of motion of the belt the force $\vec{F}_L$ and the moment $\vec{M}_L$ of the rolling contact are additionally required:

$$m_G(\ddot{\vec{r}}_{0G} + g\vec{e}_{30}) = \vec{F}_L - \vec{F}_F$$

$$\underline{J}_G^{(G)} \dot{\underline{\omega}}_G^{(G)} + \underline{\omega}_G^{(G)} \times \underline{J}_G^{(G)} \underline{\omega}_G^{(G)} = \underline{M}_L^{(G)} - \underline{M}_F^{(G)}$$

The kinematic base systems of the tyre model for calculation of the ideal centre-point of tyre contact and the speed field in the contact area are formed with the belt base $\vec{\underline{e}}_G$ and the ground normal vector in the contact area $\vec{n}_B$. For determination of the kinematic belt base $\vec{\underline{e}}_K$ and the contact patch base $\vec{\underline{e}}_L$ the following applies:

$$\vec{e}_{1K} = \frac{\vec{e}_{2K} \times \vec{n}_B}{|\vec{e}_{2K} \times \vec{n}_B|} \qquad \vec{e}_{1L} = \frac{\vec{e}_{2K} \times \vec{n}_B}{|\vec{e}_{2K} \times \vec{n}_B|}$$

$$\vec{e}_{2K} = \vec{e}_{2G} \quad \text{and} \quad \vec{e}_{2L} = \vec{n}_B \times \vec{e}_{1L}$$

$$\vec{e}_{3K} = \vec{e}_{1K} \times \vec{e}_{2K} \qquad \vec{e}_{3L} = \vec{n}_B$$

The ground normal vector can be calculated by a linearisation of the road surface in the contact area, since the ground topology in this model must be long-wave. The contact patch of the tyre is determined in an initial approximation by the cross-sectional area of the tyre torus with the road surface plane. The deformations in the massless contact patch are calculated by systems of transport equations according to Böhm /1,4/ and Schulze /2/. The contact patch is divided into several tracks over the contact area width, and the shear deformation $\vec{\Delta}$ is calculated in discrete points by an integral equation:

$$\vec{\Delta}(x,y,t) = \vec{\Delta}_0(x_E, y, t_E) + \int_{t_E}^{t} \dot{\vec{\Delta}}(\xi, y, \tau) d\tau$$

Dependent on the number of collocation points, any short-wave rolling states in the road surface plane can be calculated. The example in Figure 4 shows the calculated shear deformation in the contact patch in a simulation of steering a standing tyre. In this example 21 tracks in the discretised contact patch, each with 13 collocation points, were used.

Within the contact patch stick and slip zones are differentiated by a law of friction with memory effect. The pressure distribution in the contact patch must be simulated on the basis of measurements, due to the rigidly modelled tyre belt, and is varied according to the actual displacement of the belt with respect to the rim.

Driving manoeuvres such as standing starts or stopping are calculable without problems of unstable differential equations with this tyre model, which always occur in this case in use of classic slip definitions.

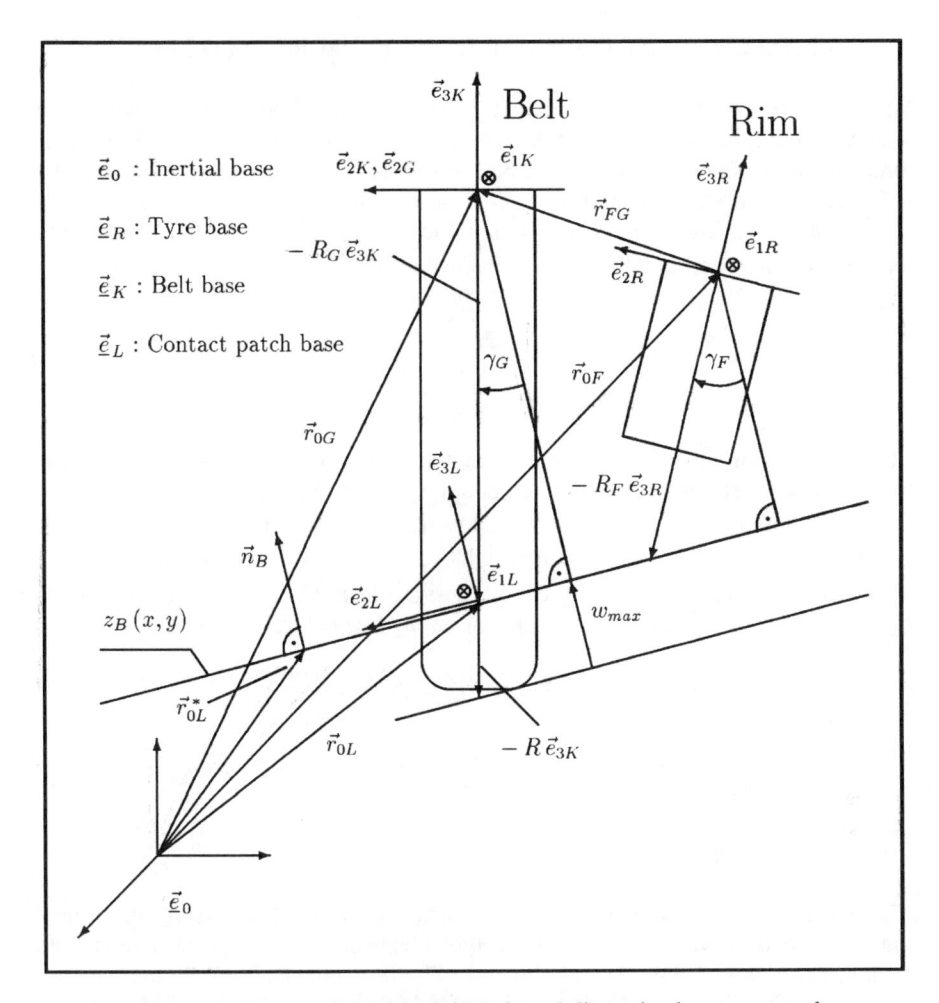

Figure 3: Tyre model with rigid belt and discretised contact patch

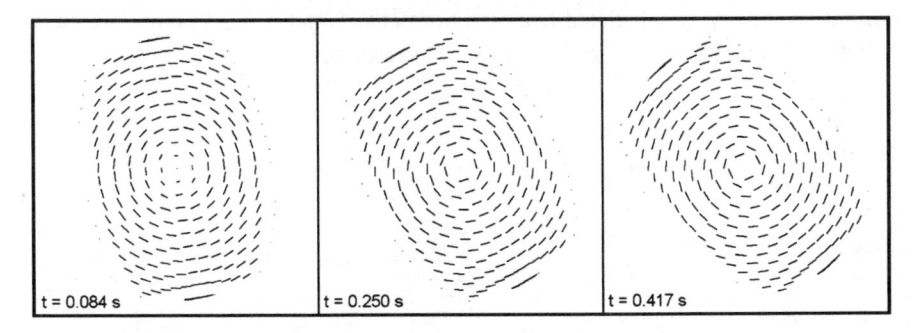

Figure 4: Shear deformation in the contact area with purely bore slip

RIDE COMFORT TYRE MODEL WITH DISCRETISED BELT

For calculation of the impact force when running over short-wave obstacles in a straight line a tyre model is available in which the tyre belt is discretised as a three-dimensional mass point system. The belt model according to Figure 5 reduces the belt assembly, in practice comprising at least three layers, two steel and one rayon cord, to a single ring of n_m mass points which are interconnected by tension and torsion springs c_t and c_α to simulate the tensile and bending stiffness EF and EI_y:

$$c_t = \frac{EF}{2\pi R_G} n_m$$

$$c_\alpha = \frac{EI_y}{2\pi R_G} n_m$$

The sidewall is assumed to be massless, and attaches the mass points of the belt by means of radial springs c_a to assigned coupling points on the rim. The stiffness of the sidewall and the resistance of the belt to transverse displacements with respect to the rim are simulated by torsion springs c_γ and c_φ in the rim coupling points.

The stiffness of the belt bedding can be derived from the elastic potential of the rigid ring and is here calculated dependent on the free natural frequencies:

$$c_a = \frac{d^2V}{da^2} = 2\pi R_G k_a \frac{1}{n_m} = 4\pi^2 (2f_u^2 - f_\varphi^2) m_G \frac{1}{n_m}$$

$$c_\gamma = \frac{d^2V}{d\gamma^2} = 2\pi R_G (R_G - R_F)^2 k_s \frac{1}{n_m} = 4\pi^2 f_s^2 (R_G - R_F)^2 m_G \frac{1}{n_m}$$

$$c_\varphi = \frac{d^2V}{d\varphi^2} = 2\pi R_G (R_G - R_F)^2 k_\tau \frac{1}{n_m} = 4\pi^2 f_\varphi^2 (R_G - R_F)^2 m_G \frac{1}{n_m}$$

The contact force is calculated from the restoring force of elastically deformed, massless bristles, with the tread of the tyre idealised as a single-track model. In each sensor point on the tread, as shown in Figure 6, contact between the tyre and the road surface is checked and the local tread normal vector calculated, so that a single unbroken contact area need no longer be taken as the basis. The local tread normal vector $\vec{e}_{3L}$ is calculated from the normalised vector $\vec{e}_{ii+}$ in circumferential belt direction and the local ground tangential vector $\vec{t}_B$:

$$\vec{e}_{1L} = -\vec{e}_{ii+}$$

$$\vec{e}_{2L} = \vec{e}_{3L} \times \vec{e}_{1L}$$

$$\vec{e}_{3L} = \frac{\vec{e}_{1L} \times \vec{t}_B}{|\vec{e}_{1L} \times \vec{t}_B|}$$

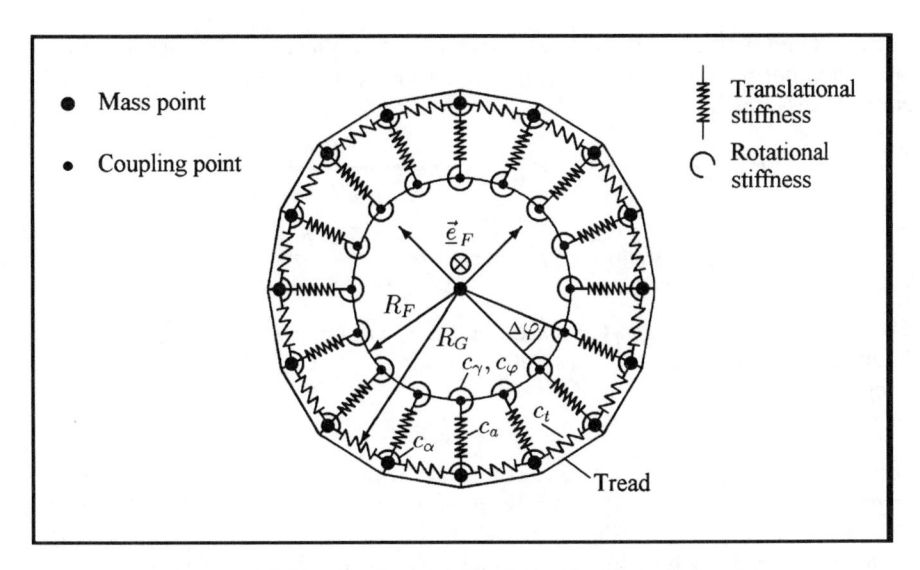

Figure 5: Ride comfort tyre model with elastic 1-ring belt

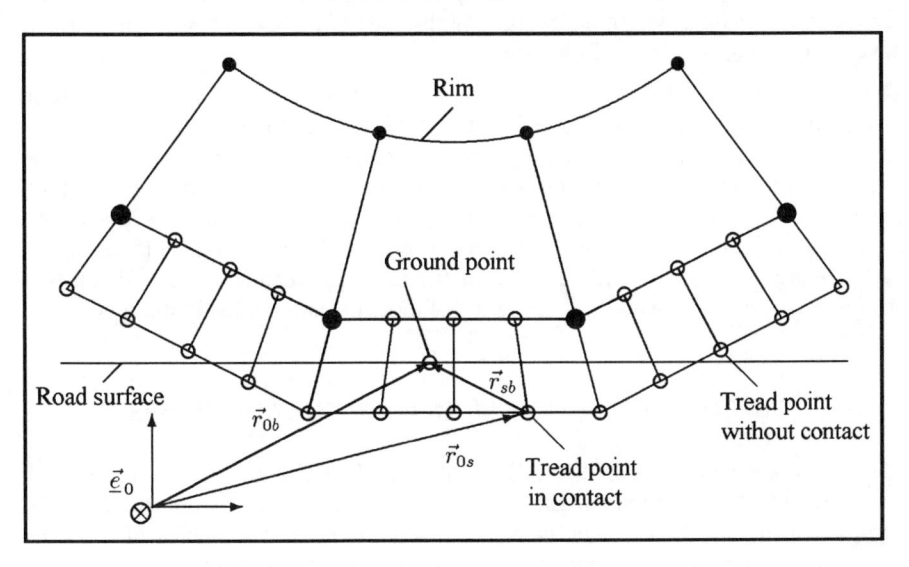

Figure 6: Tread of the 1-ring ride comfort tyre model

The restoring force of a bristle is dependent on condition of sticking or sliding. In case of sticking the restoring force is determinable from the elastic deformation of the bristle:

$$\underline{F}_{si}^{(L)} = \underline{C}_L^{(L)} \underline{T}_{0L}^T \underline{r}_{sb}^{(0)}$$

In case of sliding the components of the bristle force in local tread plane are calculated from normal force and sliding friction, dependent on sliding velocity and contact pressure:

$$\vec{F}_{si} \cdot \vec{e}_{1L} = \mu_G(p_z, v_G) \vec{F}_{si} \cdot \vec{e}_{3L} \frac{\vec{r}_{sb} \cdot \vec{e}_{1L}}{\sqrt{(\vec{r}_{sb} \cdot \vec{e}_{1L})^2 + (\vec{r}_{sb} \cdot \vec{e}_{2L})^2}}$$

$$\vec{F}_{si} \cdot \vec{e}_{2L} = \mu_G(p_z, v_G) \vec{F}_{si} \cdot \vec{e}_{3L} \frac{\vec{r}_{sb} \cdot \vec{e}_{2L}}{\sqrt{(\vec{r}_{sb} \cdot \vec{e}_{1L})^2 + (\vec{r}_{sb} \cdot \vec{e}_{2L})^2}}$$

The equations of motion of the elastic belt are very simple and efficient for high frequency systems - only the translational acceleration of the mass points has to be integrated:

$$m_i(\ddot{\vec{r}}_{0i} + g\vec{e}_{30}) = \vec{F}_i$$

For examinations in the field of tyre comfort in which running over obstacles at an angle is also to be calculated, a 4-ring mass point model with multi-track contact was developed, but not validated yet. Modelling of the load-bearing performance of the belt over the belt width is always necessary when running over an irregularity on the ground causes a change in the pressure distribution over the belt width, see Eichler /8/. Other tyre models with a high grade of complexity were previously developed by Gipser /3/ and Böhm /5/.

The use of tyre models with discretised belts is only practicable in the simulation phases of running over obstacles, due to the significantly slower computing time in comparison to models with discretised contact patches. To reduce the computing time, the tyre model with flexible belt is therefore adaptively coupled to the tyre model with rigid belt according Bannwitz and Oertel /7/. A road surface scanner checks information of the parametric ground model relating to road surface topology in a preview zone in the direction of travel of the vehicle wheels. When obstacles occur in the preview zone a change of model is automatically initiated.

COUPLING OF THE TYRE MODELS TO THE ADAMS MBS

The tyre models are addressed in ADAMS by a GFORCE statement with defined parameter input. According to Eichler and Oertel /6/ the rim is considered as a component of the MBS and delivers the kinematic control variables of the tyre modules for one integration time step of the MBS, i.e. rim position, speed and acceleration, see Figure 7. Other transfer variables include information on selection of the tyre model and assignment of the parameter files of the tyre models.

The subroutine GFOSUB supplies the above tyre models with the kinematic control variables of the rim by way of a standardised mechanical interface. The task of the numeric interface in GFOSUB is to synchronise the internal integration procedure of the tyre modules with the ADAMS integrator.

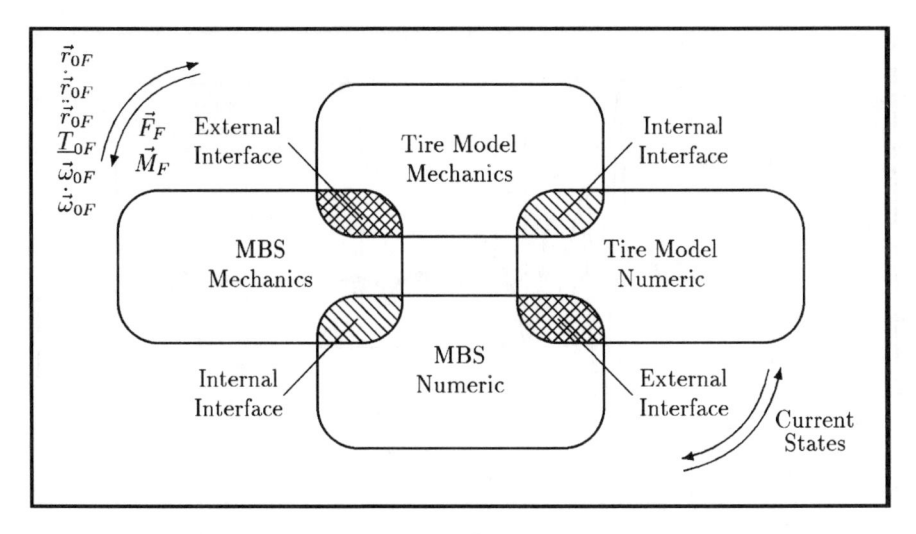

Figure 7: Internal and external interfaces between MBS and tyre model

Efficient modelling of the ground topology gains greater significance in the case of uneven road surfaces and tyre models with fine contact patch discretisation, requiring a larger number of ground scans. For this reason a parametric ground model was developed which, in addition to rapid ground scanning, also permits simple and reliable positioning of solid obstacles. Discrete parameterisable obstacle types can be selected from a catalogue and positioned as required to create the simulated landscape. Description of the ground topology by means of continuous obstacle shapes is also possible as the use of standard ADAMS road data files in a slightly modified manner.

SIMULATED RUNNING OVER CLEATS

A standard driving manoeuvre for comfort testing at Volkswagen is straight-line running over 2-15 mm high cleats at the test site. To validate the 1-ring ride comfort tyre model a vehicle with 56 degrees of freedom was modelled in ADAMS and coupled to the tyre model. The comparison of measurement and calculation of the hub carrier acceleration of a conventional twist-beam rear axle when running over the 8 mm high cleats at a speed of 30 km/h, as in Figures 8-11, shows a good match; the road surface excitation between the cleats was not taken into account in the ground model. Another validation was made with an experimental multi-link rear suspension at a speed of 25 km/h, as in Figures 12-15, which shows a good match, too.

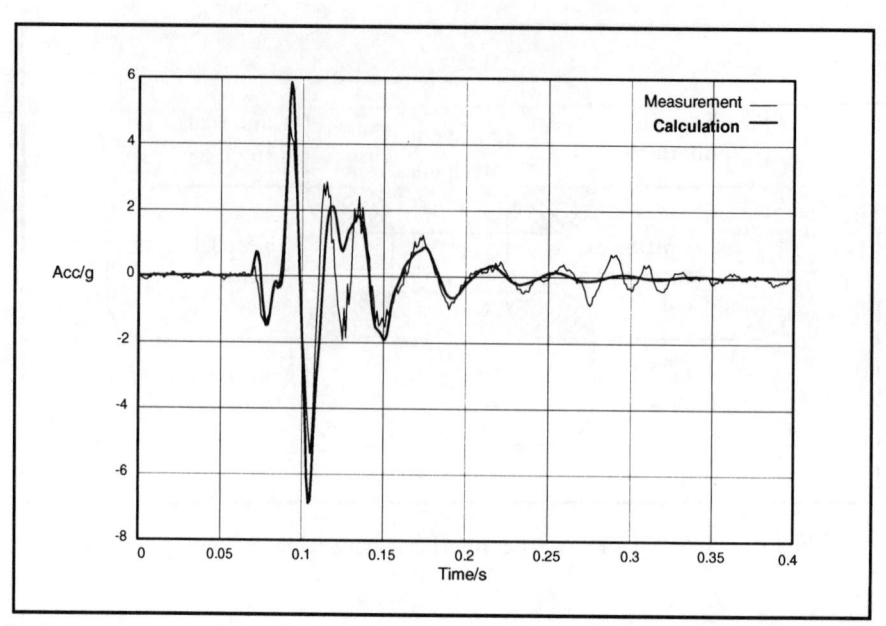

Figure 8: Horizontal hub carrier acceleration of the twist-beam rear axle in time domain

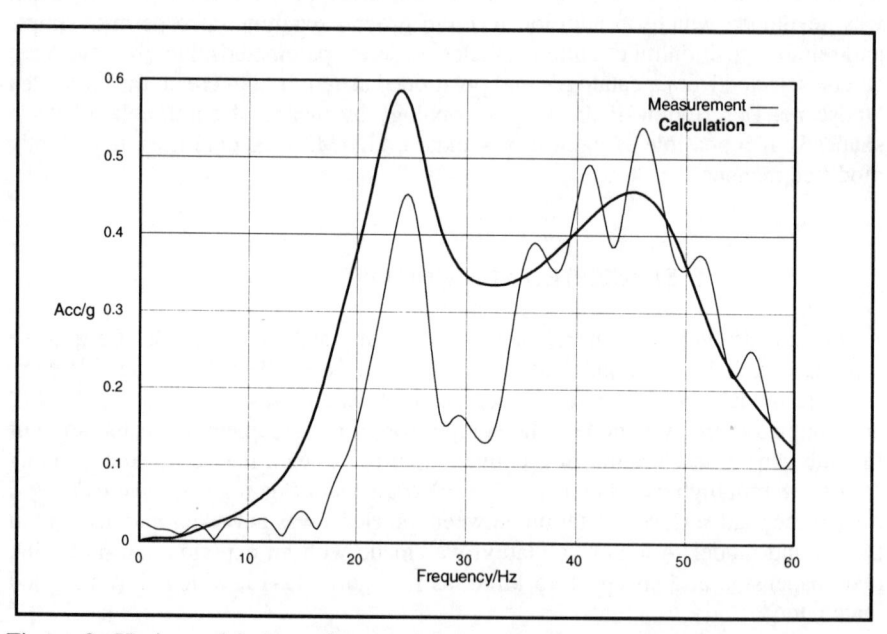

Figure 9: Horizontal hub carrier acceleration of the twist-beam rear axle in frequency domain

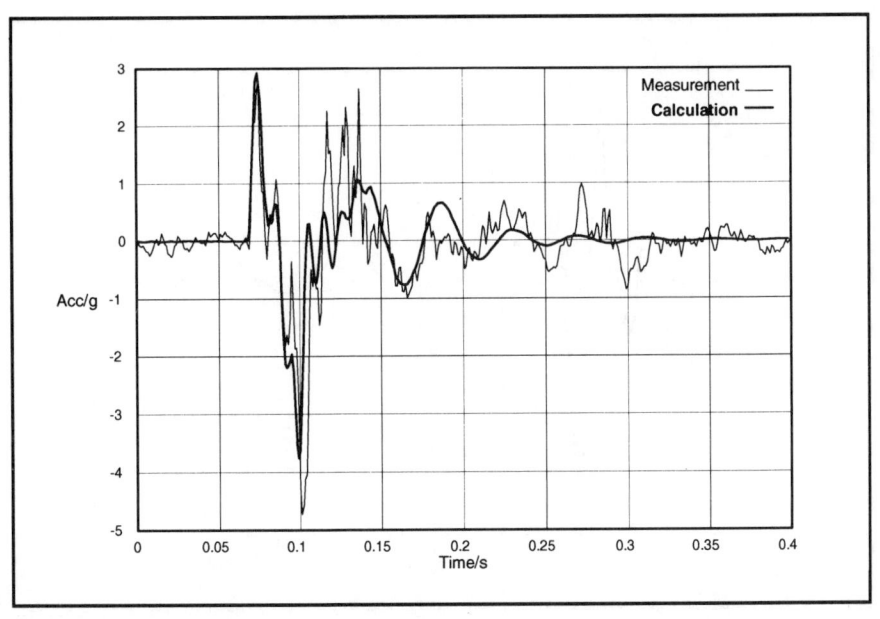

Figure 10: Vertical hub carrier acceleration of the twist-beam rear axle in time domain

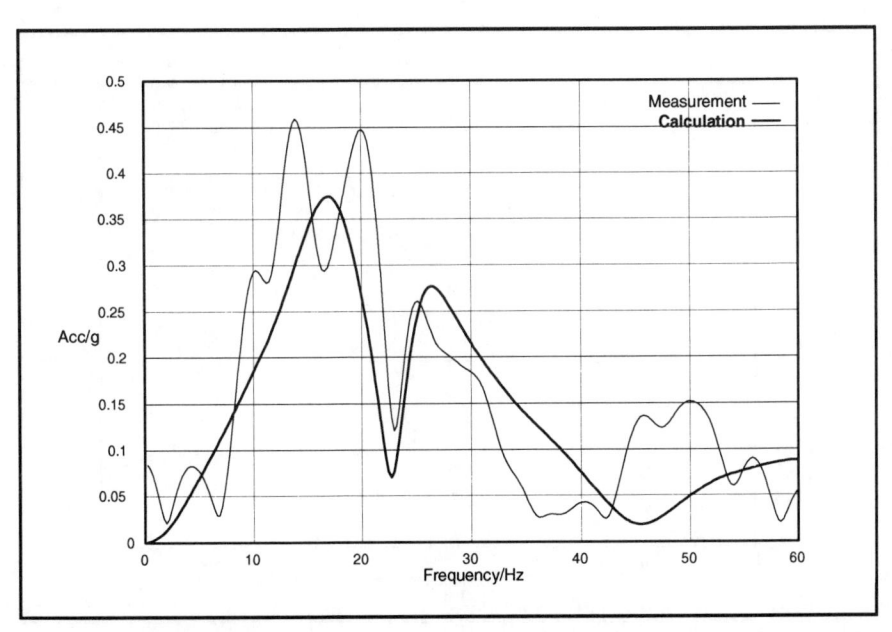

Figure 11: Vertical hub carrier acceleration of the twist-beam rear axle in frequency domain

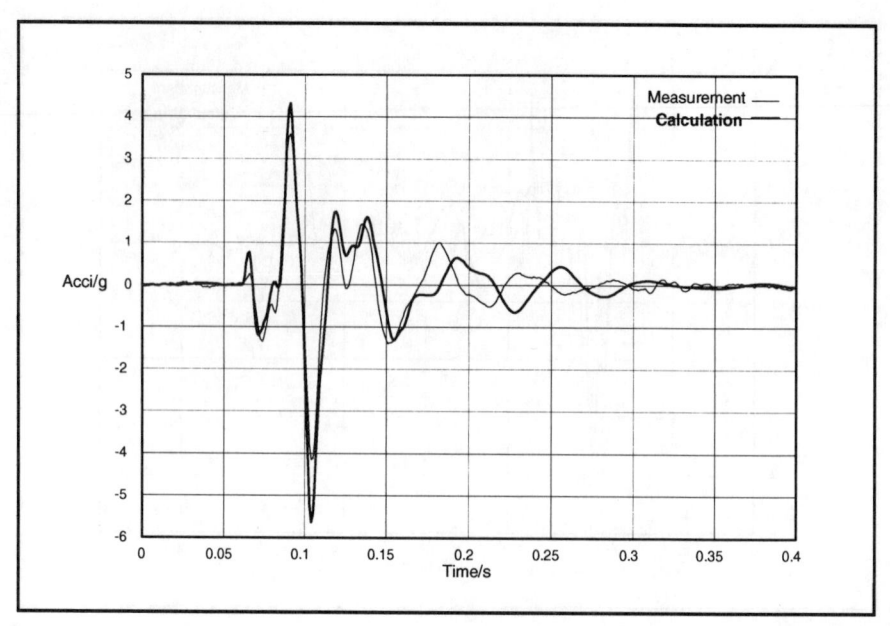

Figure 12: Horizontal hub carrier acceleration of the multi-link rear suspension in time domain

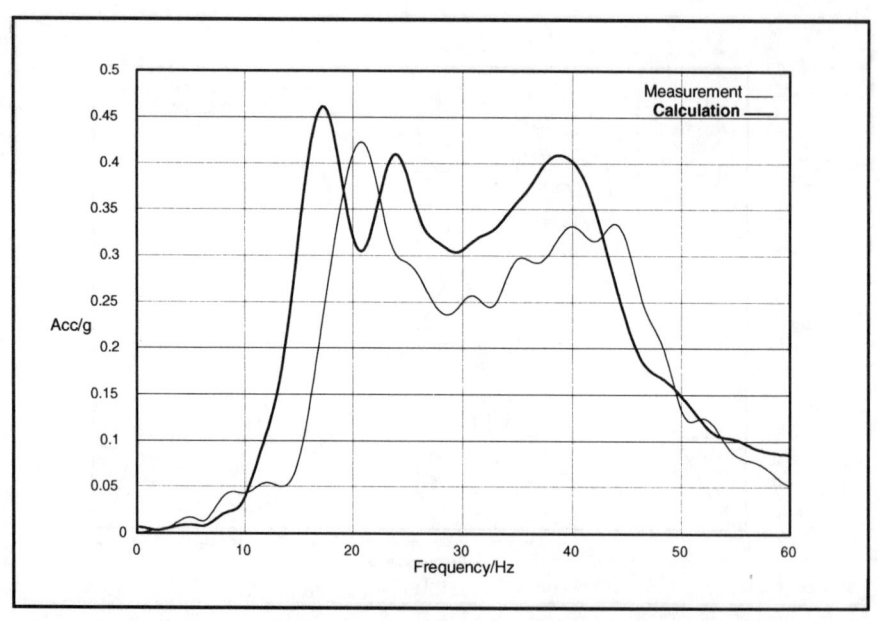

Figure 13: Horizontal hub carrier acceleration of the multi-link rear suspension in frequency domain

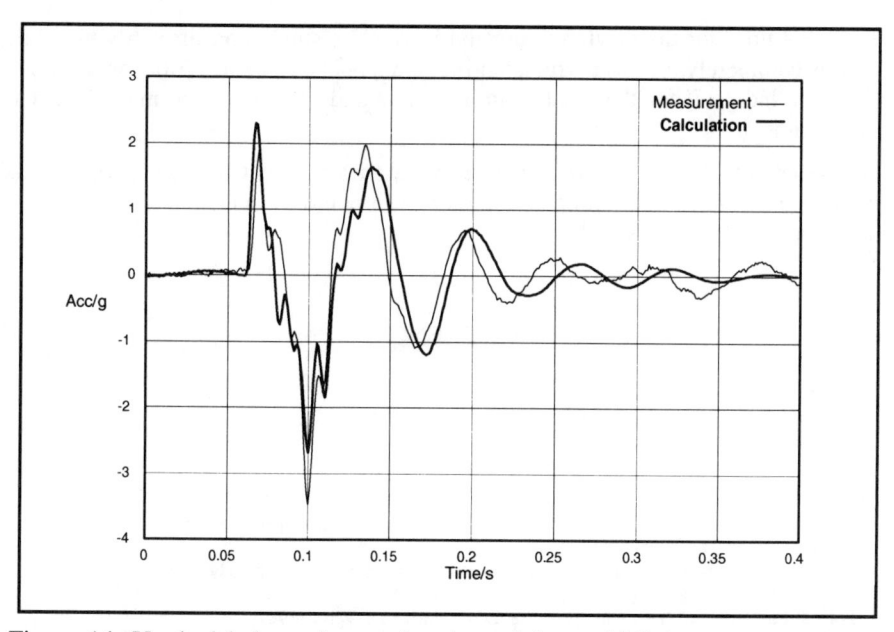

Figure 14: Vertical hub carrier acceleration of the multi-link rear suspension in time domain

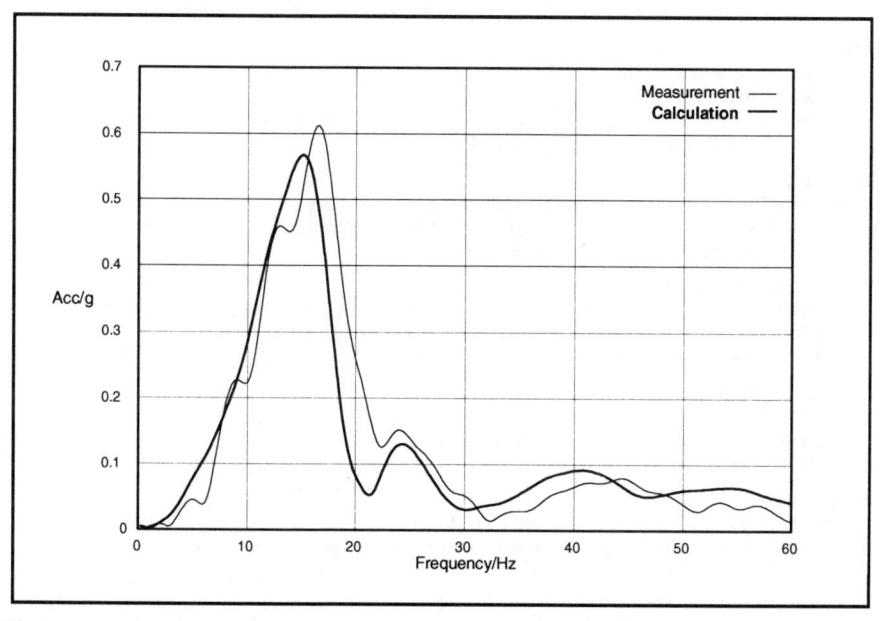

Figure 15: Vertical hub carrier acceleration of the multi-link rear suspension in frequency domain

The CPU time for the driving manoeuvre, over 2 s simulation time in which one cleat is completely run over with adaptive tyre models, is about 1200 s on a Silicon Graphics Indigo II for the twist-beam rear axle and 1500 s for the multi-link rear suspension.

To validate the 4-ring ride comfort tyre model measurements are planned, such as running over cleats at an oblique angle or deep potholes to estimate load populations for calculation of stability.

REFERENCES

/1/ Böhm, F.:
Theorie schnell veränderlicher Rollzustände für Gürtelreifen
(Theory of rapidly changing rolling states for belted tyres)
Ingenieur-Archiv 55, 1985

/2/ Schulze, D.:
Instationäre Modelle des Luftreifens als Bindungselemente in Mehrkörpersystemen für fahrdynamische Untersuchungen
(Non-stationary models of the pneumatic tyre as binding elements in multibody systems for examinations of driving dynamics)
Fortschritts-Berichte, VDI Reihe 12 Nr. 88, Düsseldorf: VDI-Verlag 1987

/3/ Gipser, M.:
DNS-Tire - ein dynamisches, räumliches, nichtlineares Reifenmodell
(DNS-Tire - a dynamic, spatial, non-linear tire model)
Reifen, Fahrwerk, Fahrbahn. VDI-Berichte Nr. 650, Düsseldorf: VDI-Verlag 1987

/4/ Böhm, F., Eichler, M., Kmoch, K.:
Grundlagen der Rolldynamik von Luftreifen
(Principles of the rolling dynamics of pneumatic tyres)
Fortschritte der Fahrzeugtechnik 1, W. Stühler (publ.): Fahrzeugdynamik,
Vieweg-Verlag, Braunschweig / Wiesbaden, 1988

/5/ Böhm, F.:
Tire Models for Computational Car Dynamics in the Frequency Range up to 1000 Hz
Tyre Models for Vehicle Dynamics Analysis, edited by H.B. Pacejka,
Swets & Zeitlinger B.V. Amsterdam/Lisse 1991

/6/ Eichler, M., and Oertel, C.:
Zur Standardisierung der Schnittstelle zwischen Reifenmodellen und Fahrzeugmodellen
(On the standardization of the interface between tyre models and vehicle models)
ATZ, 96. Jahrgang/Nr. 3, März 1994, ISSN 0001-2785

/7/ Bannwitz, P., and Oertel, C.:
Adaptive Reifenmodelle: Aufbau, Anwendung und Parameterbestimmung
(Adaptive tyre models: modelling, application and parameter identification)
Reifen, Fahrwerk, Fahrbahn. VDI-Berichte Nr. 1224, Düsseldorf: VDI-Verlag 1995

/8/ Eichler, M.:
Ride Comfort Calculations with Adaptive Tyre Models
AVEC '96, International Symposium on Advanced Vehicle Control, Aachen 1996

Elastodynamics of Cars and Tires

F. Böhm

ABSTRACT

The car as a free elastic body is moving or standing and the mechanical equations should be the same in both cases. The aim of the paper is to analyse the coupling of tire models of different complexity with the car body to get a simulation system for all possible testing conditions. So it is necessary to look for low frequency dynamics and also for high frequency behaviour of car and tires. Parameters of friction and damping have to be measured.

The tire models range from stationary rolling models to high frequency rolling up to 3 kHz for dynamics of profile elements. Models rolling on uneven road and on soft ground are also introduced. The time integration is explizite and range of numerical amount is from 114 degrees of freedom (DOF) for car body only to 1198 DOF for one elastodyamic tire with profile elements.

These complex tire models are discrete anisotropic membrane models with bending disturbance. Parameters for this models are given from material constants and from measured damping constants.

INTRODUCTION

The elastodynamic behaviour of cars on uneven road surfaces is examined using the particle method, and is compared to discrete numerical methods of continuum mechanics, Lit. [1], at the end of the paper. The usual parameters from continuum mechanics continue to be employed, cf. Lit. [2 - 4]. Regarding friction, damping, and hysteresis in the material and in the contact between the tire and the road, however, there are influences that are impossible to describe using methods of continuum mechanics, cf. Lit. [5 - 12]. Still, even by using the particle method, resolution is limited in terms of both time and space, as explicit integration is generally required without introducing constraint conditions, Lit. [13, 14]. Implicit methods of integration should be rejected as they suppress high-frequency motion components and give a wrong picture of transients during rolling and, consequently, of load collections as well, Lit. [15 - 17]. Measurements showed that the proposed method of particle systems allows a better approximation by this computational approach, Lit. [18 - 20]. Static states of equilibrium are very easy to treat by computation of dying-out oscillations, and are therefore superior to the finite element method as the latter is only suitable for examining either statics or stationary rolling conditions, Lit. [21]. As a result, if load collections up to the high-frequency range are to be computed, the only way is to apply the particle method together with an explicit method of integration.

1. ELASTODYNAMICS OF SYSTEMS WHICH ARE FREE IN SPACE AND USE ONLY FORCES FOR BOUNDARY CONDITIONS

The motorcar may be described as a rigid body system only by very rough approximation. Modern low-weight design and elasticities deliberately introduced into the design for reasons of comfort and stability cannot be described in terms of rigid bodies, but are rather controlled by the equations of motion of particle systems having a finite propagation velocity of impacts and moments of momentum. The effects of resilient elasticities to be used purposefully are not restricted to local suspension and damper systems; rather the entire car should be considered as being a mass-spring-damping system which is free in space. Moreover, the effect of internal friction expresses itself in that higher-frequency axle excitations are unable to build up resonances because the influence of impulses produced by the wheels is consumed by frictional forces in the system. Thus, measurements made by H.-P. Willumeit within the "High-frequency rolling contact of car wheels" SFB 181 project showed that the body of a middle-market car, when subjected to vibration testing on stand and excited by shakers at the axle pins, produced frequencies up to a maximum of 75 Hz and apart from that flat peaks only, Fig. 1. This result, however, is not caused by any rigid body movements, but by internal friction in the elastodynamic system preventing higher resonant frequencies from occurring so that only dynamic random movements are left. The effect of internal damping and friction is most easily illustrated by a 2-particle system moving freely in the x-y plane, Fig. 2.

In Fig. 2, the upper particle was started at an initial horizontal velocity, while the lower particle was accelerated from its position of rest by the connecting spring. If a damping-free system is used, no uniform rotational and translational motion will be caused, but a swinging of kinetic energy between translation and rotation produced by the high-frequency energy exchange between the two particles. On the other hand, if the energy exchange is impeded by damping and friction in addition to the spring, the system will smooth out after a short period of time with reference to the distance oscillation of the two particles toward each other, and a stationary translational and rotational motion occurs. The resultant trajectories of the two particles are the cycloids known from the rolling motion of a rigid wheel. Such a rotational motion corresponds to that of a rigid body, the difference to the rigid body being that additional impulses, if acting on the system, pass to the respective other non-charged particle with a time delay, which corresponds to natural behaviour, while in a rigid body this transition takes place without time delay. Consequently, the numerical integration of free elastodynamic systems has to take the highest velocity of impulse spreading in the system into account and may not, in a quasi-static fashion, impress the force effect simultaneously on all body particles under consideration. Hence, it is necessary to perform the time integration explicitly and to avoid implicit methods as the latter operate in a quasi-static way regarding high-frequency forms of motion. The resulting impulse distributions in space and time are incorrect and falsify the computation of kinetic energy, potential energy, and dissipation energy by damping. Thus, the calculated forces are also wrong, as far as they are determined by high-frequency forms of motion.

2. THE CAR BODY AS A SPACE-FRAME SYSTEM STRUCTURE

The car body in this case is modelled as a particle system held together by zero-mass elastic rods. Damping and friction during longitudinal expansion and compression of these rods are also taken into account. Here, Newton's third law "actio est reactio" must exactly be taken into account, which is done automatically by checking the combining matrix of the space-frame system. Fig. 3 shows such a particle system comprising MacPherson front struts and a longitudinal-bar axle design at the rear axle. For numerical integration it is necessary to estimate the highest frequency occurring in the system. To do so, it will be assumed that all rods acting on the selected particle act in the same direction, presuming that the oscillation point which has to be located between two adjacent particles at highest frequency must be situated at half the rod length. Accordingly, twice the total of all rod rigidities will be the total restoring effect of a selected particle. The particle itself is determined by estimation from the nodal lines produced by connecting nodes in the middle of rods. Such a field is shown in Fig. 3 in hatched lines, the rod rigidities themselves resulting from a comparison with a tubular frame car of equal overall rigidity behaviour as it is frequently used in the trial stage for newly designed cars. To simplify the numerical computation following as a one-step predictor-corrector integration, axles and rims are treated as rigid bodies because of their small overall dimensions.

3. CAR MOTION WHEN USING SIMPLE TIRE MODELS

In their simplest form, tire models are presented as tire characteristic models suitable only for road speeds exceeding 5 m/sec, their advantage being the high computational speed attainable. Inputs used to determine the tire guiding forces are just the longitudinal creep and lateral creep, while spin creep is not taken into account. The creep quantities are determined for the contact centre point resulting as the point of stick-through of the steepest descent line lying on the wheel plane perpendicular to the axle through the central point of the wheel with the road plane. The trace of the wheel plane as the straight line intersection with the road plane is also needed. A parameter required in the characteristic diagram of tire forces is the wheel load resulting from deflection, i.e., settling-down of the wheel. During the computation of steered motion of the car, the steering rod length is varied such that the axle basis formed by the shadowed triangle may turn about the central line of the front wheel spring, Fig. 4. Fig. 5 shows the corresponding numerical solution to the equations of motion for steering into a road curve. The oscillatory steering test is illustrated as a computer animation. However, the associated computation results for wheel load and force distribution in the car body do not yet involve the additional impulses produced by the high-frequency rolling contact between car wheels and short-wave road bumps occurring in contrast to the behaviour of a point spring (corresponding to the simple tire model).

4. ELASTODYNAMICS OF THE TIRE

Being a highly flexible structure, the pneumatic tire contacts the uneven road in a hand-sized contact area, and is simulated by its contact-flattened longitudinally very stiff belt founded on a very elastic membrane, the carcass. The belt model is a circular ring simulated by 25 particles, the stiffness of connecting rods between the particles resulting from measurements of belt stiffness. The carcass membrane of nearly zero mass is replaced by an elastic foundation. Such a 2D-model is on left side shown in Fig. 6, where the tread is also assumed to be an elastic foundation. At right side a 3D-model is shown. A distinction is made between static friction and sliding friction. Due to the high rigidity of the belt and its low mass the highest eigenfrequeny of this system is about 3 kHz, meaning that the numerical discretisation must be performed in time steps of 30 kHz. Fig. 7 shows the force response of a radial tire when rolling over a short cleat. The vertical wheel load was specified to be constant at 300 daN, and the time-varying damping force of the damper strut suspension was taken into account.

5. ROLLING WITH SPECIFIED WHEEL GUIDANCE ON AN UNEVEN ROAD

In section 4, the driving processes of the elastic car were calculated using a simple tire model. The resulting axle loads are not high-frequency forces but time-varying. If the vertical axle loads calculated in this way are used in the computational model of the elastodynamic tire, the dynamic contact forces on an uneven road can be determined. To do so, the rolling distance of the selected wheel is calculated and used to define road bumps the wheel passes along this known distance. In the computation, the wheel develops high additional forces during touch-down which have to die out first, Fig. 8a. Only afterwards will it have assumed a normal uniform rolling state for which a rolling distance of about 1 m is required. Then the distance may be counted until the road unevenness of interest is reached, Fig. 8b. Computations include contact force distributions along the contact line of the belt and the tread, respectively. Longitudinal force distribution, lateral force distribution and vertical force distribution follow immediately from the elastodynamic equations of tire motion. In this case, too, integration uses a predictor-corrector method. However, since time steps differ from those used during car integration, it is useful to select an integer ratio of time steps. While the car is integrated in time steps of 10 kHz, the tire uses 30 kHz time steps. Therefore, the solution obtained in the car program has to be subjected to three interpolations per step in the tire program. In addition to contact force distribution, another result is the tire's integrated force and moment effect on the rim, and due to the high rim mass and rotational mass a considerable averaging effect is caused here. This means that acceleration measurements at the axle fail to reflect the strong fluctuations of contact force distribution. Thus, the stress on the road is much higher from tire contact forces than the stress on the car from wheel rim forces. However, ride comfort investigations have always assumed the focus to be on the load collection at the car and at the driver. Introducing wear behaviour of the road in proportion to the locally acting forces in rolling contact, wear distribution patterns as caused at intersections during accelerating and braking and as found at expressway slabs are obtained. Of interest to the vehicle engineer are the impulsive longitudinal forces on uneven roads because these are transmitted in an almost unsprung manner into the chassis basis of the car body.

6. ROLLING COUPLED TO BOTH ELASTIC SYSTEMS, CAR AND TIRES ON UNEVEN GROUND

Taking the high-frequency rolling behaviour into account, systems of equation for numerical computation having a great plurality of degrees of freedom (DOF) are obtained. The case of the elastic car used with four elastic circular ring models results in 326 DOF. The case of the car combined with the elastodynamic membrane tire model produces 1114 DOF, see Fig. 9, while finally an off-road tire model used with profile element dynamics leads to 1978 DOF. These complex models are discrete particle models including anisotropy and bending stiffness. The parameters for these models are derived from material constants and measured damping/hysteresis constants. The parameters for material friction and contact friction must be estimated and are temperature-dependent, especially for tire parameters. To limit the computational effort in the examples, only the right front wheel was simulated as a complex tire model. The other three wheels were computed as tire-characteristic models.

In all examples the excitation intensity of the complex model is distinctly higher than the excitation intensity of the tire-characteristic models.

The following examples were computed:
- load intensity factor at the points on the right half of the car using the complex 2D-model compared to the left half of the car, see Table 1;
- on an expressway having defective plates, Fig. 10, and
- on surface undulations of different lengths, Fig. 11.

These examples are demonstrated by animation.

The author wishes to thank Deutsche Forschungsgemeinschaft (DFG) for the provision of support in kind.

LITERATURE

01 F. Böhm: Elastodynamik der Fahrzeugbewegung. Fahrzeugdynamik-Fachtagung Essen, 1990.
02 D. Schulze: Instationäre Modelle des Luftreifens als Bindungselemente in Mehrkörpersystemen für fahrdynamische Untersuchungen. Fortschritt-Berichte VDI, Reihe 12, Nr. 88, VDI Verlag 1987.
03 M. Kollatz: Kinematik und Kinetik von linearen Fahrzeugmodellen mit wenigen Freiheitsgraden unter Berücksichtigung von Reifen und Achsen. Fortschritt-Berichte VDI, Reihe 12, Nr. 118, VDI Verlag 1989.
04 F. Böhm: Theorie schnell veränderlicher Rollzustände für Gürtelreifen. Ingenieur-Archiv 55 (1985), 30-44

05 C. Oertel: Untersuchung von Stick-Slip-Effekten am Gürtelreifen. Fortschritt-Berichte VDI, Reihe 12, Nr. 147, VDI Verlag 1990.
06 F. Böhm: Einfluß der Laufflächeneigenschaften von Gürtelreifen auf den instationären Rollkontakt. KGK Kautschuk und Gummi Kunststoffe, 41 (1988), 359-365
07 D. H. Schulze: Zum Schwingungsverhalten des Gürtelreifens beim Überrollen kurzwelliger Bodenunebenheiten. Fortschritt-Berichte VDI, Reihe 12, Nr. 98, VDI Verlag 1988.
08 F. Böhm, M. Kollatz: Some Theoretical Models for Computation of Tire Nonuniformities. Fortschritt-Berichte VDI, Reihe 12, Nr. 124, VDI Verlag 1989.
09 F. Böhm: Reifenmodell für hochfrequente Rollvorgänge auf kurzwelligen Fahrbahnen. VDI-Tagung 1993, VDI Berichte 1088
10 F. Böhm, M. Swierczek, G. Csaki: Hochfrequente Rolldynamik des Gürtelreifens - das Kreisringmodell und seine Erweiterung. Fortschritt-Berichte VDI, Reihe 12, Nr. 135, VDI Verlag 1989.
11 F. Böhm: Über die Wirkung von Hysterese und Kontaktreibung auf das dynamische und thermische Verhalten von Luftreifen. KGK Kautschuk Gummi Kunststoffe, 47, (1994), 824-827
12 F. Böhm: Action of hysteretic and frictional forces on rolling tires. Int. Rubber Conference, Moscow (1994)
13 A. Gallrein: Berechnung hochfrequenter Stollendynamik am Gürtelreifen. Diplom-Arbeit TU Berlin, 1992.
14 U. Masenger: Dynamisches Verhalten von Ackerschlepper-Reifen auf der Straße und auf dem Acker - Erweiterung einer instationären Rolltheorie auf großvolumige Reifen im Einsatz auf visko-elastischen Böden. Dissertation TU Berlin, 1992.
15 M. Gipser: DNS-Tire - ein dynamisches, räumliches, nichtlineares Reifenmodell. VDI-Berichte 650 (1987).
16 M. Gipser: Modellbildung, Numerik und Anwendungen eines komplexen Reifenmodells. VDI-Berichte 699 (1988).
17 M. Gipser: Zur Modellierung des Reifens in CASDaDE. VDI-Berichte 816, (1990), 41-57.
18 M. Eichler: Messungen des Einrollverhaltens von Gürtelreifen am Glasplatten-Rolltisch und Vergleiche mit theoretischen Ergebnissen. Diplom-Arbeit TU Berlin, 1987.
19 F. Böhm: From Non-Holonomic Constraint Equation To Exact Transport Algorithm For Rolling Contact. Dynamics, Identification and Anomalies, TU Budapest, Nov. 1994.
20 F. Böhm, A. Duda: Verschleiß und Zerstörung von Fahrbahnoberflächen infolge hochfrequenter Rolldynamik von LKW-Reifen. Haus der Technik der RWTHA, Febr. 1996.
21 K. Feng: Statische Berechnung des Gürtelreifens unter besonderer Berücksichtigung der kordverstärkten Lagen. Fortschritt-Berichte VDI, Reihe 12, Nr. 258, VDI Verlag 1995.

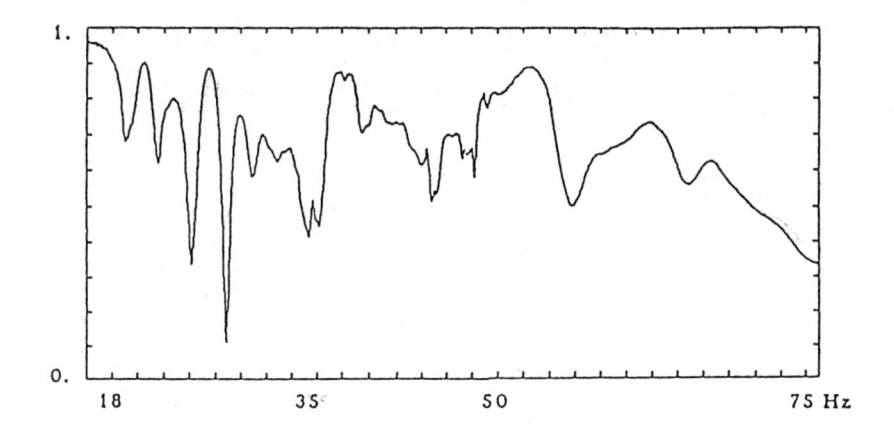

Fig. 1 Indicatorfunction of chassis

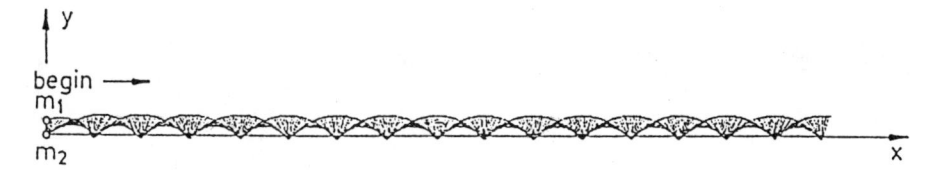

Fig. 2 Two masses connected by one spring with friction

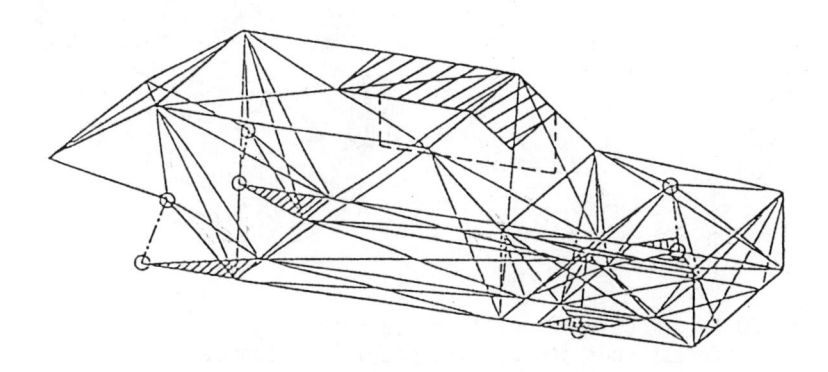

Fig. 3 Car model with 38 particles

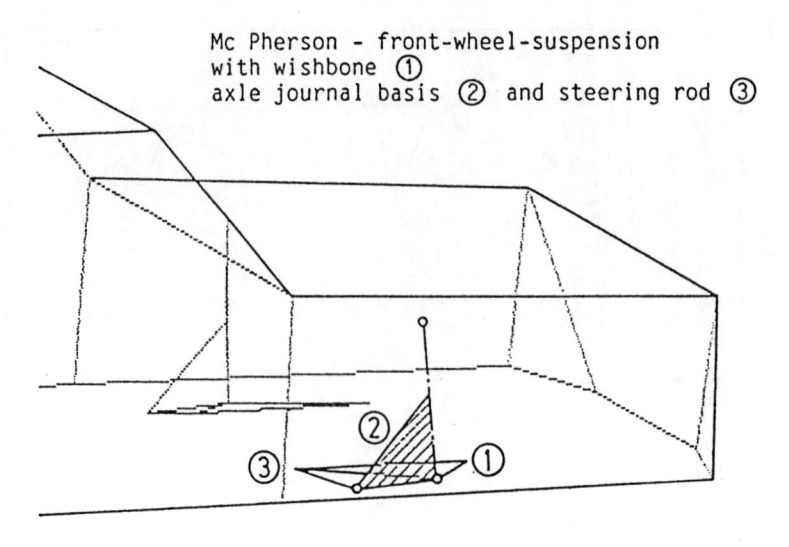

Mc Pherson - front-wheel-suspension
with wishbone ①
axle journal basis ② and steering rod ③

time in ms T=791

<u>Fig. 4</u> Mc Pherson front-wheel suspension

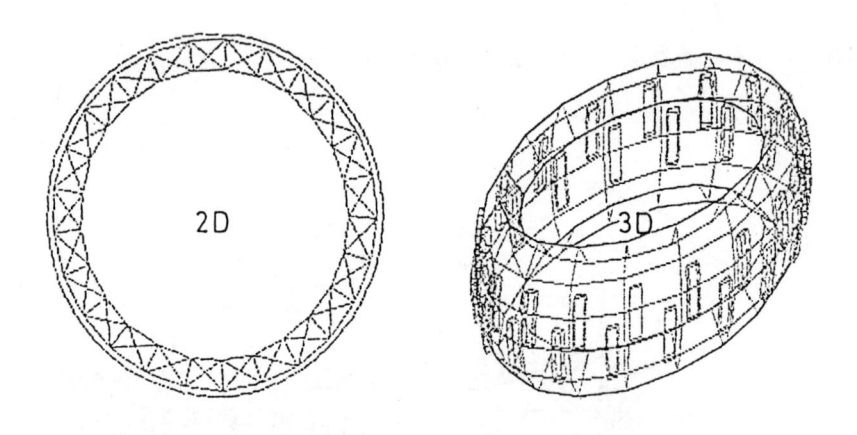

<u>Fig. 6</u> 2D-Tyre model with belt, radial and diagonal stiffness-rods and 3D-Tyre model in diagonal design for off-road applications

ELASTODYNAMICS OF CARS AND TIRES 131

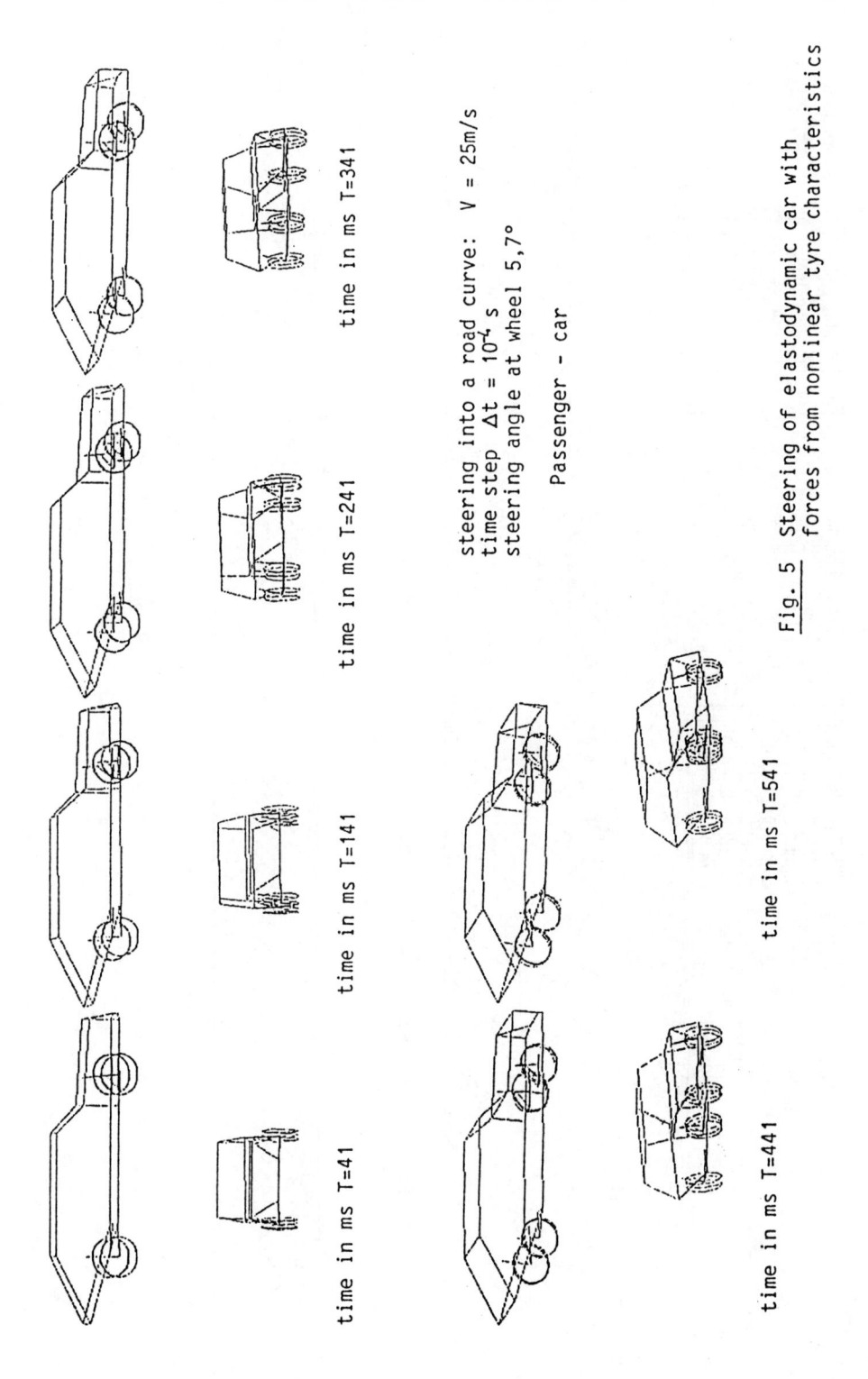

steering into a road curve: V = 25m/s
time step Δt = 10⁻⁴ s
steering angle at wheel 5,7°

Passenger - car

Fig. 5 Steering of elastodynamic car with forces from nonlinear tyre characteristics

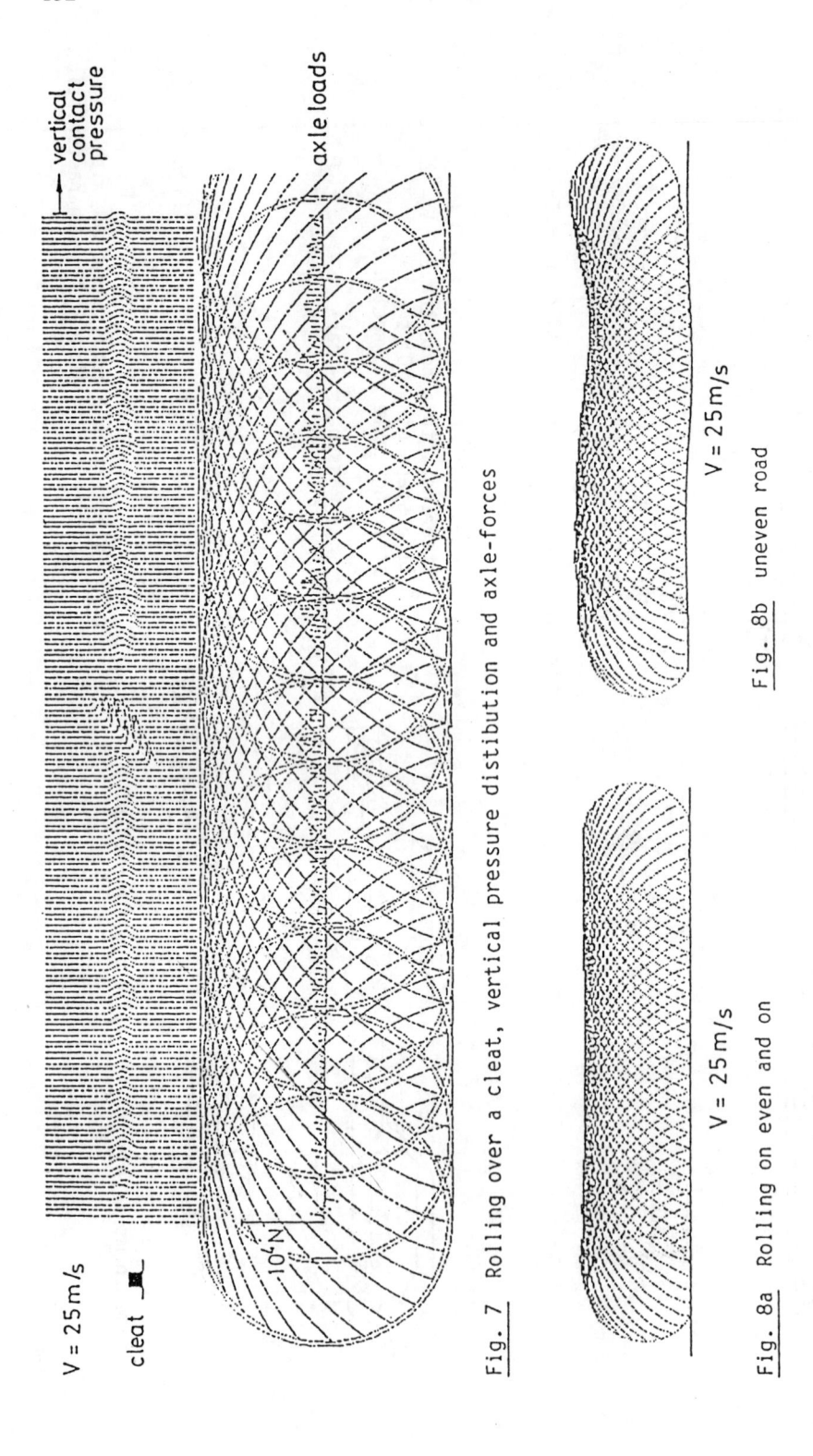

Fig. 7 Rolling over a cleat, vertical pressure distibution and axle-forces

Fig. 8a Rolling on even and on

Fig. 8b uneven road

Table 1 Accelerations at point I = 1-38 of car model running with V = 25m/s on express - way plates L =8m, Δh = 4cm

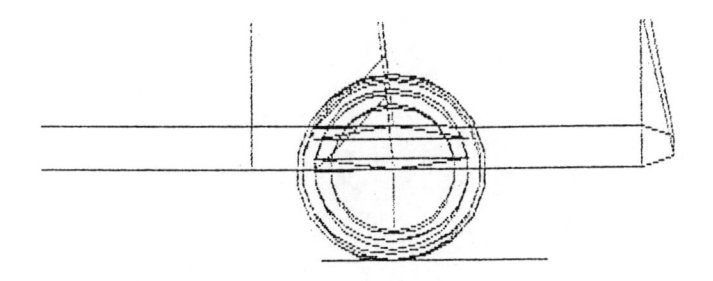

I	10g	20g	30g	40g	50g
1	2429	491	51	16	6
2	2451	512	22	9	2
3	2927	69	2	0	0
4	2989	8	1	0	0
5	2347	617	27	6	1
6	2973	24	1	0	0
7	2651	320	23	1	3
8	2903	82	9	2	2
9	2971	26	1	0	0
10	2984	13	1	0	0
11	2996	2	0	0	0
12	2998	0	0	0	0
13	2766	163	42	14	7
14	2984	14	0	0	0
15	2997	1	0	0	0
16	2644	255	51	29	8
17	2621	344	26	5	0
18	2027	768	152	31	13
19	1145	933	417	181	129
20	2296	649	46	5	1
21	2491	486	15	1	3
22	2986	10	2	0	0
23	2981	17	0	0	0
24	2898	87	9	3	1
25	2981	14	3	0	0
26	2812	184	2	0	0
27	2994	4	0	0	0
28	2998	0	0	0	0
29	2998	0	0	0	0
30	2998	0	0	0	0
31	2998	0	0	0	0
32	2958	38	2	0	0
33	2998	0	0	0	0
34	2998	0	0	0	0
35	2898	92	7	1	0
36	2836	160	2	0	0
37	2962	30	3	2	1
38	2886	76	22	12	2

Fig. 9 Car - model with 3D - Tire model

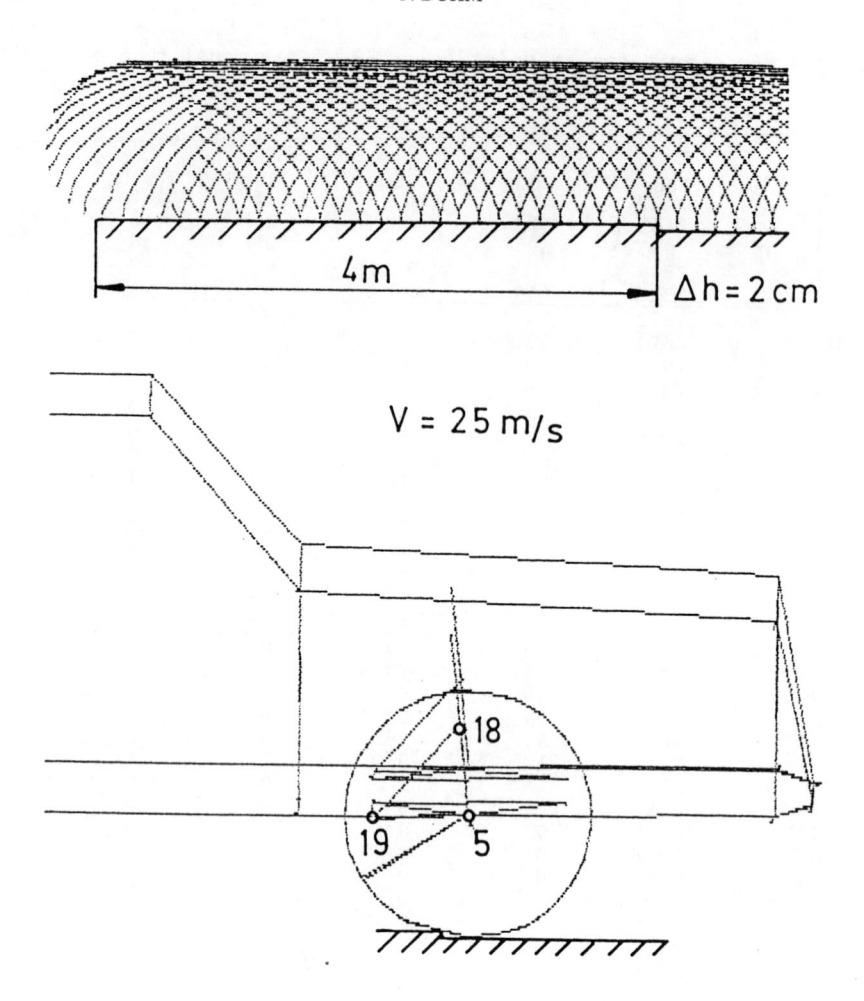

Fig. 10 Rolling on express - way plates

ELASTODYNAMICS OF CARS AND TIRES 135

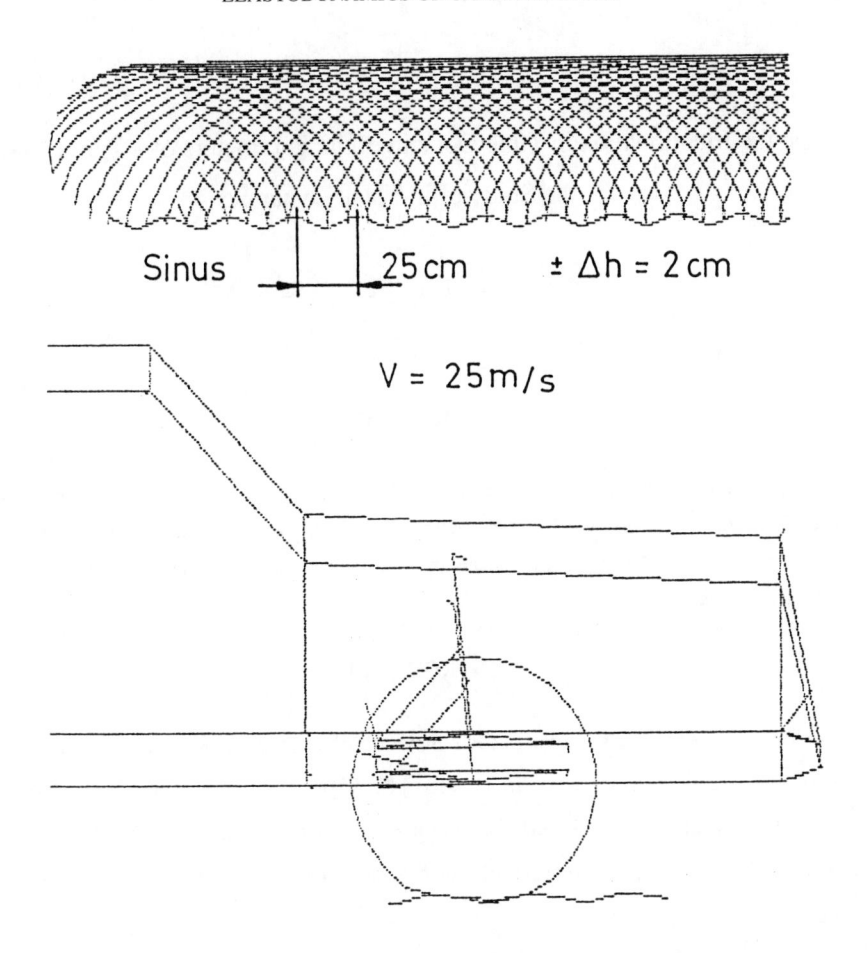

Sinus   25 cm   ± Δh = 2 cm

V = 25 m/s

Fig. 11 Rolling on weavy road

On the Identification in Time Domain of the Parameters of a Tyre Model for the Study of In-Plane Dynamics

S. BRUNI   F. CHELI   F. RESTA

ABSTRACT

The paper proposes an original approach to identify the physical parameters of a "rigid ring" tyre model for vehicle comfort, braking and driving analysis. A brief description of the standard experimental tests and the time domain identification techniques adopted is done. First experimental results are reported.

INTRODUCTION

In order to simulate the dynamical behaviour of a road vehicle, it is necessary to create an adequate model of the tyre ([1],[2],[4]), keeping into account also the wideband excitations coming from the road irregularities, from the not uniform mass and stiffness, from the tread sculpture of the tyre and from the motion impressed to the vehicle itself.

This work deals with the researches associated to the vehicle's driving and comfort performance in the medium-low frequency range (0-130 Hz), analysing the problems connected to the tyre modellisation: the tyre model can be utilized both for the vehicle in-plane behaviour simulations and as a link between the FEM models (used in the tyre design) and the forecasting of the real tyre dynamic behaviour analysis. In the work development, an existing literature model known as "rigid ring model" was used, to reproduce the tyre behaviour. An innovative identification methodology of the model parameters, which makes use of a reduced number of experimental standard tests with consequently reduced time and costs, is proposed. In the last section of the paper some experimental results are presented.

1. THE RIGID RING MODEL

The "rigid ring" tyre model, proposed in [1], describes the zeroth and first order modes of the tyre in the vertical plane and gives an accurate description of the dynamical vertical and longitudo-torsional behaviour up to the mid frequency range. The model consists of a rigid ring representing the tread band, connected elastically to the rim with radial c_b and tangential c_{bt} stiffnesses (fig.1). In the present work the rigid ring model was implemented, considering a fixed rim and introducing also the damping effects, neglected in [1]. The equations of motion are written for small perturbations of the tyre motion around the stationary conditions corresponding to the tyre rolling with constant angular speed $\Omega = V_x/R$ with no external torque applied on the rim, being V_x the longitudinal velocity of the rim and R the "effective rolling radius".

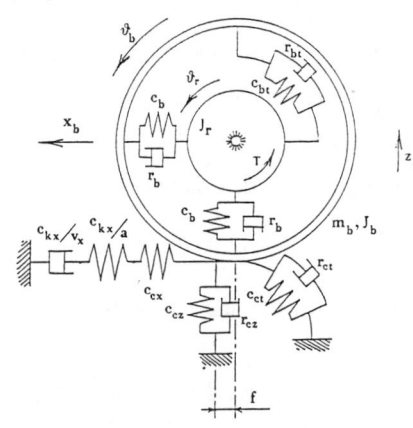

Fig.1 - The rigid ring tyre model used in the work

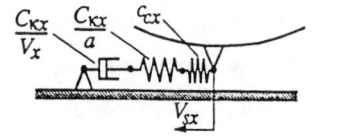

Fig. 2 - The relaxation model for the longitudinal force in the contact area (side view of the tire).

The vertical z_b and longitudinal x_b displacements of the ring and the ring θ_b and the rim θ_r angular rotations were assumed as independent coordinates.

The ring is also bonded to the ground with residual stiffnesses c_{cx}, c_{cz} and c_{ct}, introduced to account for the correct static behaviour of the model and a brush type slip model (fig.2) in order to introduce the non-stationary behaviour of the tyre-road interface considering half of the contact length "a", as a relaxation length.

The equations of motion are written in a state coordinate form:

$$[A]\,\underline{\dot{x}} = \underline{x} \qquad (1)$$

including, as $\underline{x}$ state variables:

$$\underline{x} = \{\dot{x}_b, \dot{z}_b, \dot{\vartheta}_b, \dot{\vartheta}_r, x_b, z_b, \vartheta_b, \vartheta_r, F_x\}^T \qquad (2)$$

also the longitudinal force F_x generated in the contact path, reproduced by means of the relaxation model shown in fig.2. The [A] state matrix of (1) is: (3)

$$[A] = \begin{bmatrix} -\dfrac{r_b}{m_b} & 0 & 0 & 0 & -\dfrac{c_b}{m_b} & 0 & 0 & 0 & -\dfrac{1}{m_b} \\ 0 & -\dfrac{r_b+r_{cz}}{m_b} & \dfrac{fr_{cz}}{m_b} & 0 & 0 & -\dfrac{c_b+c_{cx}}{m_b} & \dfrac{fc_{cz}}{m_b} & 0 & 0 \\ 0 & \dfrac{fr_{cz}}{J_b} & -\dfrac{f^2 r_{cz}+r_{bt}+r_{ct}}{J_b} & \dfrac{r_{bt}}{J_b} & 0 & \dfrac{fc_{cz}}{J_b} & -\dfrac{f^2 c_{cz}+c_{bt}+c_{ct}}{J_b} & \dfrac{c_{bt}}{J_b} & \dfrac{R}{J_b} \\ 0 & 0 & \dfrac{r_{bt}}{J_r} & -\dfrac{r_{bt}}{J_r} & 0 & 0 & \dfrac{c_{bt}}{J_r} & -\dfrac{c_{bt}}{J_r} & 0 \\ 1 & 0 & 0 & 0 & 0 & 0 & 0 & 0 & 0 \\ 0 & 1 & 0 & 0 & 0 & 0 & 0 & 0 & 0 \\ 0 & 0 & 1 & 0 & 0 & 0 & 0 & 0 & 0 \\ 0 & 0 & 0 & 1 & 0 & 0 & 0 & 0 & 0 \\ c_{eq} & 0 & -c_{eq}R & 0 & 0 & 0 & 0 & 0 & -\dfrac{V_x c_{eq}}{c_{kx}} \end{bmatrix}$$

where c_{eq} is the equivalent value of the series of the slip stiffness of the brush model c_{kx}/a and the residual stiffness c_{cx}. Using this model, the dynamic longitudinal F_{xm} and vertical F_{zm} forces transmitted to the hub can be simply calculated.

The aim of the paper is to identify the values of each physical parameters of the rigid tyre model (see eq.3), from few experimental standard tests.

2. THE PROPOSED IDENTIFICATION METHODOLOGIES

The whole identification strategy proposed (a more extensive description of the global procedure and of the test machines was described in [3]) consists of two steps:
- experimental tests for the direct measure of some rim-tyre physical parameters like tyre inertial parameters, torsional stiffness and damping of tyre structure, longitudinal slip stiffness of the tread in the contact area and length (2a) of the contact patch;
- experimental "passing over an obstacle with fixed hub" tests, measuring the forces transmitted to the hub during the passage of the tyre on an isolated obstacle, for the definition of the other parameters of the tyre model.

In the first step, the mass m_b and the ring rigid moment of inertia J_b and of the rim J_r are identified using a "torsional pendulum", where the tyre is connected to a rigid frame with a known torsional spring ([3]). The total tyre + rim + hub inertial moment J_{tot} can be evaluated from the measured natural frequency of the torsional system. The rigid ring inertial moment J_b can be easily evaluated keeping into account tread rubber + belts + casing. Then m_b mass can be evaluated from the relation $m_b = J_b / R^2$ being $R = r_e$ the effective rolling radius of the tyre, measurable on a "road-wheel" test machine.

Stiffness and damping parameters c_{bt} and r_{bt} are identified by minimizing the difference between the model frequency response $h^{(c)}(\omega)$ between the tread rotation θ_b and the rim rotation θ_r:

$$h^{(c)}(\omega) = \frac{\theta_{bo}(\omega)}{\theta_{ro}(\omega)} = \frac{(j\omega r_{bt} + c_{bt})}{-\omega^2 J_b + j\omega r_{bt} + c_{bt}} \qquad (4)$$

and the experimental ones $h^{(S)}(\omega)$ measured using a "torsional vibrator" ([3]). In this test, the wheel is mounted with the rim rigidly linked to a shaft alternatively rotated by means of an electrical actuator: the angular rotation θ_b of the tread and the imposed angular rotation θ_r of the rim are measured by means of accelerometers placed on the tread and on the rim, allowing the experimental transfer function $h^{(S)}(\omega)$ to be defined.

The slip longitudinal stiffness c_{kx} is identified from outdoor experimental tests by means of a dynamometric trailer allowing the measurement of the slip velocity V_{sx} and the longitudinal force transmitted to the hub F_{xm} ([3]). The test consists of a "random braking" torque, far enough from the blocking conditions: from the measured signals, it is possible to evaluate the harmonic transfer function $H_{F_{xm},V_{sx}}$ between the rim F_{xm} longitudinal force and the V_{sx} slip velocity. Such function converges to $H_{F_x,V_{sx}}$ transfer function between the longitudinal F_x force on the ground and the slip velocity V_{sx} when frequency value approaches zero. The slip stiffness is evaluated from the following property:

$$\lim_{\omega \to 0} \left| H_{F_x,V_{sx}} \right| = c_{kx} / V_x \qquad (5)$$

The second step in the identification procedure allows us to identify the remaining unknown parameters. More in details, the algorithm develops in two subsequent phases:

- identification of the modal parameters, that is natural frequencies ω_i and damping factors $(r/r_c)^{(i)}$ of the real tyre (as a function of V_x velocity), by analysing the time histories of the vertical F_{zm} and longitudinal F_{xm} hub forces in standard "passing over an obstacle with fixed hub" tests. In this phase three different time domain identification techniques were used: Ibrahim's method and two different formulations of Prony's method.
- estimation of the physical rigid tyre model parameters, by minimizing the differences between the experimental natural frequencies and damping factors, identified in the previous phase on the real system, and the same quantities evaluated, as a function of the unknown parameters, by means of the rigid ring model (see eq.1).

In the first phase we use an indoor road-wheel machine where a free wheel is loaded with fixed rim on a vertical axle drum (diameter = 2.5 m), keeping a constant and controlled inflating pressure. The drum is rotating and a generic shape obstacle can be positioned. After a warm up period, vertical and longitudinal rim forces in coast-down condition are measured, by pulling the wheel-road at the maximum velocity and then removing the torque and allowing the wheel itself to decelerate naturally. The experimental impulse response of the hub forces is registered. The experimental free oscillation of the forces can be analytically reproduced by means of the following expressions:

$$\underline{F}_m(t) = \sum_{i=1}^{2N} \begin{Bmatrix} F_{zm}^{(i)} \\ F_{xm}^{(i)} \end{Bmatrix} e^{\lambda_i t} \qquad (6)$$

where $\lambda_i = \alpha_i + j\,\omega_i$ (i=1, ... ,2N) represents the unkown damped natural frequencies of the real system in the considered range. To estimate these modal parameters, as already said, three different algorithms were used.

Ibrahim's method [6] is a time domain identification method, evaluating the modal parameters natural frequencies ω_i, the damping coefficients $(r/r_c)^{(i)} = \alpha_i/|\omega_i|$ and the vibrating modes $\underline{X}^{(i)}$ from the free response measured on the real system. The algorithm evaluates the real decay as the response of a generic N d.o.f. linear system:

$$\underline{X}(t) = \sum_{i=1}^{2N} \underline{X}^{(i)} e^{\lambda_i t} \qquad (7)$$

where λ_i is the generic frequency and $\underline{X}^{(i)}$ the corresponding mode, both unknown parameters. The problem is reduced to an eigenvalue-eigenvector problem:

$$[B][\Phi_d] = [\Phi_d][\alpha] \qquad (8)$$

where [B], opportunely organized, contains the experimental measurements $\underline{X}^{(S)}$, while the matrices $[\Phi_d]$ and $[\alpha]$ contain respectively the eigenvectors and the eigenvalues of [A]. The eigenvalues of the mechanical system λ_i depend on the $a_i = b_i + j\, c_i$ eigenvalues of [A] matrix in the following way:

$$\omega_i = \frac{1}{\Delta t_1}[\tan^{-1}\left(\frac{c_i}{b_i}\right) + k\pi] \qquad \alpha_i = \frac{1}{2\Delta t_1}\ln(b_i^2 + c_i^2) \qquad (9)$$

The complex Modal Confidence Factors (MCF = |MCF| $e^{j\phi_{MCF}}$) ([6]) can be used as indicators of the goodness of the identification: parameters estimations are assumed as correctly identified if characterized by 95 % < | M.C.F | < 105 % and -10 < $\varphi_{M.C.F.}$ < 10.

The first formulation of the Prony method ([5]), used in the work, evaluates the modal parameters by analizing the measured free system response $\underline{X}^{(S)}(t)$. The method estimates the generical discrete series $\underline{X}(m\Delta t)$ by means of the following series of exponential terms:

$$\underline{X}(m\Delta t) = \sum_{i=1}^{p} \underline{A}_i e^{[\alpha_i(m-1)\Delta t + j\omega_i(m-1)\Delta t + j\varphi_i]} \qquad (10)$$

being Δt the sample period. The unknown modal parameters are identified, with a least square procedure, by minimizing the difference between the experimental measurements $\underline{X}^{(S)}(m\Delta t)$ and the analytical values $\underline{X}(m\Delta t)$ (see eq. 10):

$$J = \frac{1}{2}\sum_m \left[\underline{X}^{(S)}(m\Delta t) - \underline{X}(m\Delta t)\right]^T \left[\underline{X}^{(S)}(m\Delta t) - \underline{X}(m\Delta t)\right] \qquad (11)$$

The impulse response can be reconstructed from the identified parameters in order to validate the results. A second formulation of the Prony Method, using a modified version (furnished by MATLAB library), was also used to evaluate the real tyre natural frequencies ω_i and damping factors $(r/r_c)^{(i)}$.

In the second phase of the second step of the identification procedure, the unknown physical parameters (that is residual stifness c_{cx}, c_{ct}, c_{cz} and damping r_{ct}, r_{cz} and radial c_b, r_b coefficients) are estimated by minimising the difference between the experimental damped natural frequencies $\lambda_i = \alpha_i + j\omega_i$ and the analitycal ones $\lambda_i^{(c)} = \alpha_i^{(c)} + j\omega_i^{(c)}$, using a mean square approach with the Nelder-Mead algorithm for each V_x velocity:

$$\sum_{V_x}\sum_i \left[\lambda_i^{(c)}(V_x,\underline{p}) - \lambda_i\right]^2 = \min \qquad (12)$$

being $\lambda_i^{(c)}$ the eigenvalues of the tyre model state matrix [A] (see eq. 3), function of the unknown parameters "$\underline{p}$".

3. FIRST EXPERIMENTAL RESULTS

The global methodology was first validated using numerically simulated pseudo-experimental data and then applied to real experimental measurements.

As an example, modal parameters identification results from experimental decay of the rim forces from the "passing over an obstacle with fixed hub" tests are shown in a real application.

A 195/60 R15 type tyre (inflating pressure 2.4 bar and vertical load 5150 N) was tested in the range 40-110 km/h using a square section obstacle (15x15 mm).

In figs. 3 the time histories of the vertical F_{zm} and longitudinal F_{xm} forces, measured at the hub, are reported: the advancing speed V_x changes the system's dynamic responses in terms of both amplitude and frequencies, as confirmed by the spectral analisys reported in the next fig. 4.

TYRE MODEL FOR THE STUDY OF IN-PLANE DYNAMICS

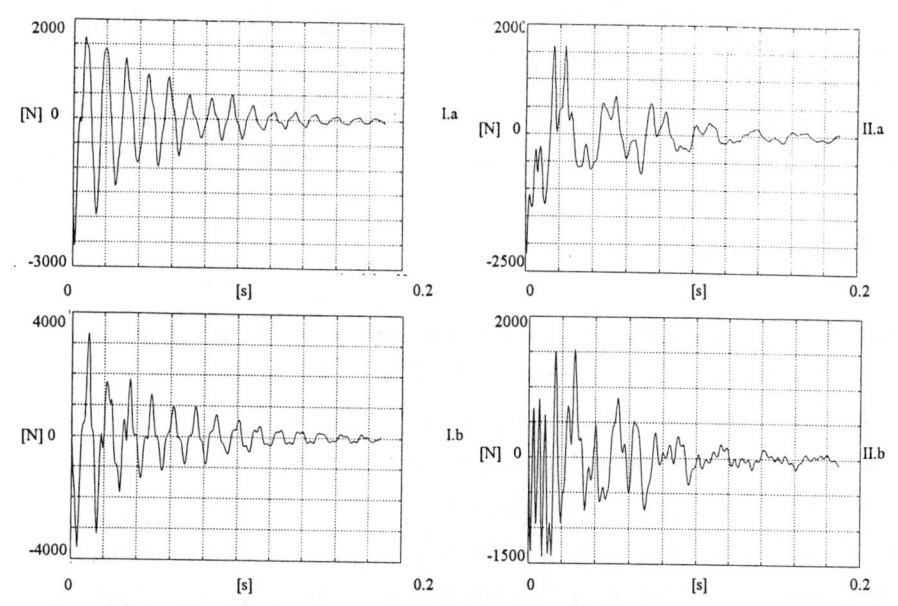

Fig. 3 - Tyre 195/60 R15 cleat test: experimental decay of the hub forces

- $V_x = 80$ km/h: (I.a) vertical force F_{zm}, (II.a) longitudinal force F_{xm}
- $V_x = 110$ km/h: (I.b) vertical force F_{zm}, (II.b) longitudinal force F_{xm}

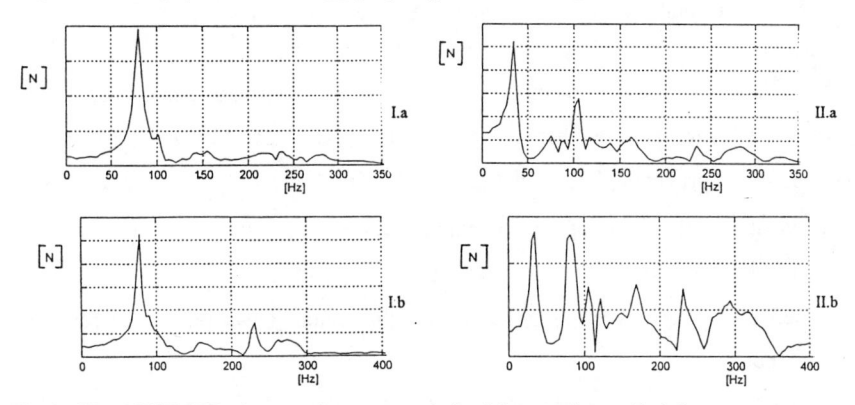

Fig. 4 - Tyre 195/60 R15: cleat test: frequency analysis of the experimental hub forces

- $V_x = 80$ km/h: (I.a) vertical force F_{zm}, (II.a) longitudinal force F_{xm}
- $V_x = 110$ km/h: (I.b) vertical force F_{zm}, (II.b) longitudinal force F_{xm}

In the range between 0 and 130 Hz the most excited modes are:

- 1st mode (natural frequency of about 34 Hz) classified as torsional-longitudinal mode, visible in the longitudinal force signal;
- 2nd mode (at about 80 Hz) with a predominant component in the vertical force signal;
- 3rd mode (torsional-longitudinal) with natural frequency of about 94 Hz, visible in the vertical force signal;
- 4th mode (torsional-longitudinal) at 103 Hz.

In figure 5 the modal parameters, which were identified with three different approaches, are reported for different V_x speed velocities. The first and the second vibrating modes have been identified with great accuracy by the three methods in the whole considered speed range, both in terms of natural frequencies and of damping factors. The third and the fourth modes have been partially identified with some difficulties, since the type of obstacle used was unable to introduce enough energy in the system. The third vibrating mode shows also a "hole" in the definition between 80 and 110 km/h.

It is possible to state that all the results obtained with the three identification techniques give similar values in the presence of adequate force signals.

The whole identification methodology is shown in a real case concerning a 185/60 R14 tyre (inflating pressure 2 bar, vertical load 1990 N) mounted on a 5 ½ J 14 rim. The first step of the procedure has permitted (see table I):

- to evaluate of the rigid ring mass m_b and the J_b and J_r inertial moment by means of the torsional-pendulum test;
- to define the ring radius $R = r_e$ (effective rolling radius measured on the road-wheel machine);
- to identify the torsional stiffness c_{bt} and damping r_{bt} by means of the torsional vibrator tests. Figure 6 reports the experimental transfer function $h^{(S)}(\omega)$ between the rotation θ_b of the tread band and the one θ_r imposed to the hub, compared with the analytical ones reconstructed with the identified parameters (see eq. 4);

TYRE MODEL FOR THE STUDY OF IN-PLANE DYNAMICS

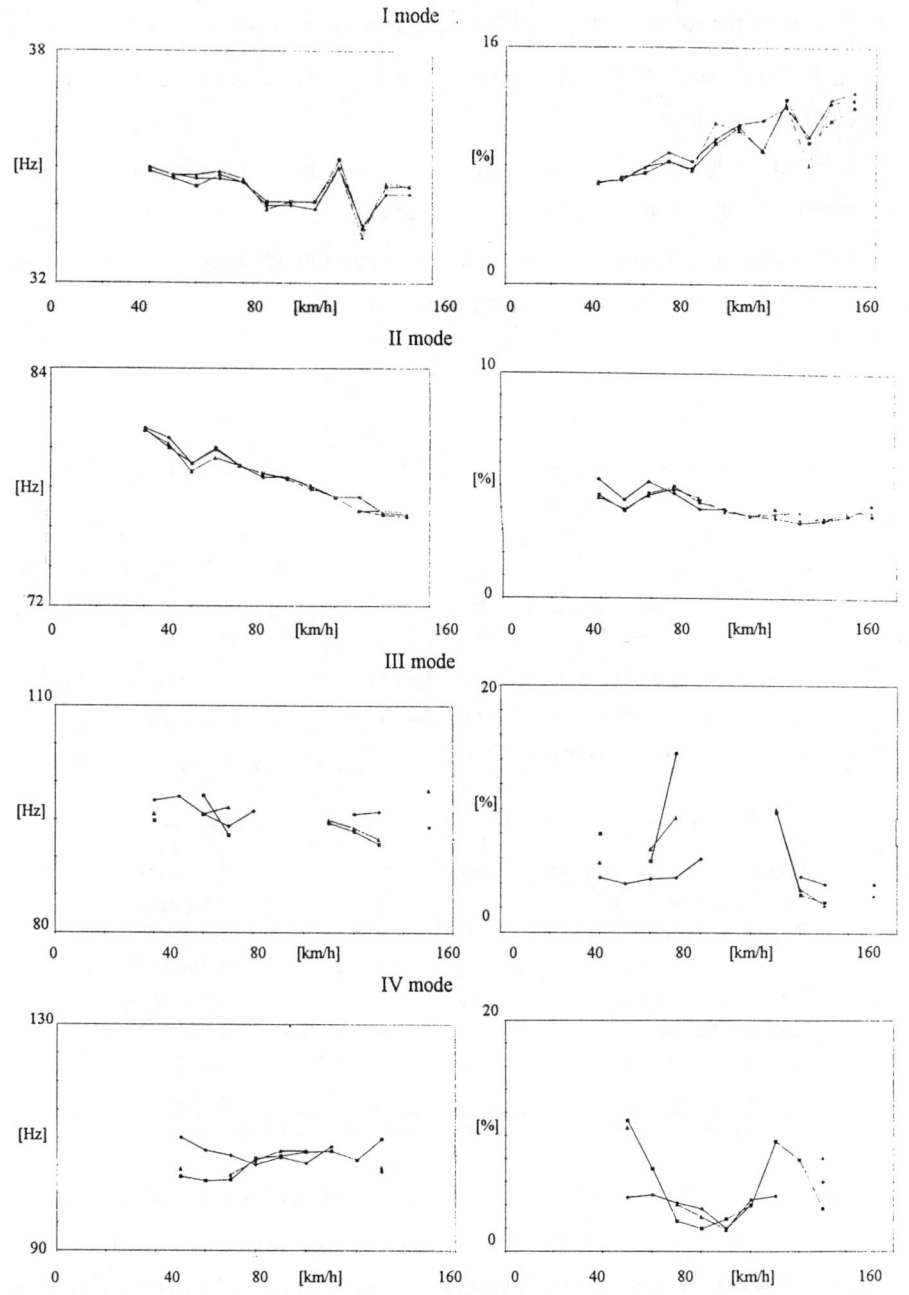

Fig 5: Tyre 195/60 R15: experimental natural frequencies [Hz] and related damping factors [%] vs velocity Vx

- to identify the longitudinal slip stiffness c_{kx} from the transfer function $H_{F_{xm}.V_{sx}}$ (see figure 7) between the longitudinal force at the rim F_{xm} and the slip speed V_{sx} measured on the dynamometer trailer;
- to evaluate the semi-length of the contact area "a" from the measurement of the dimensions of the road-wheel contact patch area ([3]);
- to evaluate the distance "f" of the vertical resultant force with respect to the wheel centre by measuring the rolling resistance moment.

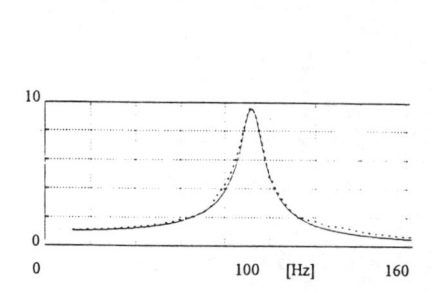

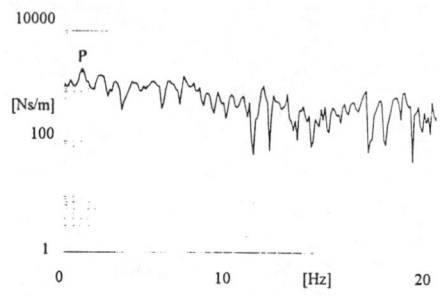

Fig 6: Tyre 185/60 R14: transfer function beetween tread band and rim rotation (dotted line: experimental $h^{(S)}$, continous line: analitycal $h^{(c)}$)

Fig 7: Tyre 185/60 R14: experimental transfer function $H_{F_{xm}.V_{sx}}$ between longitudinal rim force and slip velocity V_{sx} (see eq. 5)

Rigid ring mass	m_b	= 4.24	[kg]
Rigid ring inertial moment	J_b	= .33	[kg m²]
Rim inertial moment	J_r	= .41	[kg m²]
Rigid ring radius	R	= .279	[m]
Torsional stiffness	c_{bt}	= 110660	[Nm/rad]
Torsional damping	r_{bt}	= 20.1	[Nm/rad]
Creep stiffness	c_{kx}	= 19470	[N]
Half contact lenght	a	= .05	[m]
Vertical result arm	f	= .002	[m]

Tab. I: Tyre 185/60 R14: Identified values in the first identification step

The response of the forces transmitted to the fixed hub in the "passing over an obstacle with fixed hub" tests was measured. The tests have been conducted at different velocities between 20 and 140 km/h using both a square (15x15mm) and a triangular (base 15mm, height 15mm) obstacle to excite the tyre in a different way.

As an example, figure 8 reports the spectra of the wheel hub vertical and longitudinal forces, at the speed of 60 and 130 km/h, using the two different obstacles.

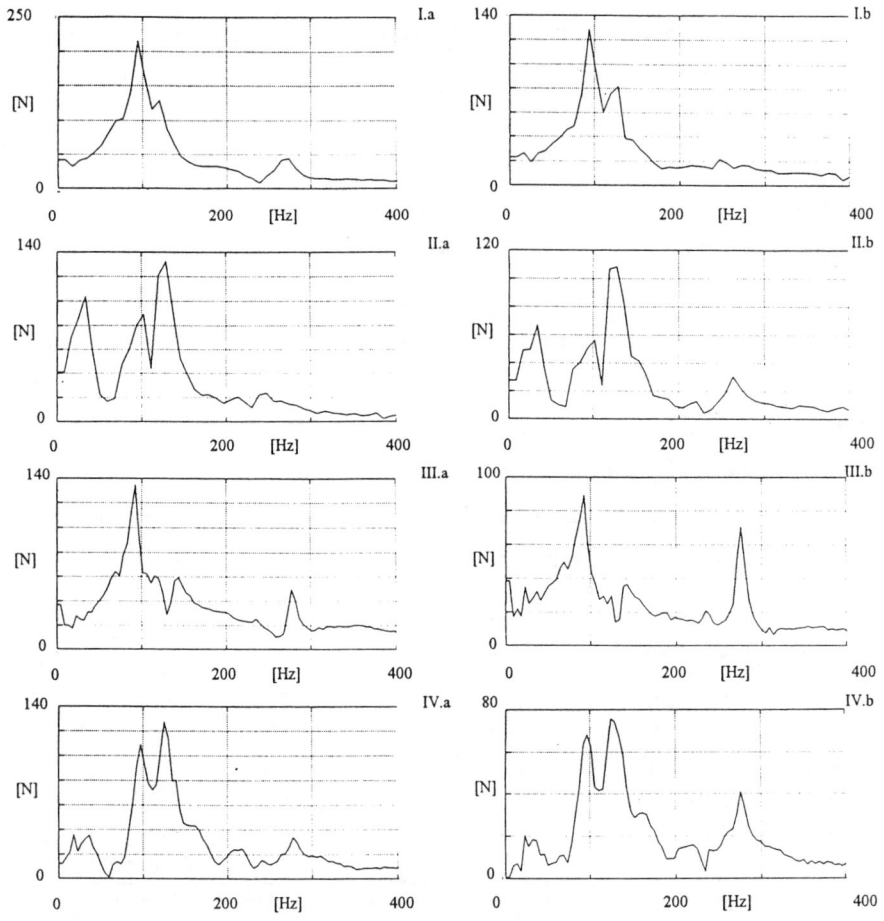

Fig 8: Tyre 185/60 R14: tyre cleat test: frequency analysis of the experimental hub forces: vertical force F_{zm} (I and III) and longitudinal force F_{xm} (II and IV) at the speed of 60 (I and II) and 130 km/h (III and IV), using the two different obstacles: the square (a) and the triangular (b)

It is possible to see quite clearly the presence of the four natural frequencies of the tyre in the range 0-130 Hz:
- 1st mode (frequency 36 Hz) classifiable as a longitudinal-torsional;
- 2nd mode (at 94 Hz) with a predominant component in the vertical force;

- 3rd mode (torsional-longitudinal) with frequency at about 102;
- 4th mode torsional-longitudinal at 125 Hz.

Comparing these spectra, it is possible to see how the different obstacles excite the vibrating modes differently, thus allowing a better identification of the parameters of the rigid ring model.

The responses at different velocities have been analyzed with the different time domain methods of identification, evaluating the experimental behaviour of the natural frequencies and the damping factor at different speeds V_x. These experimental values have been obtained as an average of the identified values with the three different methods for each vibrating mode and speed, the associated standard deviations being very limited. The 185/60 R14 tyre natural frequencies remain quite constant at any advancing speed, while the damping factor slowly increases, except for the first mode value, which increases rather sensibly with the speed.

The identification of the third mode (better identified with a triangular obstacle) is difficult since it is often confused with the second vibrating mode (very close in frequency).

Radial stiffness	c_b	=	1351200.0	[N/m]
Residual longitudinal stiffness	c_{cx}	=	449700.0	[N/m]
Residual vertical stiffness	c_{cz}	=	192900.0	[N/m]
Residual torsional stiffness	c_{ct}	=	23870.0	[Nm/rad]
Radial damping	r_b	=	276.0	[Nm/rad]
Vertical damping	r_{cz}	=	123.5	[Nm/rad]
Torsional damping	r_{ct}	=	41.5	[Nm/rad]

Tab. II: Tyre 185/60 R14: identified values in the second identification step

From these results, the identification of the physical characteristics of the remaining unknown parameters of the "rigid ring" tyre model (reported in Tab. II) was performed, minimizing the difference between the modal parameters, calculated with the model, and the ones experimentally identified.

The analytical natural frequencies and damping factors, reconstruced with the reference model using all the identified parameters (reported in Tabs. I and II) are compared with experimental values. It is possible to see that the error concerning the

TYRE MODEL FOR THE STUDY OF IN-PLANE DYNAMICS 149

first three modes is little, while it increases for the fourth vibrating mode, where a significant underestimation of the damping exists.

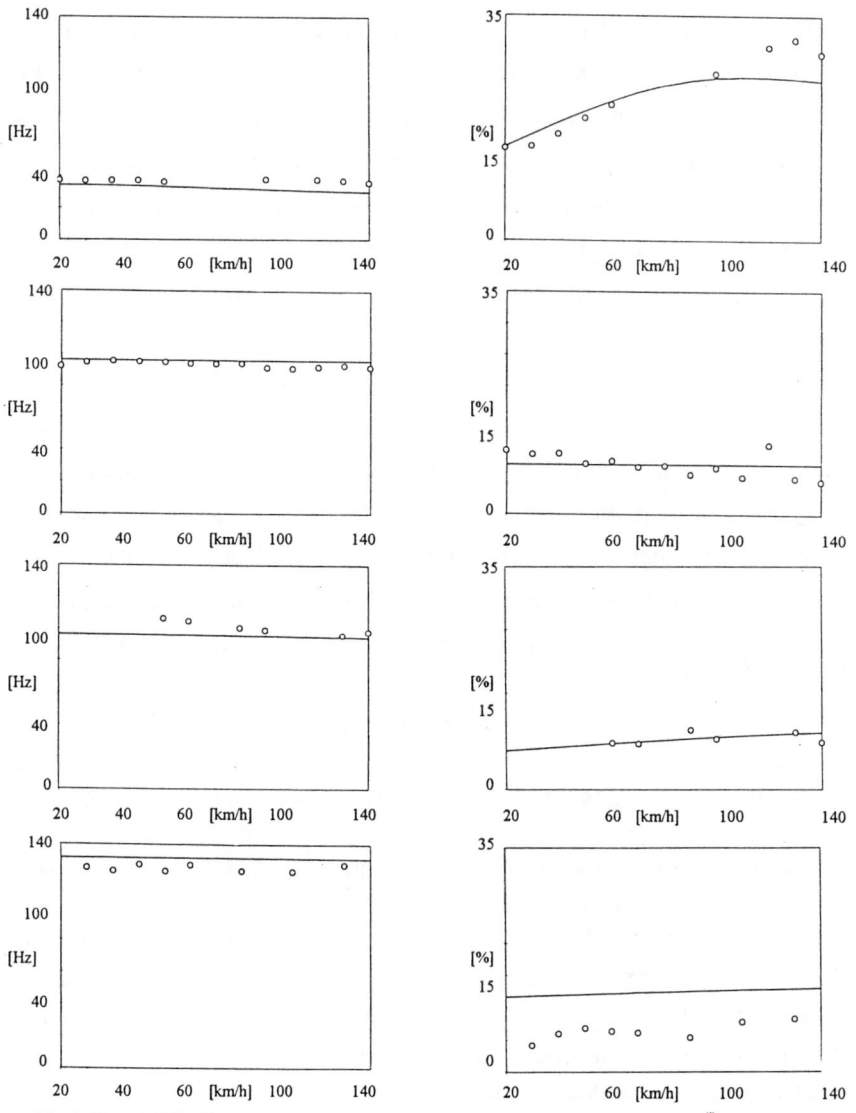

Fig 9: Tyre 185/60 R14: natural frequencies (I) and related damping factor $(r/r_c)^{(i)}$ (II) vs velocity V_x
(ooo experimental data; --- reconstructed behaviour with identified parameters)

4. CONCLUSIONS

The rigid ring tyre model is good for the reproduction of the tyre dynamics in the frequency range of interest in order to achieve a good modelization of the driving comfort and of the phenomena linked to the torque transmission to the wheels (30 - 130 Hz), reproducing both vertical and longitudinal-torsional motions. In this work, with respect to the Zegelaar-Gong-Pacejka model, also the damping parameters have been introduced. A new methodology to identify the physical parameters of the rigid ring model has been studied from a limited number of tests, among which the braking and the obstacle overtaking test, already used by the tyre producing factories.

The obtained results, applied to a real case, are as a whole largely satisfactory, remembering that the main goal of this research phase was to propose the methodology, to evaluate the feasibility and to validate the results.

Surely this model must be perfected: by introducing the frequency dependance of the damping and stiffness physical parameters, by implementing the identification problem with statistical methods and by analizing possible modifications of the obstacle overtaking test, for instance by means of road models to better excite the tyre vibrating modes and to solve the problems for some frequencies at certain advancing speeds.

REFERENCES

1. Zegelaar, P.W.A., Gong, S., and Pacejka, H.B., "Tyre models for the Study of Inplane Dynamics", The Dynamic of Vehicle on Roads and Tracks, Vehicle System Dynamics, Swet & Zetlinger, 1994.
2. Pacejka, H.B. "Modelling of Pneumatic Tyre and its Impact on Vehicle Dynamic Behaviour", Technische Universiteit Delft, 1988.
3. Mancosu, F., Matrascia, G., Cheli, F., "Techniques for Determining the Parameters of a Two-Dimensional Tyre Model for the Study of Comfort", 15th Meeting of the Tyre Society, Ohio, 1997.
4. Clark, S.K., "Mechanics of Pneumatic Tires", Ed. U.S. Department of Transportation. National Highway Traffic Safety Administration. Washington, 1982.
5. Marple, S., "Digital Spectral Analysis, with Applications", Ed. Prentice Hall.
6. Pappa, R.S., Ibrahim, S.R., "Large Modal Survey Testing using of the Ibrahim Time Domain Identification Tecnique", Spacecraft and Rockets (Vol. 19), N. 5, September 1992.

ACKNOWLEDGEMENTS: This study received a grant by MURST and CNR (Italy).

On the low frequency in-plane forced vibrations of pneumatic tyre / wheel / suspension assemblies

D.J. ALLISON and R.S.SHARP

ABSTRACT

The paper contains an interim report on a research project into the longitudinal vibrations of vehicles. The aim is to improve understanding of the process by which tyres come to vibrate by rolling along surfaces which may be smooth or profiled and the transmission of the vibration into the vehicle cabin. A description of a specially designed rig is given. This allows control of the boundary conditions of a tyre / wheel assembly which runs on a rotating drum, and allows forcing to be applied by an electromagnetic shaker to the hub in the vertical or longitudinal direction. A measurement scheme has been implemented to record forces and translational and spin accelerations of the hub. A mathematical model of the rig has been built using the multibody modelling software Autosim. The tyre is represented as a hub and rigid circular ring with translational and rotational freedoms between them. The hub is mounted in a suspension system allowing it translational and spin freedoms. The model and the measurements are used to generate frequency responses, and a parameter identification scheme is being followed to obtain a best fit of the theoretical results to the experimental results in the frequency range up to about 100 Hz. Vibration spectra and comparative experimental and theoretical frequency response functions are included.

1. INTRODUCTION

Pneumatic tyres are resonant structures making area contact with generally profiled and textured road surfaces. In normal operation, the tyre rolls along, so that the vibrating structure rotates and the contact involves different material parts of the tyre tread from instant to instant. Excitation of the tyre structure into vibration is also contributed by the tyre tread pattern, necessary for drainage from the tyre / road interface when standing water is on the roadway. It seems inevitable that pneumatic tyre structural vibrations will be excited by roads in the future. Adding vibration damping mechanisms to the structures seems to be diametrically opposed to the need to minimise rolling resistance and is not a practical option. Therefore good vibration engineering in the tyre area would seem to depend on tuning tyre and mounting system, at least to avoid the worst of possible interaction problems.

The vibrations of tyres mounted on wheels with a fixed hub have been studied quite extensively [1-9]. Gong's work, in particular, reveals that considerable progress can be made with the understanding of tyre vibration properties up to 100 Hz through relatively simple structural models, in which the circumferential ring of the tyre is considered rigid and non-sliding tread "bristle" interactions with the ground occur. Beyond about 100 Hz, bending vibrations of the tread ring beam and above about 1000 Hz, extensional vibrations of the ring become important and the complexity of the theory increases markedly.

A tyre mounted on a modern vehicle does not have a fixed hub. The hub is normally compliant both vertically and longitudinally. Consequently, the possibility arises that the hub motions occurring as a part of the system vibrations, which involve the tyre structure and tyre ground interaction forces, will influence those forces and thereby affect the vibrations. Then, the tyre has become part of a system, the vibrations of which must be considered as a whole.

The research described subsequently is based on this systems approach. A test rig allowing in-plane wheel hub and tyre vibrations, control of the wheel mounting conditions and the application of shaking forces from an electromagnetic vibrator, together with tyre / rotating drum interactions, has been designed, built and instrumented. A mathematical model of the rig has been constructed and an identification scheme for matching model behaviour of rig behaviour has been devised. Vibration test results from the rig for longitudinal, sinusoidal excitation have been obtained and signal processing of the results has indicated good reliability of most of the measurements. Frequency responses have been derived and are currently being used to identify tyre and rig parameters via an optimiser.

The rig and its instrumentation system are described in the next section, following which a mathematical model of the complete rig is described. Experimental results are discussed next and this is followed by an explanation of the procedure used for the identification of rig and tyre parameters. Conclusions drawn are of an interim nature, since the work is not yet complete.

2. EXPERIMENTAL RIG

2.1 Design overview.

The rig is located within the Dynamics Laboratory of SP Tyres (UK) Limited. It utilises one drum of a four drum rolling road test facility. The ground location of any rig operating at this facility is achieved by fixing the base structure to large test bed plates positioned beside each drum. The large diameter smooth drums can be fitted with a variety of regular profiles, such as cleats and sinusoids, and various types of replica road surface textures.

The rig design is illustrated in Fig. 1. Shown in addition are an electromagnetic vibrator, the wheel assembly, and the measurement transducers.

The rig consists of two base block assemblies, constructed from square section mild steel. These are bolted to the test bed plates to provide a rigid base for the hub suspension. The vertical suspension uses a standard Range Rover front air spring assembly. The load is transmitted between the hub and spring by a pullrod and bell crank arrangement. The hub is supported by spherical joint ended tubular steel suspension members. By reference to Blevins [10], member dimensions were chosen to avoid any rig resonances at frequencies within the test range 0-100 Hz. The hub is from the rear of a Rover 600 series car, fitted with an anti-lock braking system (ABS) sender ring and pick-up. The construction allows a range of tyre and wheel assemblies to be fitted.

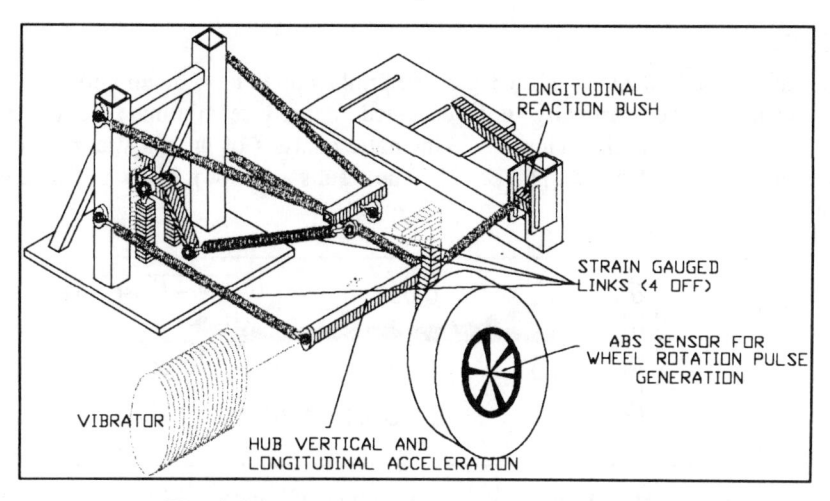

Fig. 1. Diagrammatic representation of test rig.

2.2 Instrumentation.

The positions of the various transducers can be seen in Fig. 1. The tubular suspension links are fitted with full Wheatstone bridge strain gauge assemblies. This allows the measurement of the tensile or compressive loads within the suspension. A commercial piezo-resistive load cell is used to measure the force applied to the hub by the electromagnetic shaker. The hub longitudinal and vertical accelerations are measured by commercial piezo-electric accelerometers with built in pre-amplifiers, excited by constant current supplies.

The hub angular acceleration is measured by utilising the signal from the ABS system fitted to the hub. The system fitted consists of a swaged tooth ring attached to the rotating brake disc, and a statically mounted inductive coil pick-up. The pick-up generates a periodic voltage with frequency equal to the tooth pass frequency. This signal is processed off-line, by specially written software, to extract the hub angular velocity. Artificial data has been manufactured to describe various wheel motion conditions and the software processing has successfully reconstructed the wheel velocity and acceleration profiles.

An optical pick-up, (not shown in Fig. 1), is used to generate 'a once per hub revolution' TTL pulse. To output a high signal state at a single hub position per revolution, a section of reflective tape is used to trigger the optical pick-up. This signal is used to correlate time series measurements from test runs performed at the same drum and wheel speeds, to allow for time domain averaging and, in particular, subtraction from shaker, tyre and drum forced vibration responses, those due to the tyre and drum imperfections.

2.3 Design Features.

The suspension kinematics of the rig constrain the hub to move nominally in-plane. As the hub moves vertically, it is designed to produce minimal lateral scrub, camber, and steer angle change. A kinematic analysis of the rig was performed using the software Autosim, [11,12,13]. The results are shown below in Fig. 2 and Fig. 3.

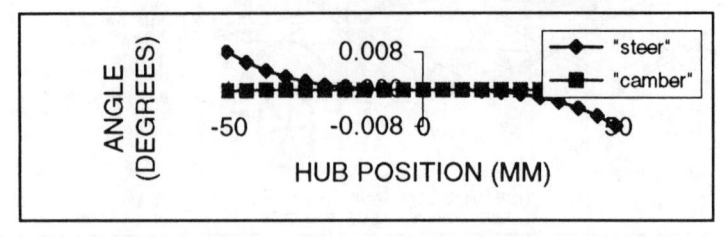

Fig. 2. Steer and camber change for range of hub vertical position.

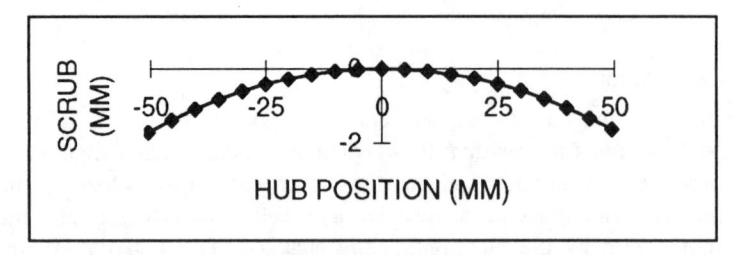

Fig. 3. Lateral scrub for range of hub vertical position.

The load of the tyre against the rolling road drum can be adjusted by varying the pressure within the airspring. The anti-dive characteristics of the suspension can be varied by adjusting the angle of the longitudinal reaction rod with respect to the ground plane. The longitudinal kinetics are adjustable by using bushes of differing stiffness at the end of the longitudinal suspension rod.

2.4 External forcing.

The rig can be forced into vibration by attaching road surface profile or texture to the rolling road drum. It can also be forced by the electromagnetic shaker, Fig. 1. This is suspended by elastic straps from a frame, and allows application of force to the hub in either vertical or longitudinal direction. Forcing applied has been both sinusoidal and random. However random forcing resulted in a very poor signal to noise ratio; therefore it is considered disadvantageous for this work.

2.5 Measurement system - data capture.

The data capture system is a portable PC based system, which can be completely battery operated. The system contains strain gauge excitation and signal conditioning modules, and differential voltage measurement. All the recorded channels can be low pass filtered at hardware alterable cut-off frequencies. Currently, filters are configured to roll off at 1000 Hz. Matlab software is used to analyse the raw data. This allows very flexible off-line processing and graphing.

2.6 Proving - impulse hammer testing.

The rig must have no resonances within the range of frequencies of tyre and suspension vibrations being studied. This was investigated by using an instrumented impulse hammer with a soft tip to excite the rig, and measuring the response with a magnetically attached accelerometer. Different pairs of forcing and response measurement positions were used and tests of each pair were repeated several times. Force time histories were checked to exclude records showing double hits. By using standard Fast Fourier Transform based methods in Matlab, frequency responses and coherence functions were computed for each pair. It was discovered that the longitudinal reaction assembly resonated at the lowest frequency, but that this was above the highest frequency of interest. Plots of the frequency response gain and coherence over 16 trials for the longitudinal reaction block acceleration response to longitudinal force input are shown in Fig. 4 and Fig. 5.

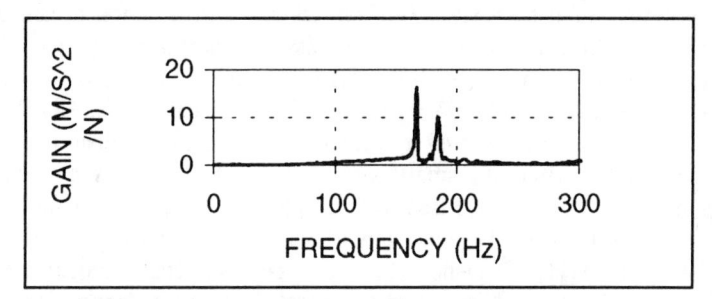

Fig. 4. Longitudinal reaction block acceleration to force frequency response function gain.

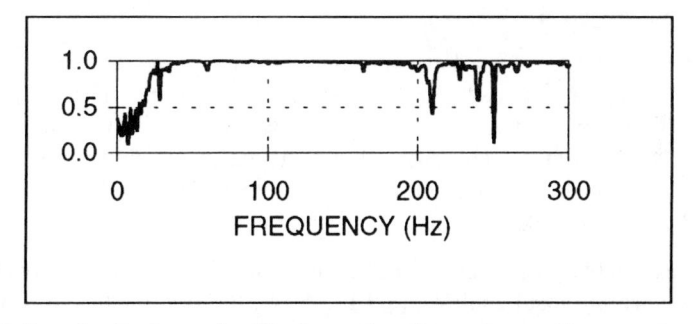

Fig. 5. Longitudinal reaction block acceleration to force coherence function.

3. MODEL

The tyre / wheel / suspension system model is shown diagrammatically in Fig. 6. For clarity it is divided into two parts.

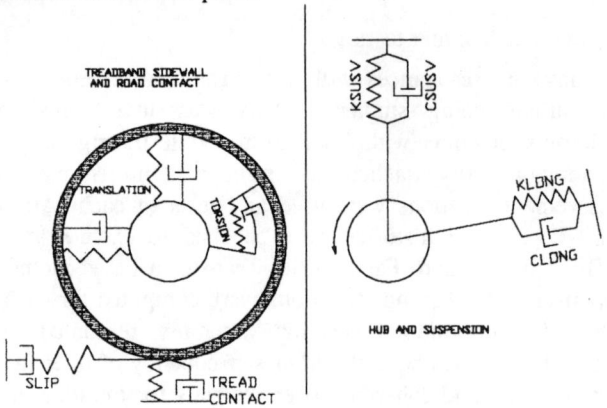

Fig. 6. Tyre and wheel model.

The tyre is represented by a rigid ring on an elastic foundation attached to the wheel hub. The ring is considered to have mass and inertia, and is given freedom to translate in-plane and rotate with respect to the hub. The tread / road contact patch is represented longitudinally by a non-sliding "bristle" model [14], and the tread vertical contact is represented by a force applied to the treadband centre. The bristle model is linearised for small perturbations of slip about an equilibrium operating state and the effective free rolling radius is represented as an empirical function of the vertical tread contact force [14]. Specifically the shear force law used is $\{(Vr-Vx) / |Vr|\}*Cx$, where Vr is the constant road velocity, Vx is the unsteady linear velocity of the point of contact of the tyre treadband, and Cx is the longitudinal slip stiffness of the tread. Also the rolling radius is related to the vertical hub deflection by $RRad = Rad - \xi \cdot (1-\eta) * \delta$, where $RRad$ is the effective rolling radius, Rad is the free rolling radius, ξ and η are constants and δ is the vertical hub deflection.

The hub is also considered to have mass and inertia. It has freedom to rotate and is supported by a suspension system allowing in-plane motion. The planar motion is resisted by a spring and viscous damper arrangement in both vertical and longitudinal directions. The suspension kinematics are adjustable to represent different rig anti-dive configurations. The suspension members themselves are assumed to be rigid links, justified by the rig proving work.

From in extensible ring theory of a tyre treadband on an elastic foundation [1-9], and experimental modal analysis, [8,9,15,16,17], the first treadband flexural mode occurs at about 120 Hz for family car sized tyres. This mode of vibration is lightly damped, and will not be excited much by forcing below 100 Hz. The model is therefore considered suitable for a frequency range of tyre vibration up to about 100 Hz.

Autosim is used to build the model. It generates linearised state space equations of motion in a Matlab M-file. Subsequent processing has been done within Matlab involving, in particular, the calculation of frequency responses with the shaker force taken to be the input to the system.

4. EXPERIMENTAL RESULTS.

The experimental results presented were obtained by forcing the rig using the electro-magnetic shaker. A tyre was loaded against the smooth drum which was rotating at constant velocity. The shaker applied a single frequency sinusoid to the hub in the longitudinal direction. Time history measurements were recorded for a sequence of forcing frequencies from 10 to 100 Hz in steps of 1 Hz. In order to maximise signal to noise ratios, the shaker was set to deliver its maximum output without mechanical limiting for each frequency. Records of the hub vertical and longitudinal accelerations and reaction forces, the input shaker force, and the output from the ABS sensor, were captured for an eight second period at each forcing frequency. Each record was divided into seven overlapping segments, the mean values removed, and hanning windowed to prevent leakage. Using standard Fast Fourier Transform techniques, the power spectra of each signal, and the cross spectra between the input force and resultant hub motions and forces were

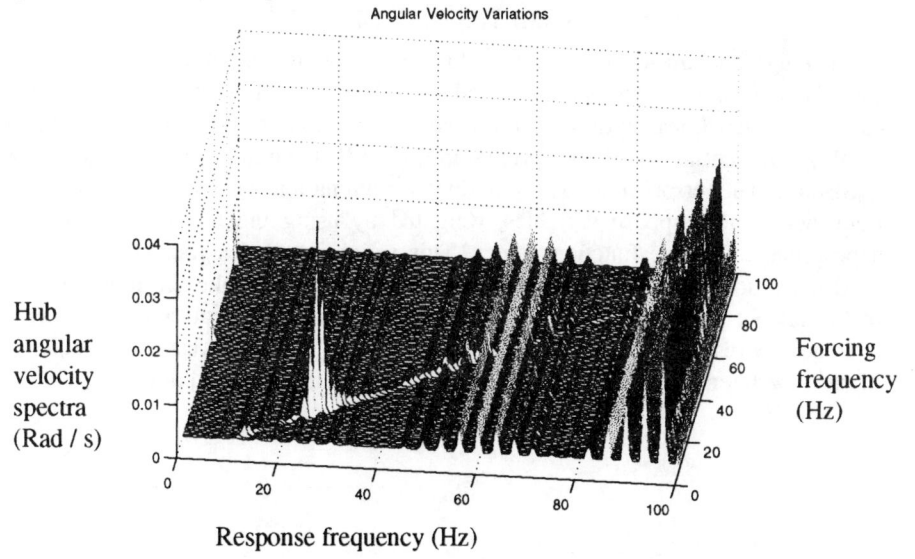

Fig. 7. Hub angular velocity spectra for forcing in the range 10 - 100 Hz.

calculated. This data was normalised using an asymptotically unbiased scaling factor to remove the effect of the window, and the overlapping spectral estimates ensemble averaged. Hub angular velocity line spectra (showing vibration amplitudes, not mean square values) from tests employing the full range of forcing, 10 - 100 Hz, with one tyre at a constant speed, are shown in Fig. 7.

The wheel rotation frequency was 3.6 Hz. The lines parallel to the forcing frequency axis correspond to fixed response frequencies which are harmonics of wheel rotation. Responses at the forcing frequency applied to the rig can be seen as a diagonal line passing through the origin, with the largest peaks indicating the most vigorous resonant conditions. It is significant that the greatest response is at a frequency of 28 Hz. This is around the natural frequency of the torsional mode of the treadband contra-rotating against a free hub.

Coherence functions were calculated, to test the reliability of the recorded data. Results have confirmed that coherence approaching unity is obtainable for most of the response variables at the forcing frequency. Where the coherence is significantly worse than this, it is associated either with smallness of the response signal (vertical hub force) or with a wheel rotation harmonic coinciding with the forcing frequency. Coherence values have been employed as weighting parameters in the optimisation (see below), to reflect their physical significance.

5. PARAMETER IDENTIFICATION

The parameter identification scheme is to match experimental and model results by suitably adjusting model parameter values. The algorithm used is one of the constrained non-linear optimisation procedures available in the Matlab toolbox. It employs sequential quadratic programming [18] to minimise a scalar error function. The error function is a double summation of terms over all test frequencies and response variables, weighted by using the fourth power of the appropriate coherence function value. This weighting is designed to relieve the need in the identification to find parameter values which make the model match real behaviour known only with low confidence. The terms themselves are squares of frequency response phasor differences normalised by the corresponding squares of the experimental phasor magnitudes. Such a term is illustrated in Fig. 8.

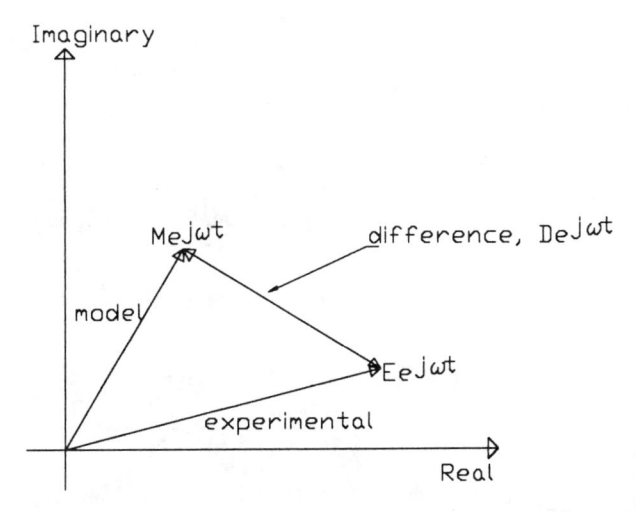

Fig. 8. Experimental and model frequency response phasors for a particular variable at a particular frequency.

The contribution of this variable at this frequency to the error function would be $\{(D/E)^2 \gamma^4\}$ where γ is the relevant coherence function value. In total the error function is given by:

$$\sum_{n=1}^{N}\left\{\sum_{\omega_1}^{\omega_2}\frac{\text{Re}(Expr_n(\omega)-Modl_n(\omega))^2+\text{Im}(Expr_n(\omega)-Modl_n(\omega))^2}{(Expr_n(\omega))^2}\right\}*\left(\gamma_{xy_n}(\omega)\right)^4$$

in which N variables contribute and the frequency range used is $\omega_1 \rightarrow \omega_2$.

After identification, model frequency responses should match those experimental ones which are known with certainty, and the parameters of the rig and tyre should be known. Rig development work is not yet complete and, so far, convergence of the theoretical results on the experimental ones has been by no means perfect.

Frequency response function results for the longitudinal rod force output, after optimisation, are shown in Fig. 9. The low frequency matching is better than that at higher frequencies and this is a general feature of results obtained so far.

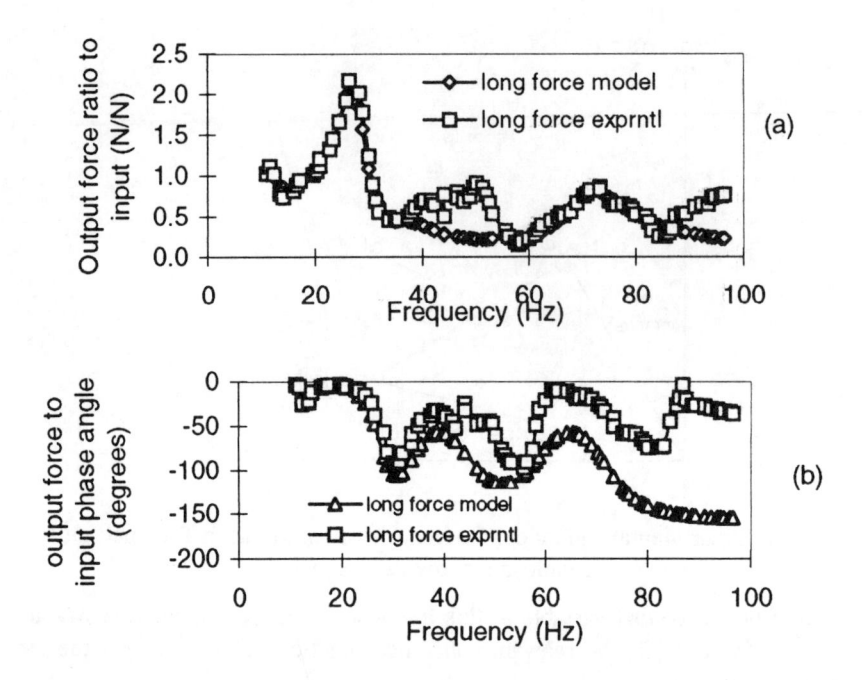

Fig. 9. Longitudinal rod axial force frequency response function (a) gain (b) phase.

From the same test series, the longitudinal hub acceleration frequency response gain has been computed and this is shown in Fig. 10.

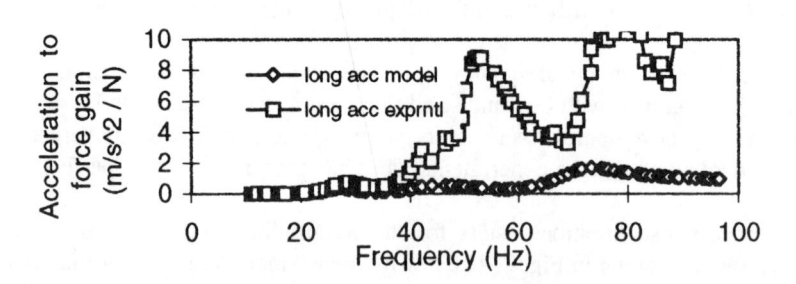

Fig. 10. Longitudinal hub acceleration frequency response gain.

The figure illustrates the main point of difference between the theory and the experiments at present. The rig has some higher frequency resonant behaviour, involving longitudinal hub motions, which the theory is not able to reproduce. The problem has been traced to the anchorage of the longitudinal rod support and it is thought that attention to detail in this area will remove the major differences.

6. CONCLUSIONS

An instrumented test rig providing for in-plane motions of wheel and tyre and variations in the kinematic and compliance hub boundary conditions has been designed and built. Shaker access to the wheel hub is provided, so that forced vibration responses can be recorded. Sinusoidal forcing has been shown to provide repeatable results and data processing has yielded frequency response functions, including those for wheel angular velocity.

A rig model has been constructed and a parameter identification scheme put in place to allow optimisation of tyre and rig parameters, minimising significant differences between theory and experiment. Only partial success with parameter determination has so far been achieved. Work is continuing to improve the match between theory and testing.

ACKNOWLEDGEMENT

The authors are pleased to acknowledge the financial support of the U.K. Engineering and Physical Sciences Research Council and the technical and resource contributions of their colleagues at SP Tyres and Rover Group.

REFERENCES

1. Allaei, D., Soedel, W. and Yang, T.Y., Eigenvalues of rings with radial spring attachments, Journal of Sound and Vibration, Vol. 121, No. 3, 1988, 547 - 561.
2. Huang, S.C. and Soedel, W., Effects of Coriolis acceleration on the free and forced vibrations of rotating rings on elastic foundations, Journal of Sound and Vibration, Vol. 115, No. 2, 1987, 253 - 274.
3. Endo, M., Hatamura, K., Sakata, M., and Taniguchi, O., Flexural vibration of a thin rotating ring, Journal of Sound and Vibration, Vol. 96, No. 2, 1984, 261-272.
4. Huang, S.C. and Su, C.K., In-plane dynamics of the tyre on the road based on an experimentally verified rolling ring model, Vehicle System Dynamics, Vol. 121, 1987, 247 - 267.
5. Allaei, D. and Soedel, W., Natural frequencies and mode shapes of rings that deviate from perfect axis symmetry, Journal of Sound and Vibration, Vol. 120, No. 1, 1986, 9 - 27.
6. Bickford, W.B. and Reddy, E.S., On the in-plane vibrations of rotating rings, Journal of Sound and Vibration, Vol. 101, No. 1, 1985, 13 - 22.
7. Huang, Shyh-Chin and Su, Chen-Kai, Vibrations of spinning rings with radial and circumferential displacement constraints, ASAE Design Division, Vol. 36, 1991, 141 - 146.
8. Potts, G.R., Bell, C.A., Charek, L.T. and Roy, T.K., Tire vibrations, Tire Science and Technology, Vol. 5, No. 4, 1977, 202 - 225.
9. Gong S., A study of in-plane dynamics of tyres, Doctoral Dissertation, Delft University of Technology, 1993.
10. Blevins, R.D., Formulas for natural frequency and mode shape, Van Nostrand Reinhold, New York, 1979.
11. Sharp, R.S., Vehicle dynamics modelling with the aid of a symbolic multibody code, Proc. AVEC '96 Int. Symp. on Advanced Vehicle Control, Vol. 2, Aachen, June 1996, 971-984.
12. Sayers, M.W., Automated formulation of efficient vehicle simulation codes by symbolic formulation (AUTOSIM), The Dynamics of Vehicles on Roads and Tracks (Anderson R. Ed.), 1990, Swets and Zeitlinger, Lisse, 474 - 487.
13. Mousseau, C. W., Sayers, M.W. and Fagan D.J., Symbolic quasi-static and dynamic analyses of complex automobile models, The Dynamics of Vehicles on Roads and Tracks (Sauvage G. Ed.), 1992, Swets and Zeitlinger, Lisse, 446-459.

14. Pacejka, H.B., Tyre factors and front wheel vibrations, Int. Journal of Vehicle Design, Vol. 1 No. 2, 1980, 97 - 120.
15. Barone, M.R., Impact vibration of rolling tyres, SAE 770612, 1977.
16. Richards, T.R., Brown, J.E., Hohman, R.L. and Sundarum, S., Modal analysis of tyre relevant to vehicle system analysis, 3rd Int. Modal Analysis Conf., Orlando, TSTCA Vol. 21, 1985, 23 - 39.
17. Richards, T.R., Charek, L.T. and Scavusso, R.W., The effects of spindle and patch boundary conditions on tyre vibration modes, SAE 860243, 1986.
18. Gill, P.E., Murray, W. and Wright, M.H., Practical optimisation, Academic Press, London, 1981.

3D MEMBRANE SHELL MODEL IN APPLICATION OF A TRACTOR AND PKW TYRE

DIETMAR ZACHOW

ABSTRACT

Tyre models as interface between vehicle and track are developed and used with different target. Generally one differentiates between vehicle dynamics and vehicle comfort. To fullfil the standards of the vehicle comfort, tyre models must be able to calculate high frequency and geometrical nonlinearity, caused by steps or ground waves. For the dynamic modelling of the roll characteristics of tyres three types of models are developed in SFB181.

The discrete 3D-membrane-shell-model for the tyre can be used for vehicle comfort calculation with frequencies even beyond 1000Hz.

The discretisation of the ribs as a mass-point-system enables to investigate the driving quality on deep ground. The modelling of the membrane as a mass-point-system to calculate geometrical nonlinearity.

This investigation presents the 3D-membrane shell model applied to a tractor tyre and a pkw tyre on rigid ground and deep soil.

INTRODUCTION

The dynamical calculation of the rolling process of a tyre need the equilibrium configuration of the tyre under pressure. Various experiments have shown, that a different choice of the geometry in comparisation to the equilibrium configuration causes instability during the inflating process of the tyre. Local buckling of the mass point system, caused by the stiffness of the ribs, are not taken into account. With the help of a laser rig, the outside shape will be take from the inflated tyre. Than we have a cross section measured by 0.0 bar and the inflated cross section by 1.6 bar.

A programm calculates the elastic equilibrium configuration under pressure. The belting function assumed as a quadratic function and the belting constant will be determinated in this way, that the belt curvature of the tyre will aggree with the measurement. In cause of the spatial dynamik of this mass point system, to simulate the rolling process of a tyre, the number of the degrees of freedom is limited.

DESCRIPTION OF THE TYRE MODELL

The 3D-membrane shell modell as a tyre modell is represented in his simpliest way by 8 tracks and 18 cross section. Two tracks are used for the rim, which is rigid, four tracks are used for the belt included the tyre shoulder and two tracks are used for the sidewall. The placement of this 8 points is given by the refered equilibrium configuration with 19 points for the cross section. Fig.1

shows the equilibrium shape as a mass point-system represents the belt and the sidewall of the tyre Fig.1.

The belt is calculated as a rope stiffened membrane. The belt area is modelled by a radial carcass layer and two cord layers with a small angle to the circumferential direction. The stiffness of one layer is given by the E-modul of one plyline, multiplied by the area of one rope, divided by the length of the eqivalence rod of the point system and multiplied by the density for the ropes along the perimeter. The sidewall has the stiffness of the carcass.

A single rib is represented by four masspoints, which will be coupled by basic points on the membrane. The position of the basic points is equal to the ribform. Additional ribpoints can be used to generate complicate ribshapes. The stiffness is given by the E-modul of rubber. The contactforces are acting on the ribpoints and proportionally to their position on the membrane surface, the forces summing up on the membrane points. As contact algorithm is used plastic impact on rigid road and Bekker-Theory on deep ground. Friction is included specially.

The timestep for integration of the equation system is determinated by the highest eigenvalue of this system. Various calculation have shown, that the Shannon theorem is fine enough to integrate explizite this system with a constant timestep.

The inflation acts as non conservative force in normal direction to the tyre surface and acts as prestress to the sidewall and the belt.

The stiffness in bending of the belt will be replaced by additional force couples. These forces are determinated by an angle of bending in circumferential and meridional direction.

THE DETERMINATION OF THE PARAMETER

The next step is to control this description for a tyre. For this, the dilatation in circumference direction of the inflated tyre will be compared with the dilatation of the inflated membrane model. The dilatation of the tyre is measured on a laser rig Fig.2. The dynamical calculation for the inflating process for the membrane is shown in Fig.3.

In comparision to the measurement on the laser rig is the dilatation of the model in the middlest track has the same magnitude. The calculated value is 0.8cm and the measured value is 0.9cm. Because of these simple discretisation the description of the dilatation in the tyre shoulder is not exact. The sidewall is correctly shown.

These results show, that already a simply description for the tyre as a membrane, described by a mass point system with two points for the sidewall and four points for the belt, are enough to calculate the geometry and the behaviour of an inflated tyre. For more information and a better approximation in some areas of the tyre, more mass points will be necassary. See the outlook.

APPLICATIONS

First, we will investigate the rolling process for this model. Before the model begins to roll, it gets a vertical deflection by 3.0 cm. These deflection grows in 0.2s from 0.0 to 3.0 cm. After 0.2s the tyre begins to roll and rolls than with a constant velocity at 18Km/h. After the rolling up, the model rolls stationary.

Fig.3 indicates the radial moving of one membrane point near the equatorline of the tyre. The course of the vertical deflection and a structure vibration about 45 Hz is shown. At a velocity of 36km/h the same behaviour is observed. The difference is the smaller wheel load fluctuation and the driving through a tyre resonance. That means, that the contact frequency of the rolling tyre is in the near of one eigenvalue of the tyre at a velocity of 18 km/h. The contact frequency of this rolling tyre is 42 Hz. This is recognizable at the growing wheel load during the roll up process and the reduction of this load after the roll up.

Next we will compare the rolling characteristics on rigid ground without cornering angle and with a cornering angle by 5 degree. The typical characteristics of cornering drive are the changing of the vertical load and a lateral deflection of the tyre, caused by a lateral shifting of the ribs. Both characteristic are included in this modelling. Fig.5 shows the comparision of the vertical load. The decrease of the load is visible.

We see a similar behaviour at a velocity of 18km/h and 36Km/h (Fig.6).
Fig.4 shows the lateral deflection of one rib point during three rotations. Fig.7/8 reproduce only the contact area and the typical curve for lateral deformation under the influence of a cornering angle. The cornering move of the ribpoint without cornering angle is caused by the touchdown-process of the ribs. In [5] is written, that the ribs move away in lateral direction from the equatorline of the tyre during vertical deflection and then they slip back to their old position. This means, that the ribs will move during contact in lateral direction, which grow up under the influence of a cornering angle.

For the rolling process on deep ground, the Becker theory for sandy clay soil is used. It is a 2D model for the soil pressure caused by the wheel load of the rolling tyre. Shear-forces are not included in this theory but coulombfriction. Fig.9 shows the radial moving of one membrane point near the equatorline of the tyre with its radial deflection and his structure vibration. A stationary rolling after two seconds is recognized. The vertical deflection of the tyre is smaller as on rigid ground.

During the rollup-process, after vertical deflection of the tyre, beginning at 0.5 seconds, the tyre climbed up from his starting position, begins once more to subside into the soil and climbs up to a nearly constant subside deep. These examples show the possibilities of the 3D membrane mass point system supposed by contact algorithm for rigid ground and deep soil.

2D MASSPOINT TYRE MODEL

Another application of the masspoint system is the generating of a particle system as 2D Tyre model [1]. This model is an elastic circularring model with a fine sensor layer to discribe the tyre tread. The determination of the parameter for this model is discribed in [2],[3] and [4]. This model scans ground waves or steps and gives information about the deformation included contact forces of a single tread element. Measuring of the deformation of single ribs with a Hall sensor, which where performed in the Institute for Vehicle Technology in Darmstadt [6], used as base to compare both results. Fig.10 shows the vertical and longitudinal deformation of one sensorpoint.

The difference between calculation and measuring is seen in the high of the amplitude of the vertical deformation. The reason is, that the Hall sensor with his magnet lays above the belt inside the tread. The calculation of one sensor point comprised the whole deformation between contact area and the belt. As these figure show the difference can be corrected by a factor. This example shows the use of a simple tyre model, to describe the behavior of the tread as single ribs during the driving.

OUTLOOK

The motivation for this paper was the description of the function of the the 3D-Membrane model and the presentation of some application. For more information about the behaviour of the tyre belt, tyre shoulder an the bead area is another discretisation nessary. The next discretisation step will be added by two points for the shoulder and two points for bead area.

The inflation pressure will be decreased at 0.4 [bar] on deep soil and another soil theory with shear will be used for the rolling on deep soil. The next investigation will be to improve safe driving and vehicle stability in a curve and on a slope. For this we will compare a normal wheel with a wheel with elavated wheelrim on the inside of the wheel. Further, we couple tyre models with vehicle models.

In this examples are used radial tyre. The description of bias tyre with their special theory is also necessary. In this context the 3D-membrane model will be supplied by a layered structure [7].

Bibliography

[1] Böhm, F. :
Reifenmodell für hochfrequente Rollvorgänge
auf kurzwelligen Fahrbahnen
VDI Berichte Nr.1088, 1993

[2] Böhm, F. :
Beanspruchung und Verformung
des Gürtelreifens unter Innendruck
Automobil-Industrie 3/85 Reifenmeßtechnik 317

[3] Böhm, F. :
Mechanik des Gürtelreifens
Ingenieur-Archiv 35.Band, 2.Heft, 1966, S.82-101

[4] Bannwitz.P,Oertel.Ch:
Adaptive Reifenmodell
Aufbau, Anwendung und Parameterbestimmung
VDI Berichte Nr.1224, 1995

[5] Siefkes Tjark :
Die Dynamik in der Kontaktfläche von Reifen
und der Fahrbahn und ihr Einfluß
auf das Verschleißverhalten
von Traktor-Triebradreifen
VDI Verlag Reihe 14, Nr.67

[6] Breuer Bert, Prof.Dr.-Ing:
Darmstädter Reifenkolloqium
VDI Verlag Reihe 12, Nr.285

[7] Böhm, F. :
Dynamic rolling process of tires as layered structures
Riga, Oktober 1995 NITH INTERNATIONL CONFERENC ON
MECHANICS OF COMPOSITE MATERIALS

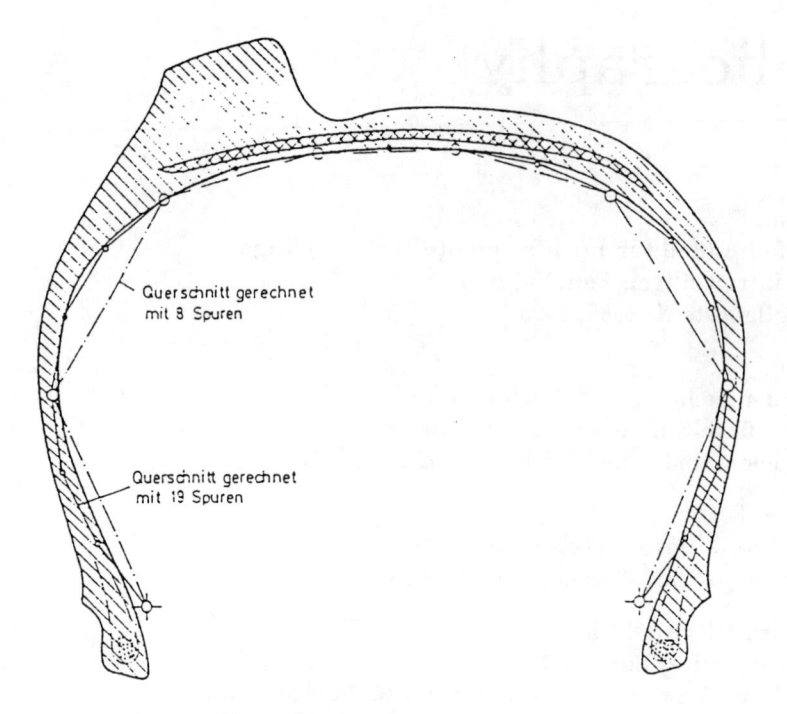

Fig.1: Cross section of tractor tyre

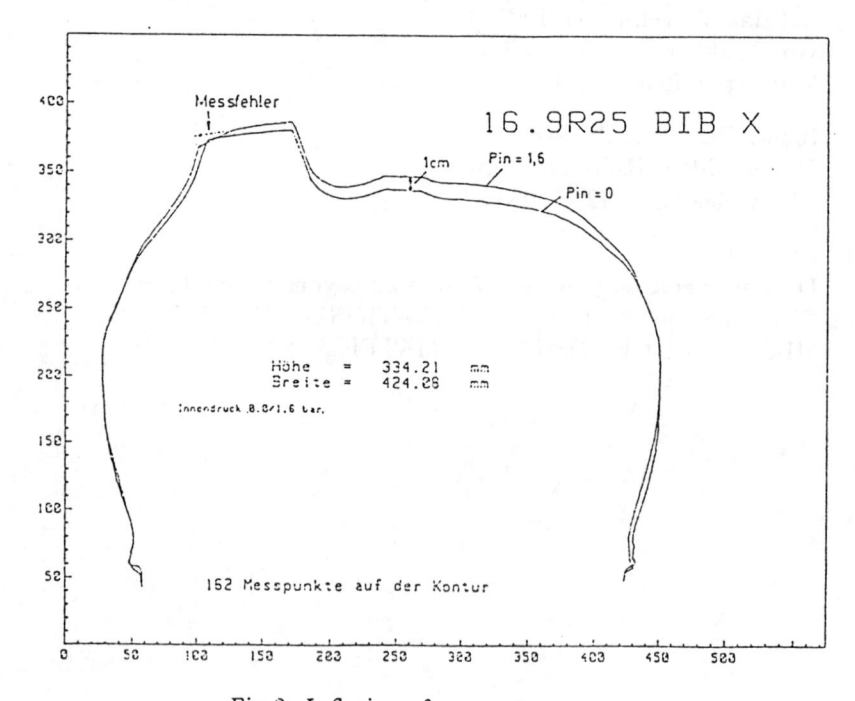

Fig.2: Inflation of tractor tyre

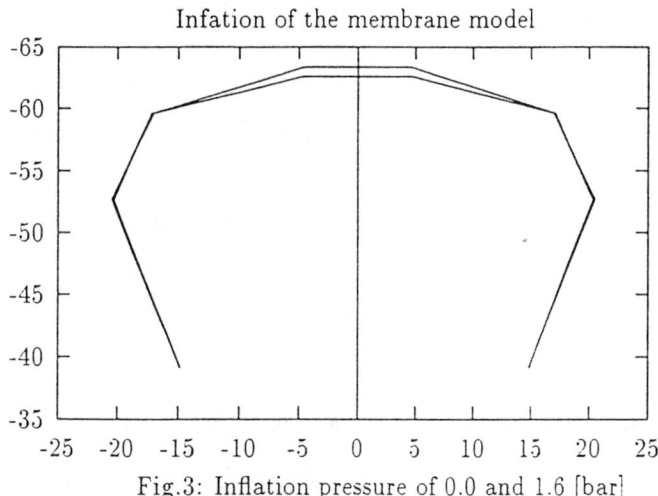

Fig.3: Inflation pressure of 0.0 and 1.6 [bar]

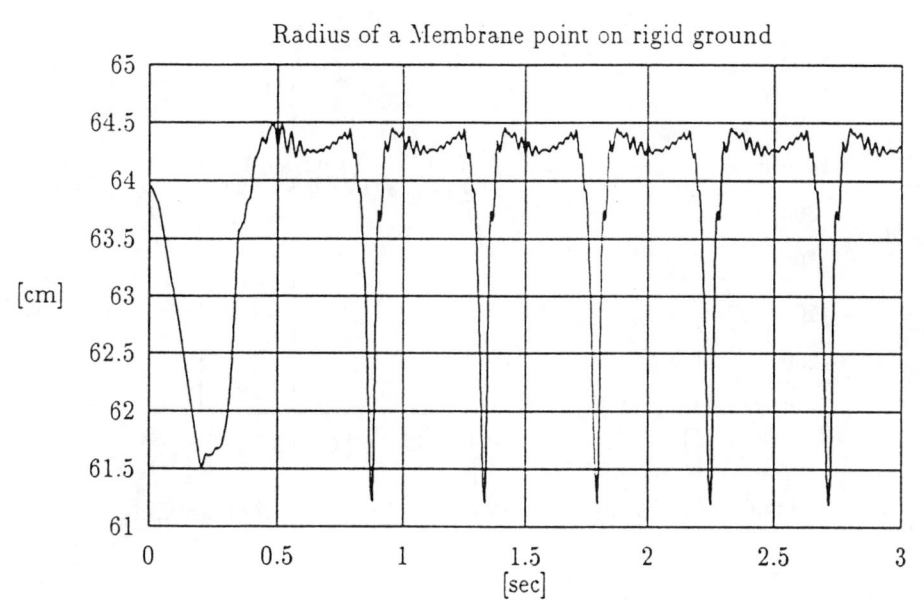

Fig.4: Calculated at a velocity of 18 km/h

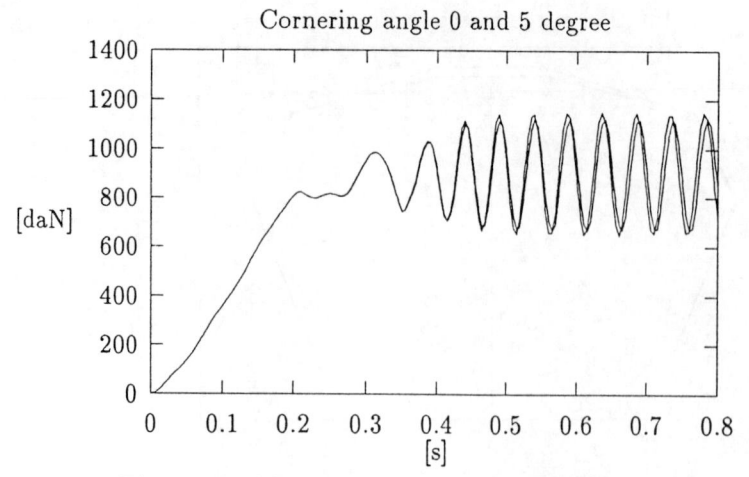

Fig.5: Wheel load at a velocity of 18 km/h on rigid ground

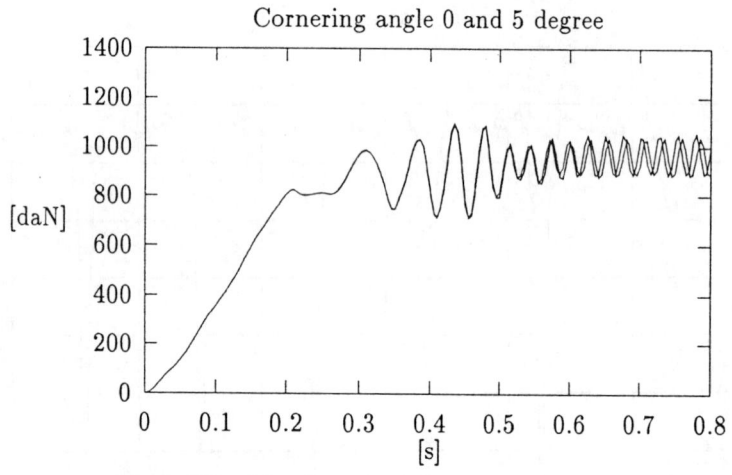

Fig.6: Wheel load at a velocity of 36 km/h on rigid ground

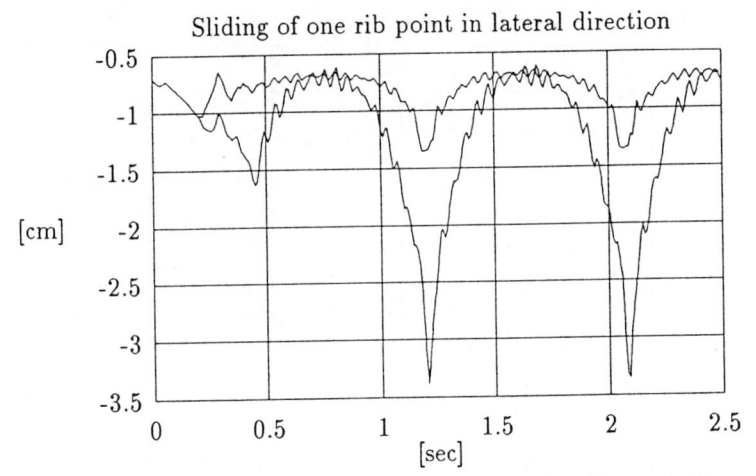

Fig.7: Wheel load at a velocity of 18 km/h on rigid ground

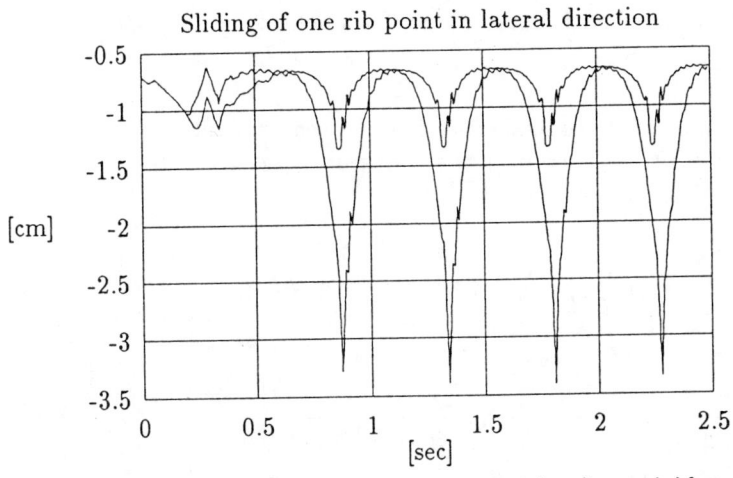

Fig.8: Wheel load at a velocity of 36 km/h on rigid ground

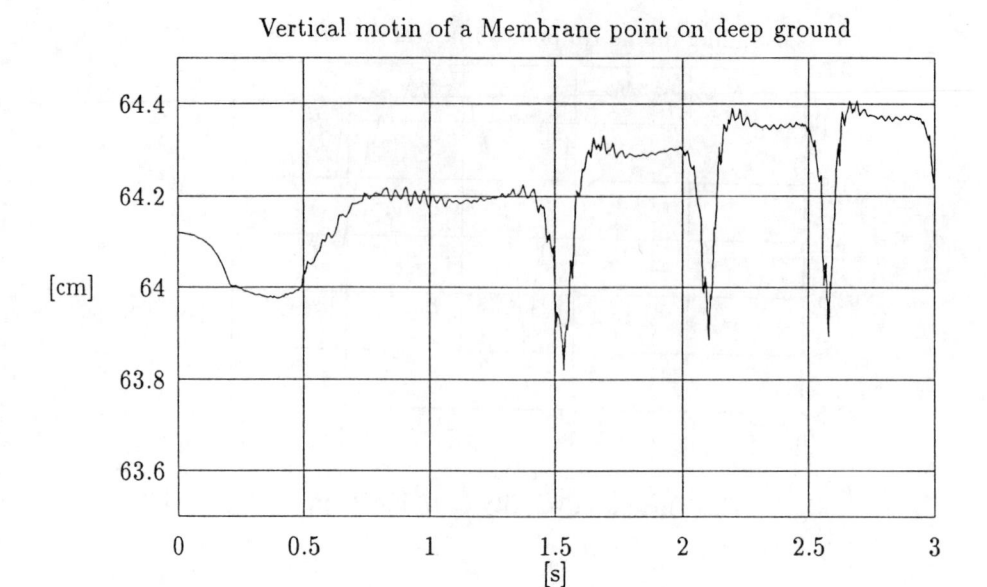

Vertical motin of a Membrane point on deep ground

Fig.9: Calculated with Becker theory at a velocity of 18 km/h

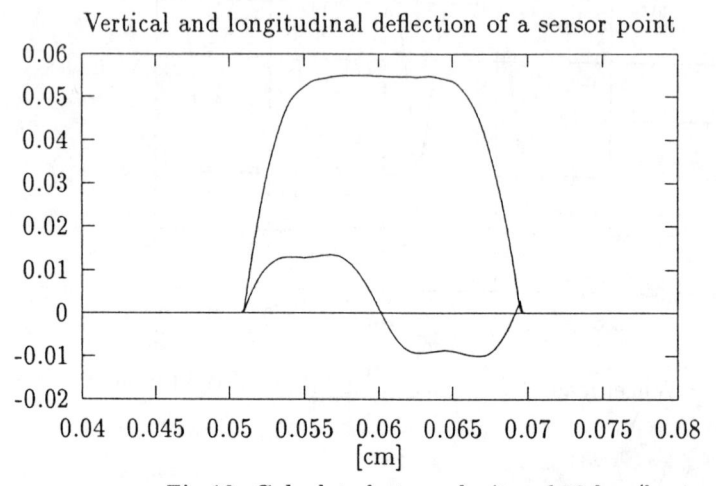

Vertical and longitudinal deflection of a sensor point

Fig.10: Calculated at a velocity of 10 km/h

APPROACHES TO PREDICT THE VEHICLE DYNAMICS ON SOFT SOIL

F.R. FASSBENDER, C.W. FERVERS, C. HARNISCH

ABSTRACT

This paper deals with the dynamic behavior of vehicles, especially on soft ground. Two different ways of simulation are described. One is based on the rigid body simulation program *ADAMS*, the other approach uses the Finite-Element-Method in conjunction with a dynamic solving algorithm. By the presented simulation results it is pointed out, that on soft soil a self excitation of the wheel exists which is caused by the deformation of the ground. The influence of this effect becomes critical when the second wheel of the vehicle is excited by the left back deformations of the first wheel. Additional measurements, investigating the dynamic influence on the behavior of the soil are used in a third, simple simulation model, based on the rigid wheel approach of BEKKER. The results show that the critical effect of self excitation is reduced by the dynamic damping behavior of the soil.

INTRODUCTION

Vehicle dynamics, especially the vertical vibrations, have a great influence on safety and ride comfort. For the layout of vehicles the prediction of ride dynamics is an important feature. While the ride dynamics on rigid roads have been investigated in detail, the dynamic behavior of a wheel on soft ground is fairly unknown. Here the vertical dynamics are strongly influenced by the reaction of the soft ground. As most obvious effect the excitation from the road bumps is modified by the vibration of the vehicle. The feedback of this modification changes as well the vibrational behavior of the vehicle. Especially for terrain, which mainly operate off the road, it is desirable to have an opportunity for the prediction of ride dynamics on soft ground. The extension of existing simulation models seems to be a suitable way for the investigation of wheel-soil dynamics.

THE SIMULATION PROGRAM *ADAMS*

The simulation program *ADAMS* (*A*utomatic *D*ynamic *A*nalysis of *M*echanical *S*ystems) [1,2] is a commercially available program for the investigation of the kinematic and dynamic behavior of three-dimensional mechanical systems. In order to run a simulation process in *ADAMS*, a mechanical model has to be translated into *ADAMS* program code. Each part of the system is handled as a rigid body, which is considered with its geometrical and physical properties. The rigid bodies are connected by joints, which are provided with translational or rotational degrees of freedom. Springs and dampers may be positioned between the bodies. By applying

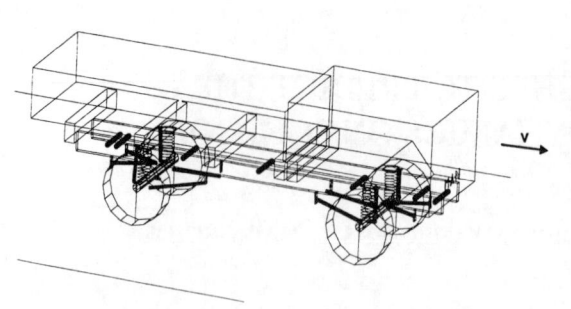

Fig. 1: Equivalent of a 4x4-truck MAN type 451, modelled in ADAMS

external forces, torques and motions, a realistic sequence of movements can be achieved for the model. The simulation is done by building the differential equations and solving them numerically. An example of an *ADAMS*-model is shown in **Fig. 1**.
This diagram shows the equivalent of a 4x4-truck MAN type 451 while driving on a solid road. The complex vehicle structure is reproduced in a true-to-life model. All its components, such as driver cabin, loading space, side rails, crossbeams, axles, steering rods, tyres, suspension and damping systems are represented in the model.

A detailed simulation requires the tyre behavior to be described with an appropriate model. The forces and torques affecting traction, lateral guidance and oscillation behavior are transmitted in the contact surface between tyre and roadway. The *TINA*-module (*T*yre *In*terface to *A*DAMS) [3] can be used to meet these demands.

THE TIRE MODULE TINA

The tire module *TINA* allows the user to consider roadways with different surfaces in *ADAMS*. However, the roadways are always rigid. In addition to this, *TINA* offers various mechanical models to describe the tire behavior. With these models it is possible to determine the forces and torques between tire and ground, that are involved in traction and lateral guidance as well as in vehicle oscillation behavior. *TINA* works as a subroutine to *ADAMS*. The interaction between *ADAMS* and *TINA* is shown in **Fig. 2**.

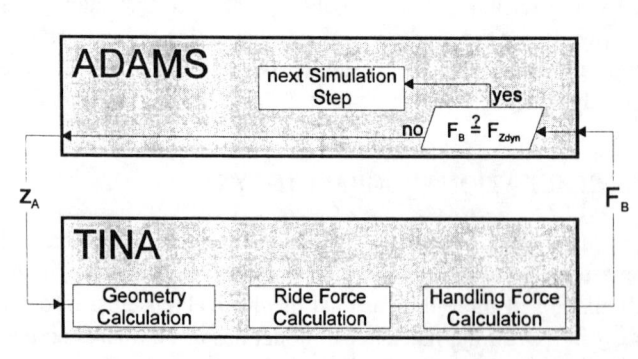

Fig. 2: Structure of *TINA* and its interaction with *ADAMS*

During the simulation *ADAMS* first calculates the subsidence z_A, related to the stationary roadway, and puts it into *TINA*. From this, the force of ground reaction F_B is determined by a chosen tire model and returned to *ADAMS*. This force is compared with the dynamic wheel force F_{zdyn}. If they are equal in tolerance, the computation of this simulation step is finished. The system then calculates the values of the next wheel. If the forces are not equal, *ADAMS* will calculate the interaction between the wheel and the ground by iteration until a solution is found. If the system does not converge after a defined number of iteration steps, the simulation will be continued with a new size of time interval. In *TINA* there are three steps to calculate the interactions:

-'*Geometry Calculation*' to calculate the geometric tire data
-'*Ride Force Calculation*' to calculate the stroke and damping of tire
-'*Handling Force Calculation*' to calculate the lateral and vertical tire forces.

THE NEW TIRE MODULE STINA

In order to solve the special formulation of the question: "*simulation of vehicle motion on soft ground*", the new module **STINA** (**S**oil **T**ire **In**terface to **A**DAMS) was developed at the IKK. As can be seen in **Fig. 3**, *TINA* consists of five separate subroutines. These subroutines are unchangeable. However, the structure of *TINA* allows the user to interlace his own subroutines parallel to the existing subroutines. This possibility formed the basis for the development of **STINA** [4].

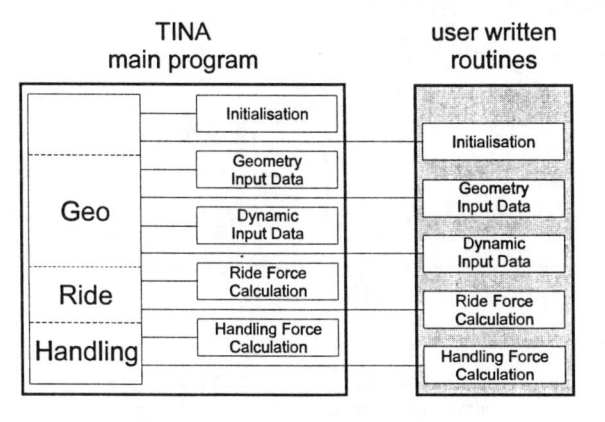

Fig. 3: Program structure of the *TINA* Software

Since the question of vertical dynamics should be discussed first, the *Ride Force Calculation* had to be substituted by a *New Ride Force Calculation*. The structure of the other subroutines remain unchanged. Only the input parameters, for instance the rolling resistance coefficient, have to be adapted to the conditions on soft ground. The resulting structure and the interfaces to *ADAMS* are shown in **Fig. 4**. In order to compute the *New Ride Force*, an interaction model is needed for the calculation of the ground reaction force F_B, based on a given subsidence z_A. The current version of **STINA** offers two different ways: the utilisation of analytical models which describe the phenomena by means of a system of formulae,

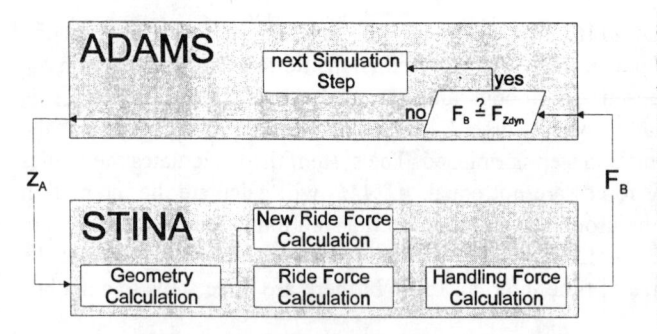

Fig. 4: Structure of *STINA* and its interaction with *ADAMS*

and the utilisation of characteristic maps, which can be gained from measurements or from other simulations (**Fig. 5**).

The contact area, which on soft ground is more complex than on solid ground, is crucial for the investigation of these phenomena. The contact area is obtained from the equilibrium of forces while taking into account the tire deflection and the sinkage, considering the resistance of soil deformation and the wheel load. In the current version of *STINA* two alternative analytical models are implemented: the rigid wheel approach of BEKKER and the parabola model of the elastic tire [7].

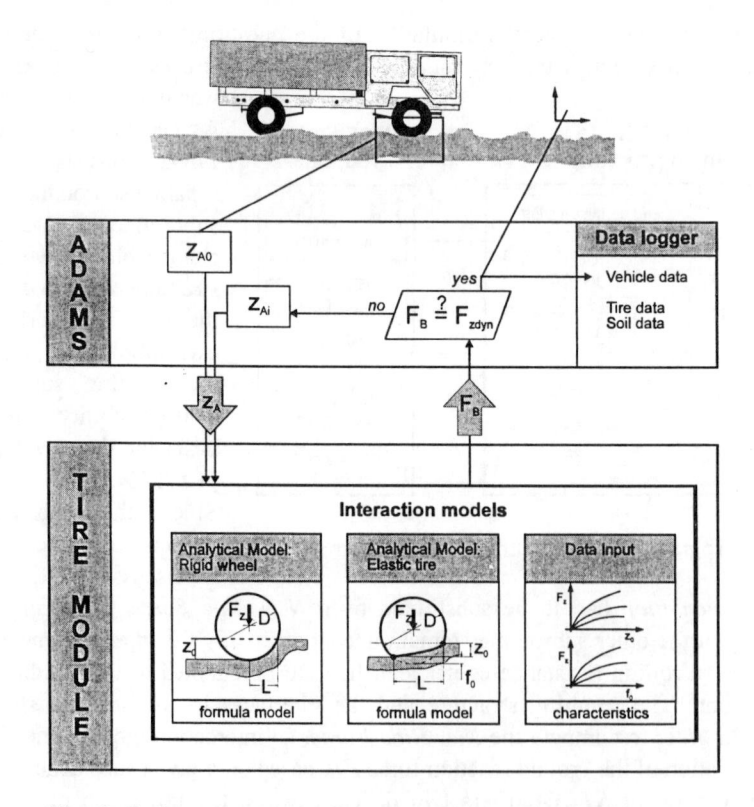

Fig. 5: Software structure of the new Tire Module *STINA*

THE PARABOLA MODEL OF THE ELASTIC TIRE

Compared to the rigid wheel approach the model of the elastic tire leads to a more realistic description of wheel soil interaction. In this model the contact curve between tire (diameter D) and soil is described by a larger rigid wheel with diameter D^*, which is approximated by a parabolic contact line [7,8].
The parabolic geometry provides the advantage of an easy mathematical treatment. The geometry is shown in **Fig. 6**. The initially unknown tire deflection f_0 on soft ground is derived from the tire deflection on solid ground, measured on a loading test stand as a function of wheel load F_z and the inflation pressure p_i. The relation between wheel load F_z and sinkage z_0 is calculated as follows:

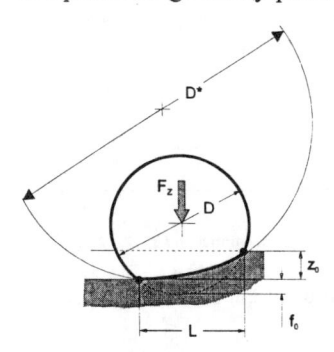

Fig. 6: Contact contour between tire and soft soil of an elastic tire

$$F_z \approx b \cdot k \cdot z_0^{n+0.5} \cdot \frac{\sqrt{1+\frac{f_0}{z_0}} + \sqrt{\frac{f_0}{z_0}}}{1+\frac{n}{2}}$$

The relation of tire deflection f_0 and sinkage z_0 has to be calculated by iteration. Considering that the sum of tire deflection f_0 and sinkage z_0 is equal to the subsidence z_A, the tire deflection on soft ground can be assumed as

$$f_0 \approx 2 \cdot \sqrt{\frac{f_0}{z_0}} \cdot \sqrt{\frac{D^*}{D}-1} \cdot f_k \cdot \frac{D}{D^*}$$

The wheel diameter ratio D^*/D can be calculated as follows:

$$\sqrt{\frac{D^*}{D}} \approx \sqrt{1+\frac{f_0}{z_0}} + \sqrt{\frac{f_0}{z_0}}$$

SIMULATION RESULTS

Several simulations with the ***STINA*** module were performed with the model of a wheel-mass-system. The simulated model was equipped with the tire of type 14.0R-20 mil. The wheel-soil interaction model during the simulation was the model with parabolic geometry. **Fig. 7** represents the influence of different inflation pressures on sinkage z_0 and tire deflection f_0 of the wheel on soft loam.
Fig. 7a shows the conditions with an inflation pressure of p_i = 4,5 bar. As can be seen, the sinkage z_0 is really high in comparison to a relatively low tire deflection f_0.
Fig. 7b shows the conditions on the same loam with an inflation pressure of

$p_i = 1.0$ bar. The softer tire has a higher tire deflection, f_0, while the sinkage, z_0, becomes smaller due to the enlargement of the tire-soil contact area.

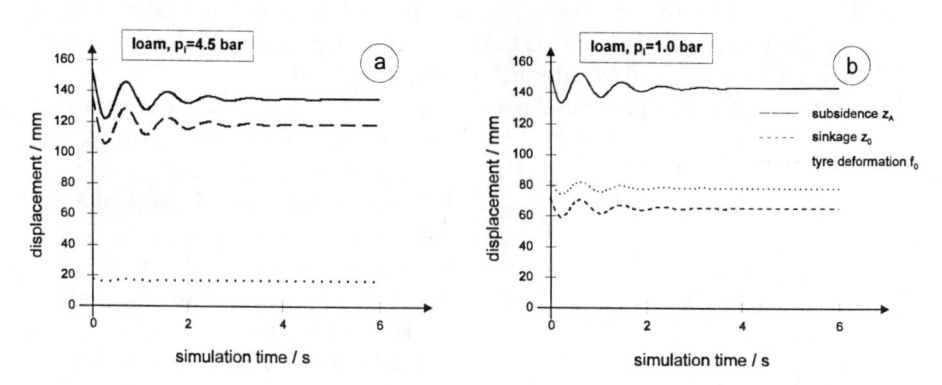

Fig. 7: Simulation result of *STINA* - vehicle drive with different inflation pressures on loam

This result corresponds to practical experience. The phenomena discussed are results of the static state on even ground. As can be seen in **Fig. 7**, these static states are reached after a transient oscillation. The reason for the overshoot at the beginning of the simulation is a user given excitation.

The most important field of application for *STINA* is the simulation of vehicle dynamics in order to investigate their vibrational behavior on soft ground. In the following the simulation of a quarter vehicle that is reproduced by a two-mass-system is presented. The simulation results demonstrate the vibrational behavior of the wheel and the vehicle body caused by a roadbump. **Fig. 8a** refers to the results of a simulation on solid ground and **Fig. 8b** to those on soft ground.

In both cases the vehicle crosses an obstacle with a height of *0.6 m* and a length of *1.0 m*, driving with a speed of *v=18 km/h*. Comparing the amplitude of the body oscillation it comes into sight that the vehicle vibration on soft ground differs significantly from these on solid ground. While the vehicle vibrations, indicated on solid ground, decrease rapidly, these on soft ground take more than three times longer to decay. This makes it obvious that the behavior of the ground, thus the parameters of the soil take a great influence on vehicle vibrations.

Another aspect to regard is the ground-surface left behind. While the surface-contour on solid ground remains unchanged, the wheel-soil interaction on soft ground implies a modification of the surface contour. This effect has an important influence on the vibrational behavior of following wheels. The rear wheel discovers the modified surface-contour that generates an excitation, influenced by the front wheel and totally different from that on solid ground. The interference of the front wheels vibration on the rear wheels excitation and vice versa, the influence of the rear wheels vibration on the dynamic load thus on the excitation of the front wheel is the most important aspect to regard with vehicle vibration on soft ground.

VEHICLE DYNAMICS ON SOFT SOIL 179

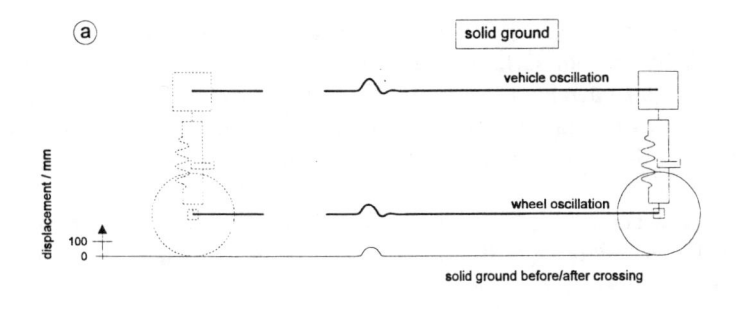

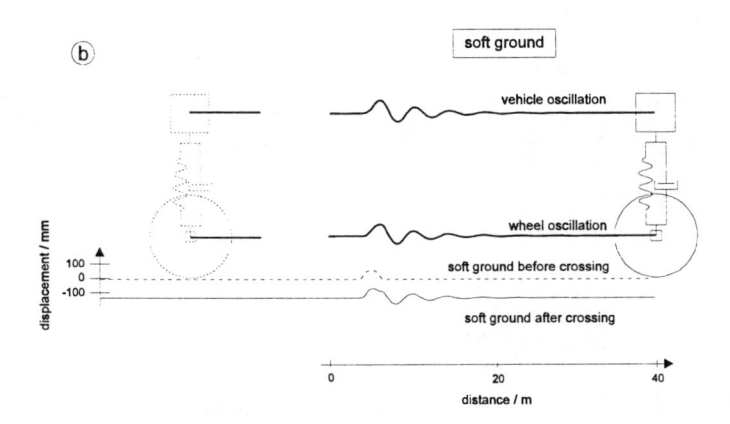

Fig 8: Simulation results of *STINA* - oscillation of a two-mass- system on solid (a) and on soft ground (b), v=18 km/h

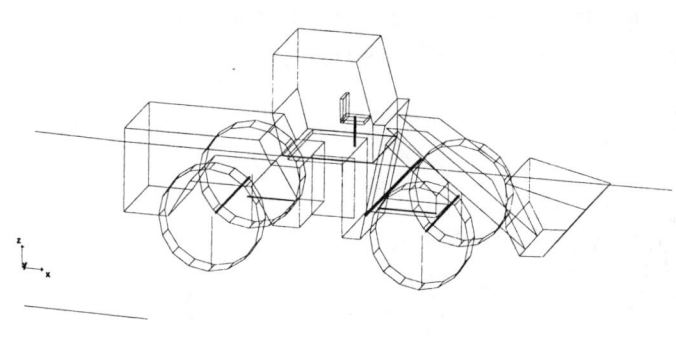

Fig 9: Equivalent of a *Zettelmeyer type ZL 3002*, modelled in *ADAMS*

To demonstrate this by an example the model of a wheeled loader was build up. The model represents the *Zettelmeyer Loader* type *ZL 3002* with a weight of 18.100 kg, a wheel base of 3550 mm and a wheel diameter of 1696 mm. This wheel loader has an unsprung body but is equipped with an airspring supported loader used as damper weight. The *ADAMS*-model of this wheel loader is shown in **Fig. 9**.

The simulation was carried out on a soil, representing a loam of 14% water-content with an initially smooth surface. The excitation was done by dropping down the wheel loader from a height of 100 mm. A controlled travelling speed of 18 km/h was established in this simulation. **Fig. 10** shows the oscillation of the vehicles body. It is to be seen that the oscillation does not decay within a time of 10 s respectively a distance of 50 m.

Looking at the bodies oscillation on solid ground, **Fig. 10**, where the oscillation decays within 6 seconds ($\hat{=}$ 30 m) the crucial point of vehicle vibration on soft ground becomes even more obvious.

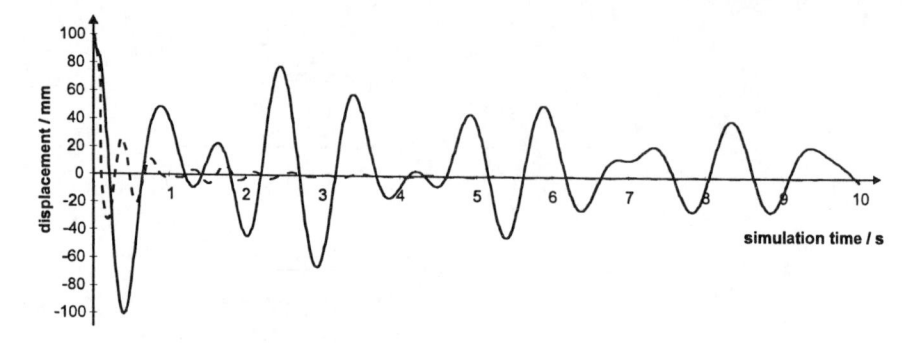

Fig 10: Simulation result of *STINA* - body oscillation on rigid and on soft ground

With respect to the conditions in field operation the riding comfort is a measure for the achievable operational speed. Here the acceleration of the drivers seat is of prior interest. As to be seen in **Fig. 11** for the simulated conditions on loam the acceleration of the drivers seat exceeds the limit of 2.5 m/s² even after a time of 3 seconds. After 6 seconds the acceleration reaches still values of 2.0 m/s². That means, that the driver of the loader is not able to drive this speed on soft ground for a longer time.

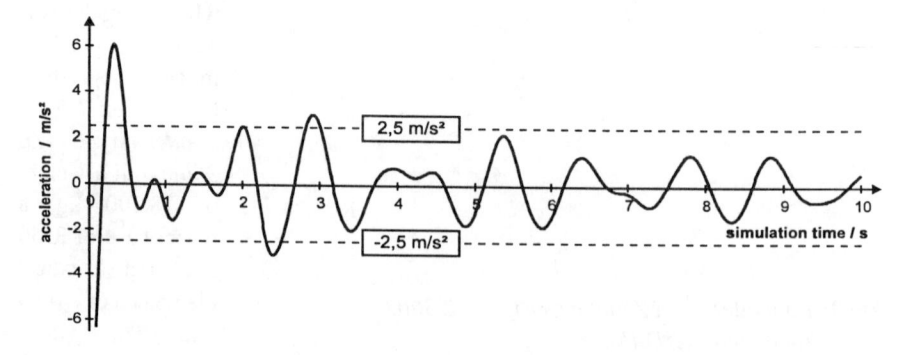

Fig 11: Simulation result of *STINA* - vertical acceleration at the driver seat

FEM-SIMULATION

Another approach for the investigation of vehicle soil dynamics by simulation is the use of the FINITE-ELEMENT-METHOD (FEM). FEM is a well known tool for the analysis of mechanical structures. It is as well used for the calculation and development of tyres [10]. Highly elaborated models exist which represent the behavior of a deformable, air-filled tyre in a detailed way [10] (**Fig. 12**). However, most of these models regard the interaction of the tyre with the ground only for rigid surfaces.

As a soil can be described as mechanical structure, too, the FEM is as well valuable for the simulation of soft soil. The IKK has used this simulation method to develop the FEM based model VENUS [9] which is used to simulate the interaction of a deformable tyre with soft soil. This model consists of a tyre model and a soil model as separate parts. The soil is described by an elasto-plastic material law on the basis of DRUCKER-PRAGER's constitutive model. A realistic representation of the soil behavior requires a detailed modelling of the soft soil and therefore takes a high amount of computer capacity. In order to reach a time effective simulation it is necessary to build the tyre with a simple model. The tyre is modelled by a structure of three concentric rings, representing the rim, the carcass and the tread (**Fig. 13**). This model has proved to simulate the wheel soil interaction in a most realistic way [9]

Fig 12: Detailed Model of a deformable tyre [10]

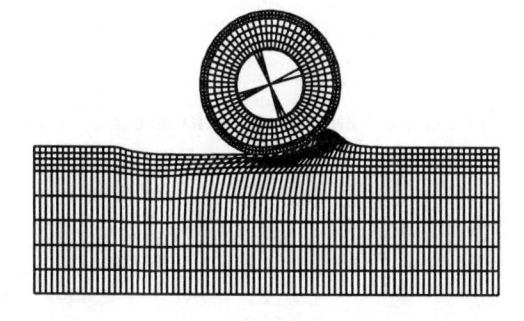

Fig 13: Wheel-soil interaction with VENUS [9]

THE DYNAMIC FEM APPROACH

The above described simulation was done with a usual FEM-solver which is based on the static equilibrium. Therefore the movement of the wheel and the deformation of the soil are regarded as quasi-static, so that dynamic effects cannot be included in this simulation. To regard dynamic effects it is necessary to calculate not only the displacements, but as well the velocities and accelerations. One possibility to consider the acceleration is the use of NEWTON's dynamic equilibrium,

$$F = m \cdot a$$

where **F** is the forces
       m is the mass
and   **a** is the acceleration

or, with regard to the FEM-Formulation,

$$\bar{F} - \bar{I} = \bar{M} \cdot \bar{A}$$

where $\bar{A}$ is the matrix of acceleration
       $\bar{F}$ is the matrix of external forces
       $\bar{I}$ is the matrix of internal forces
and $\bar{M}$ is the mass matrix.

Beside the matrices of external and internal forces, this calculation needs a discrete matrix of point masses referring to the nodes. This mass matrix can be obtained by distributing the volume mass of each element onto its nodes. The partial mass of all elements belonging to one node is summed up as the total point mass at this node. With a defined state of the model and a given value of external forces the acceleration of each node can be calculated by

$$\bar{A} = \frac{\bar{F} - \bar{I}}{\bar{M}}$$

With this acceleration the velocity $\bar{V}$ can be integrated over the time by

$$\bar{V}(t + \Delta t/2) = \bar{V}(t - \Delta t/2) + \bar{A}(t) \cdot \Delta t$$

where t is the actual time
and Δt is the incremental time interval

In accordance to this the displacement $\bar{S}$ can be calculated by

$$\bar{S}(t + \Delta t) = \bar{S}(t) + \bar{V}(t + \Delta t/2) \cdot \Delta t$$

THE SIMULATION MODEL

With this simulation method a model of a quarter vehicle on soft ground (**Fig. 14**) was built up to investigate the fundamental effects of dynamic vehicle soil interaction. As the investigation of the global effects was emphasised, the wheel in this simulation is regarded as rigid. The consideration of a deformable tire is as well possible but needs extremely high computer capacity. The parameters of the soil, namely the elasticity (8 N/mm²), the cohesion (0.02 N/mm²), the internal friction

(27.5°) and the density (1.9 t/m³) were taken from quasi static soil tests. Therefore in this simulation the dynamic behavior of the soil is concentrated on the effect of mass-inertia and the damping by the energy loss of the plastic soil deformation.

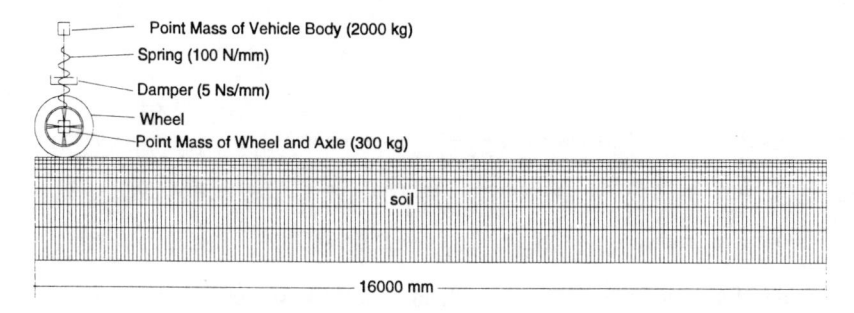

Fig 14: FEM simulation model for a quarter vehicle on soft ground

SIMULATION RESULTS

In the following the results of two simulation runs on loam are presented. In one simulation run the damper was deactivated while the other parameters of the model were held constant. The initial state of the model was built up without any forces. Thus the model was excited by applying the gravity loads suddenly at the beginning of the simulation. **Fig. 15** depicts the resulting oscillations of the vehicle body and the wheel for the vehicle with no damping.

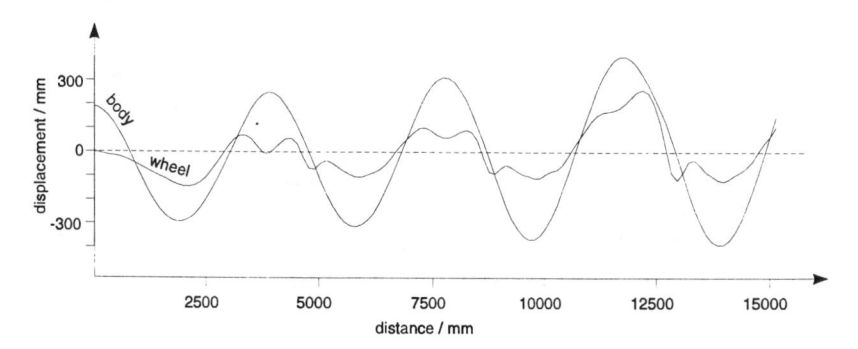

Fig 15: Oscillation of vehicle body and wheel for the vehicle with no damping

It comes into sight that the amplitude of the oscillation does not fade away. It even rises with further distance of travel. This phenomenon can be explained by the soil deformation, depicted in **Fig. 16**. It is to be seen that the soil is periodically deformed with the eigenfrequency, respectively the wave length of the bodies oscillation. Here the bow wave in of the front wheel, known as "bulldozing effect" [9], is intensified by the increasing wheel load resulting from the downward acceleration of the body. By this the soil is accumulated to a kind of "ramp" in front of the wheel. With the following upward acceleration of the body the wheel load decreases and the wheel is able to run over this "ramp", leading to an additional excitation of the body.

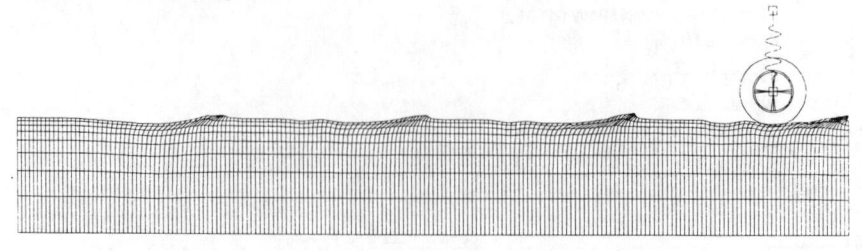

Fig 16: Soil deformation after one pass with an undamped vehicle

This effect of self excitation shows the importance of damping for terrain vehicles. According to this **Fig. 17** shows the oscillation of the body and the wheel for the same conditions but with a vibration damper. It can be observed that the damper is able to balance the excitation, so that the amplitude of the oscillation decays.

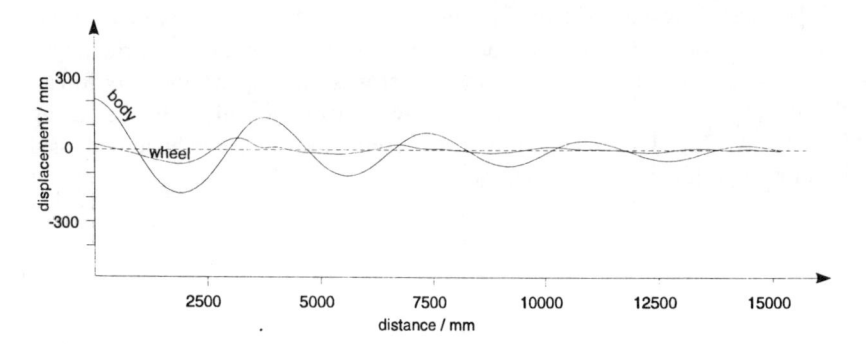

Fig 17: Oscillation of vehicle body and wheel for the vehicle with damping

The advantages of the damper are as well visible from the soil deformation (**Fig. 18**). It can be seen that only the initial excitation leads to a notable soil deformation.

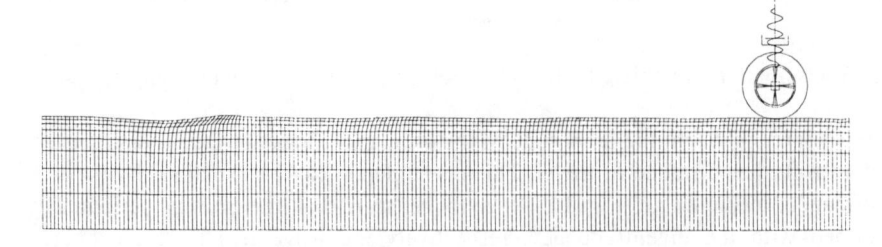

Fig 18: Soil deformation after one pass with a vehicle with damper

MEASUREMENTS

In order to get some more knowledge about the dynamic behavior of soils, measurements were carried out. According to the well known calculation-model of BEKKER, the most famous way of soil measurement, concerning wheel soil interaction, is the plate sinkage test. This test assumes a quasistatic soil penetration with low velocities. In order to regard as well the influence of the penetration speed a modified test method was elaborated by GRAHN[12]. This method is analogue to the plate sinkage test, but in contrast to the original way, here the penetration speed can be varied from 0 m/s up to 0.8 m/s. With this test method measurements were carried out on a loam with 14 % and 16 % moisture content. The tests were done by penetrating these soils with a round plate of 600 cm² area. Penetration speed was set to 2 cm/s and from 20 cm/s up to 80 cm/s in steps of 20 cm/s. **Fig. 19** and **Fig. 20** depict the corresponding pressure-sinkage curves. It is visible that the pressure increases with higher penetration velocity. As can be seen by comparison of **Fig. 19** and **20** this effect is amplified by a higher moisture content. That means, that in contrast to the static soil strength, which normally decreases with higher moisture content, the dynamic soil strength grows. The depicted test results make it obvious that there is an influence of penetration speed on soil strength. In order to consider these dynamic effects in the calculation, dynamic parameters characterising the soil strength had to be derived. Therefore the well known pressure-sinkage-relation

$$p = k \cdot z^n$$

was extended by the following approach

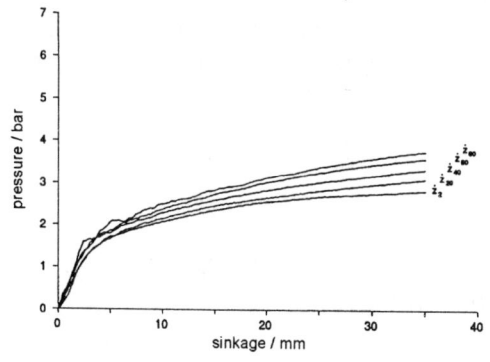

Fig. 19: Pressure sinkage curves on loam 14 % for different penetration speeds

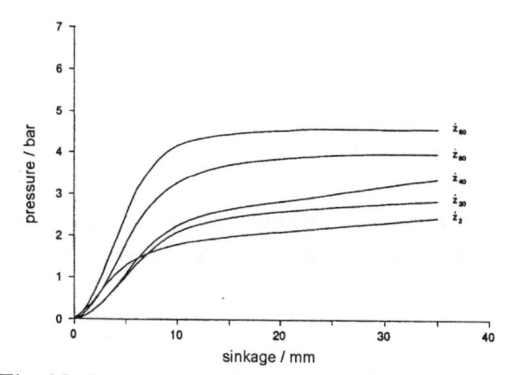

Fig. 20: Pressure sinkage curves on loam 16 % for different penetration speeds

$$p = p_{stat} + p_{dyn}(\dot{z})$$

whereby

$$p_{stat} = k_{stat} \cdot z^{n_{stat}}$$

is the original formula for the quasistatic soil penetration. In **Fig 19** and **20** it can be seen that the shape of the dynamic pressure sinkage curves is similar to the shape of the quasistatic one. Therefore the pressure sinkage relation for different velocities $\dot{z}$ can be expressed by

$$p_{dyn}(\dot{z}) = k_{dyn}(\dot{z}) \cdot z^{n_{dyn}(\dot{z})}$$

In the presented test results the sinkage exponent $n_{dyn}(\dot{z})$ is independent from the penetration speed. Thus "n" is assumed to be constant.
With

$$n_{dyn} = n_{stat} = n$$

the pressure sinkage equation can be written as

$$\begin{aligned} p &= p_{stat} + p_{dyn}(\dot{z}) \\ &= k_{stat} \cdot z^n + k_{dyn}(\dot{z}) \cdot z^n \\ &= \left(k_{stat} + k_{dyn}(\dot{z})\right) \cdot z^n \end{aligned}$$

In this equation the k_{dyn} is a function of $\dot{z}$. It could be found out that the following equation for k_{dyn} provides a good approximation for the measured pressure sinkage curves.

$$k_{dyn}(\dot{z}) = k_{dyn1} \cdot \dot{z}^m$$

With this, the dynamic pressure sinkage relation can be written as:

$$p = \left(k_{stat} + k_{dyn1} \cdot \dot{z}^m\right) \cdot z^n$$

In the formula for the rigid wheel approach of BEKKER.

$$F_Z = \frac{b * D}{2} \cdot \int_{\Theta_{in}}^{\Theta_{out}} p(\Theta) d\Theta$$

$p(\Theta)$ can be replaced by the dynamic equation, so that

$$F_Z = \frac{b*D}{2} \cdot \int_{\Theta_{in}}^{\Theta_{out}} (k_{stat} + k_{dyn1} \cdot \dot{z}^m) \cdot z^n d\Theta$$

with

$$\Theta_{in} = \arcsin\left(\frac{\frac{D}{2} - z_0}{\frac{D}{2}}\right)$$

$$\Theta_{out} = \frac{\pi}{2}$$

$$z(\Theta) = \frac{D}{2}(\sin(\Theta) - \sin(\Theta_{in}))$$

Θ = contact angle tyre / soil
F_Z = wheel load
b = width of the wheel
D = diameter of the wheel

This equation was used to calculate the ground reaction force in relation to sinkage the sinkage speed for a two mass model of a quarter vehicle. Hereby a RUNGE-KUTTA algorithm was used to solve the equations. **Fig. 21** depicts the resulting oscillation of the vehicle body and wheel on loam 14 % and the deformation of the soil. The original surface contour was smooth, the model was excited by applying the gravity loads at the beginning of the simulation. **Fig. 2** depicts the according oscillations neglecting the dynamic soil reaction. It can be seen that without regarding the dynamic soil behavior ($k_{dyn1} = 0$) the oscillation decays within a time of 6 seconds. In contrast to this, the oscillation decays within 4 seconds when considering the dynamic soil behavior.

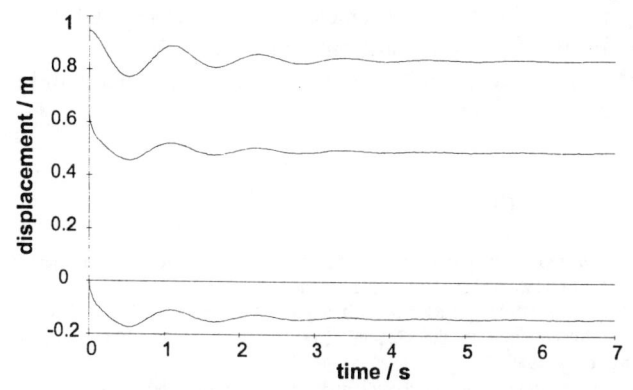

Fig. 21: Oscillation including dynamic soil behavior

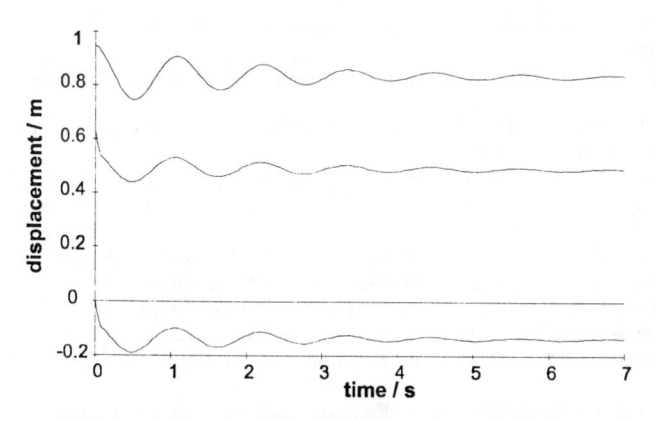

Fig. 22: Oscillation excluding dynamic soil behavior

From this it can be derived that the effect of dynamic soil damping is in opposite to the effect of self excitation. The effect of dynamic soil damping reduces the oscillation on soft ground.

CONCLUSION

In the presented simulation results of dynamic wheel-soil interaction, three different effects could be observed. First, there is the effect of self excitation due to the soil deformation. By the increasing dynamic wheel load the soil is deformed to a ramp that leads to an additional excitation of the wheel.
The second effect occurs with two wheels following each other in the track. Here the second wheel is excited by the left back soil deformation of the first wheel. In combination with the pitching of the vehicle this effect leads to critical vibrations.
The third effect is the dynamic damping behavior of the soil, which could be observed in experiments. A simple simulation model considering this effect showed, that the dynamic damping of the soil reduces the vehicle vibrations. Compared to the first and second one this effect takes a contrary influence on vehicle vibration.
For the future it is planed to integrate all effects in one model. Other soils will be regarded as well in measurement as in simulation.

REFERENCES

[1]   Bartels, M., Fischer,E.: ADAMS - Ein universelles Programm zur Berechnung der Dynamik großer Bewegungen. Automobiltechnische Zeitschrift, Bd. 86, H.9, Stuttgart, 1986
[2]   Mechanical Dynamics Inc.: ADAMS User's Manual, Version 8.1. Ann Arbor Michigan, 1995
[3]   Fischer, E.: TINA - An Introduction to the General Tire Interface to ADAMS. Seminar-Unterlagen der Firma TEDAS, Marburg, 1989
[4]   Aubel, T., Ludewig, J.: Simulation von Fahrzeugschwingungen, angeregt durch plastische Bodenunebenheiten. Forschungsbericht. IKK-Nr. 94-04,
Uni Bw Hamburg, 1994
[5]   Bekker, M. G.: Theory of Land Locomotion : The Mechanics of Vehicle Mobility. Ann Arbor: The University of Michigan Press, 1956
[6]   Bekker, M. G.: Introduction to Terrain-Vehicle System. Ann Arbor: The University of Michigan Press, 1969
[7]   Schmid, I.; Ludewig, J.: Improved Calculation of Sinkage of a Wheel on Soft Ground. Proc. 5th European Conference of the ISTVS, Budapest, 1991
[8]   Schmid, I.; Aubel, T.: Der elastische Reifen auf nachgiebiger Fahrbahn - Rechenmodell im Hinblick auf Reifendruckregelung. VDI Berichte Nr. 916. Düsseldorf: VDI-Verlag, 1991 - ISBN 3-18-090916-1
[9]   Aubel, T.: Simulationsverfahren zur Untersuchung der Wechselwirkung zwischen Reifen und nachgiebiger Fahrbahn auf der Basis der Finite Elemente Methode. Diss. Uni Bw Hamburg, 1994
[10]   Feng, K.: Statische Berechnung des Gürtelreifens unter besonderer Berücksichtigung der kordverstärkten Lagen. Fortschrittsbericht VDI Reihe 12 Nr.258, Düsseldorf, 1995
[11]   Hibitt, KARLSSON & SORENSEN, INC.: Manuals for the program "ABAQUS" Vers. 5.4 , Pawtucket, 1993
[12]   Grahn, M.: Einfluß der Fahrgeschwindigkeit auf die Einsinkung und den Rollwiderstand von Radfahrzeugen auf Geländeböden, Diss. Uni Bw Hamburg, 1996

The Contact Problem of In-Plane Rolling of Tires on a Flat Road

Son-Joo Kim and Arvin R. Savkoor

ABSTRACT

The paper describes an analysis of contact problem of free-rolling of pneumatic tires on a flat road. The tire is modelled as an elastic ring supported on a viscoelastic foundation. Additional elastic spring elements on the outer surface of the ring are included to model the compliance of the tread rubber. The contact problem of a free-rolling tire is formulated for prescribed normal deflection and subjected to constraints of both normal contact and friction. The equations of motion for the treadband displacements are expressed in terms of modal expansion and non-linear boundary conditions of rolling contact are solved numerically to determine the region of contact, the distribution of normal and shear tractions and the effective rolling radius of the free-rolling tire. The contributions of tire hysteresis and frictional sliding in the contact region to the rolling resistance are determined. The influence of damping and Coriolis acceleration at higher rolling speeds on the asymmetric distribution of normal and tangential traction in the contact patch is considered. The analysis also considers how some of the typical damping models influence the variation of rolling resistance with the rolling speed.

1. INTRODUCTION

A detailed study of the distribution of slip and traction within the footprint of a rolling tire is important, because in addition to its design functions of supporting loads and cushioning road irregularities, it has to provide sufficiently large tractive, braking and cornering forces. The other equally important design requirement is that the rolling resistance and tire wear of a free-rolling tire are required to be minimal. The present work deals with the in-plane rolling contact and concentrates on the latter aspect. The forces in the contact patch are influenced by the construction of the tire and also by the operating conditions such as the normal load, the inflation pressure, the speed of travel and the torque or force applied to the wheel to sustain the rolling motion.

It is well known that material hysteresis which occurs during cyclical deformation of tires is the main cause of the resistance to rolling. Another cause of the rolling resistance of free-rolling tires is the energy dissipation due to any frictional sliding within the contact. Generally the latter contribution is taken to be insignificant. However, this assumption has not been well substantiated and merits reexamination based on a detailed analysis of the contact problem. The study also addresses the modelling of damping and its effect on the rolling resistance of tires.

Steady rolling requires the application of a horizontal force F_{axle} in the case of a driven wheel or a torque T in the case of a driving wheel to maintain a uniform rolling motion [1],[2]. The two basically different states of free-rolling of driven and driving wheels are shown in Fig.1.1 and 1.2. The inelastic deformation of the tire is responsible for the asymmetric positioning of the contact patch with respect to the wheel axis and the asymmetric distribution of normal pressure in the contact patch. These two effects manifest in a rolling resistance moment acting on the tire, which must be overcome either by a torque or a couple applied to the wheel.

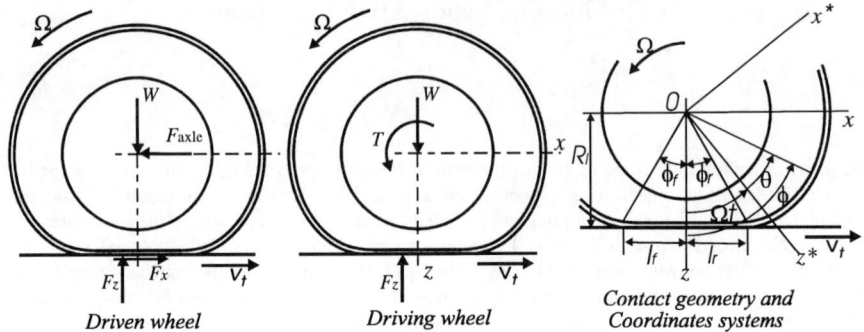

Fig.1.1,1.2,1.3: Two cases of free-rolling; Driven and Driving wheel, and Notation for contact geometry of tire and coordinates systems

2. TIRE RING MODEL

One of the simple models which represents all the essential features of in-plane tire dynamics is the model of a ring supported on two elastic foundations as shown in Fig.2.

w : radial displacement of ring
v : tangential displacement of ring
R : mean radius of ring
k_w : radial stiffness of sidewall
k_v : tangential stiffness of sidewall
k_{Et} : normal stiffness of tread rubber
k_{Gt} : horizontal stiffness of tread rubber
c_w : radial damping coefficient of sidewall
c_v : tangential damping coefficient of sidewall
q_w : external force acting on ring radially
q_v : external force acting on ring tangentially
q_β : external moment acting on ring
E : Young's modulus of ring
G : shear modulus of ring
A : cross section area of ring
I : inertia moment of cross section of ring

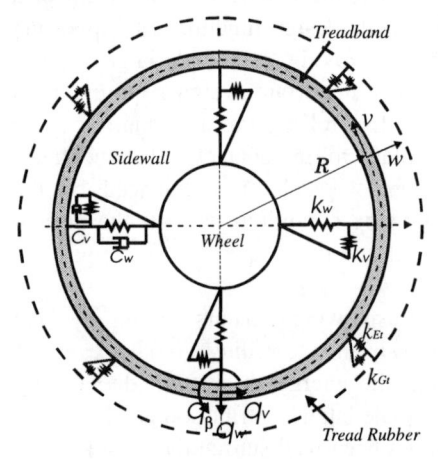

Fig.2 : The tire ring model

The treadband structure of a typical radial tire is modelled as a thin circular elastic ring which is restrained at its inner surface both in the radial and circumferential directions by a viscoelastic foundation connecting a rigid wheel-hub. The viscoelastic foundation may be represented by a continuous annulus of spring-damper elements which act both in the radial and tangential directions. The spring elements represent the membrane stiffness of the inflated torus enclosed by the carcass and the stiffness of the sidewall structure of the tire. An auxiliary elastic foundation with spring elements is attached to the outer ring surface to represent the radial and tangential flexibility of tread rubber elements of the tire. When the tire is loaded with a prescribed normal load or deflection these elements come into contact with a flat road and they are subjected to normal compression. Furthermore they undergo shear deformation due to friction depending upon the overall motion of the tire.

3. EQUATIONS OF MOTION

Coordinate systems

In order to describe the dynamics of the rotating treadband it is convenient to have a coordinate system which rotates with the wheel body. In contrast, the analysis of the contact deformations and forces is more conveniently performed using a separate coordinate system which translates with the contact patch of the tire. The origins of both coordinate systems are located at the center of the wheel. The location of an element of the tire is described using polar coordinates (r,ϕ) in the non-rotating coordinate system, or (r,θ) in the rotating coordinate system. As seen in Fig.1.3, the angular coordinates originate from the vertical axis pointing downward through the wheel center and are taken positive in the counter-clockwise direction. Sometimes the locations will be expressed by Cartesian coordinates (x, z) in the non-rotating coordinate system, or (x^*, z^*) in the rotating coordinate system. Its x-axis is positive in the direction of the forward velocity and the z-axis is positive in the downward direction.

Equations of motion

The rolling of tires on a flat road surface at a constant forward speed may be viewed as a problem of rolling contact of a stationary tire with a uniformly translating flat road in the opposite direction (see Fig.1). In this analysis we consider the wheel-body to be fixed in space and the wheel is only allowed to rotate at a constant speed Ω.

The treadband is assumed to behave as an inextensibile curved beam which can bend according to the Bernoulli-Euler assumption. The radial displacement w and the tangential displacement v at any point on the middle surface of the inextensible ring are related by:

$$w = -\frac{\partial v}{\partial \theta} \qquad (1)$$

The rotation angle β of the treadband cross section is:

$$\beta = \frac{1}{R}\left(v - \frac{\partial w}{\partial \theta}\right) \qquad (2)$$

The equations of motion of the treadband and the wheel expressed in terms of v are obtained [3]:

$$-\frac{EI}{R^4}\left(\frac{\partial^2 v}{\partial \theta^2}+2\frac{\partial^4 v}{\partial \theta^4}+\frac{\partial^6 v}{\partial \theta^6}\right)+\frac{\sigma_{\theta\theta}^0 A}{R^2}\left(v+2\frac{\partial^2 v}{\partial \theta^2}+\frac{\partial^4 v}{\partial \theta^4}\right)-k_w\frac{\partial^2 v}{\partial \theta^2}+k_v(v-R\theta_r)$$

$$-c_w\frac{\partial^2 \dot{v}}{\partial \theta^2}+c_v\dot{v}+\rho A\left(\ddot{v}-\frac{\partial^2 \ddot{v}}{\partial \theta^2}-4\Omega\frac{\partial \dot{v}}{\partial \theta}+\Omega^2\left(\frac{\partial^2 v}{\partial \theta^2}-v\right)\right)-\frac{p_0 b}{R}\left(v+\frac{\partial^2 v}{\partial \theta^2}\right)$$

$$=q_v+\frac{\partial q_w}{\partial \theta}+\frac{1}{R}\left(q_\beta+\frac{\partial^2 q_\beta}{\partial \theta^2}\right)$$

$$2\pi k_v R^3 \theta_r - R^2 \int_0^{2\pi} k_v v d\theta = T$$

$$R\int_0^{2\pi}(q_v\cos\theta+q_w\sin\theta)d\theta - F^*_{axle} = 0 \qquad \text{(3a,b,c,d)}$$

$$R\int_0^{2\pi}(-q_v\sin\theta+q_w\cos\theta)d\theta + W^* = 0$$

Here, the dot (.) denotes differentiation with respect to time; $\sigma_{\theta\theta}^0$ is the initial stress in the treadband due to the action of the centrifugal force and inflation pressure p_0; θ_r is the mean angular displacement of the treadband with respect to the wheel hub which may be described as a windup rotation resulting from the application of a torque.

4. EQUATION OF MOTION USING MODAL EXPANSION METHOD

The sixth order partial differential equation of motion of the treadband reduces to a set of second order ordinary differential equations in time by expressing the solution as a weighted summation of the natural modes of the system.

Equations in the rotating coordinate system
The response of the tangential displacement of the ring to any arbitrary forcing function is expressed in terms of a modal expansion [3],[4]:

$$v(\theta,t) = \sum_{n=0}^{\infty} \varsigma_n(t)\sin(n\theta+\omega_n t) \qquad (4)$$

Here, ω_n are the natural frequencies; n is mode number, $n=0,1,2,\cdots,\infty$; $\varsigma_n(t)$ are the modal participation factors which are unknown and which depend on the applied forces. The above expression can be rearranged as:

$$v(\theta,t) = \sum_{n=0}^{\infty}[a_n(t)\cos(n\theta)+b_n(t)\sin(n\theta)] \qquad (5)$$

where the generalised coordinates $a_n(t)$ and $b_n(t)$ to be determined are
$$a_n(t) = \varsigma_n(t)\sin(\omega_n t); \qquad b_n(t) = \varsigma_n(t)\cos(\omega_n t) \qquad (6)$$
In the case of free-rolling at constant speed, equation (3b) reduces to a simple relation between the relative angular displacement and the zeroth mode of tangential displacement of the treadband; $\theta_r = Rv_0$. Here v_0 is the zeroth mode component of the ring displacement which may be conceived as a windup rotation θ_r of the ring relative to the wheel.

Substituting equation (5) into equation (3a) and considering modes ($n \geq 1$), the motion of the tire ring model in the generalised coordinates $a_n(t), b_n(t)$ reduces to a set of linear second order ordinary differential equations.

$$\begin{bmatrix} m_n & 0 \\ 0 & m_n \end{bmatrix} \begin{Bmatrix} \ddot{a}_n \\ \ddot{b}_n \end{Bmatrix} + \begin{bmatrix} c_n & g_n \\ -g_n & c_n \end{bmatrix} \begin{Bmatrix} \dot{a}_n \\ \dot{b}_n \end{Bmatrix} + \begin{bmatrix} k_n & 0 \\ 0 & k_n \end{bmatrix} \begin{Bmatrix} a_n \\ b_n \end{Bmatrix} = \begin{Bmatrix} \xi_n \\ \zeta_n \end{Bmatrix} \qquad (n \geq 1) \quad (7)$$

The elements in the above matrices are as follows:

$$m_n = \rho A(1+n^2) \; ; \quad g_n = -4\rho A n \Omega \; ; \quad c_n = c_v + c_w n^2$$

$$k_n = \left(\frac{EI}{R^4} n^2 + \frac{\sigma_{\theta\theta}^0}{R^2} \right)(1-n^2)^2 - \frac{p_0 b}{R}(1-n^2) + k_v + k_w n^2 - \rho A(1+n^2)\Omega^2 \qquad (7.1)$$

These matrix elements represent the modal parameters, namely the mass, gyroscopic contribution, damping and stiffness of the nth mode. It will be seen that the rotation of the ring introduces a coupling through the gyroscopic terms arising from Coriolis effects.

The generalised forces corresponding to the physical forces and moment are

$$\xi_n = \frac{1}{\pi} \int_0^{2\pi} \left(q_v + q_w' + \frac{1}{R}(q_\beta + q_\beta'') \right) \cos(n\theta) d\theta$$

$$\zeta_n = \frac{1}{\pi} \int_0^{2\pi} \left(q_v + q_w' + \frac{1}{R}(q_\beta + q_\beta'') \right) \sin(n\theta) d\theta \qquad (n \geq 1) \qquad (8a,b)$$

Response to concentrated line forces

Consider at first the case of concentrated line forces and moment acting at a point on the treadband as shown in Fig.2:

$$q_w(\theta,t) = Q_w \delta(\phi - \phi_0) = Q_w \delta(\theta - (\phi_0 - \Omega t))$$
$$q_v(\theta,t) = Q_v \delta(\phi - \phi_0) = Q_v \delta(\theta - (\phi_0 - \Omega t)) \qquad (9a,b,c)$$
$$\frac{q_\beta}{R}(\theta,t) = Q_\beta \delta(\phi - \phi_0) = Q_\beta \delta(\theta - (\phi_0 - \Omega t))$$

where Q_w, Q_v and Q_β are the magnitudes of radial and tangential line forces and the moment acting at specified point ϕ_0 in the non-rotating coordinates or the corresponding point $\theta - \theta_0$ where $\theta_0 = \phi_0 - \Omega t$ in the rotating coordinates. Substituting Equations (9) into equations (8), we obtain the generalised forces:

$$\xi_n = \frac{1}{\pi} \left[\left(Q_v + (1-n^2) Q_\beta \right) \cos n(\phi_0 - \Omega t) + n Q_w \sin n(\phi_0 - \Omega t) \right]$$

$$\zeta_n = \frac{1}{\pi} \left[\left(Q_v + (1-n^2) Q_\beta \right) \sin n(\phi_0 - \Omega t) - n Q_w \cos n(\phi_0 - \Omega t) \right] \qquad (10a,b)$$

The steady state response to these generalised forces is obtained using *the undetermined coefficient method*. Using the expressions for a_n and b_n from Equation (5), the tangential displacement and the corresponding radial displacement of the tire treadband for the concentrated line forces are finally given by:

$$v(\phi) = \sum_{n=1}^{\infty} \left[\begin{array}{l} \overline{A}_{n1}\left(Q_v + (1-n^2)Q_\beta\right)\cos n(\phi_0 - \phi + \gamma_n) \\ +\overline{A}_{n2}Q_w \sin n(\phi_0 - \phi + \gamma_n) \end{array} \right] \quad (11)$$

where

$$\overline{A}_{n1} = \frac{1}{\pi\sqrt{(\overline{M}_n - \overline{G}_n)^2 + (\overline{C}_n)^2}} \; ; \quad \overline{A}_{n2} = n\overline{A}_{n1}; \quad n\gamma_n = \tan^{-1}(\frac{\overline{C}_n}{\overline{M}_n - \overline{G}_n})$$

$$\overline{M}_n = k_n - (n\Omega)^2 m_n \; ; \quad \overline{G}_n = (n\Omega)g_n \; ; \quad \overline{C}_n = (n\Omega)c_n$$

We can see from Equation (11) that the tire responses are time independent and also that the total phase lag which affects the response is influenced by two factors. The first arises from the term $n\gamma_n$ which is a function of both Coriolis and damping effects associated with the rotation of the wheel. The second arises from the external forces Q_v, Q_w and Q_β.

Response to distributed forces in the contact patch

We now consider the finite contact with the road which extends from the angular coordinate ϕ_f at the front edge of the contact patch to ϕ_r at the rear edge as seen in Fig.1c. For the case of distributed forces $q_w(\phi), q_v(\phi), q_\beta(\phi)$ $(\phi_f \leq \phi_0 \leq \phi_r)$, the total displacements of the tire treadband are obtained by integrating the basic response to concentrated forces along the contact length.

$$v(\phi) = \sum_{n=1}^{\infty} \left[\begin{array}{l} (\alpha_{n1}\overline{A}_{n1} - \alpha_{n3}\overline{A}_{n2})\sin n(\phi - \gamma_n) \\ +(\alpha_{n2}\overline{A}_{n1} + \alpha_{n4}\overline{A}_{n2})\cos n(\phi - \gamma_n) \end{array} \right] \quad (12a,b,c)$$

$$w(\phi) = -v'; \quad \beta(\phi) = \frac{1}{R}(v - w') = \frac{1}{R}(v + v'')$$

where

$$\alpha_{n1} = \int_{\phi_f}^{\phi_r} \left(q_v(\phi) + (1-n^2)\frac{q_\beta(\phi)}{R} \right) \sin(n\phi)d\phi$$

$$\alpha_{n2} = \int_{\phi_f}^{\phi_r} \left(q_v(\phi) + (1-n^2)\frac{q_\beta(\phi)}{R} \right) \cos(n\phi)d\phi$$

$$\alpha_{n3} = \int_{\phi_f}^{\phi_r} q_w(\phi)\cos(n\phi)d\phi \quad (13a,b,c,d)$$

$$\alpha_{n4} = \int_{\phi_f}^{\phi_r} q_w(\phi)\sin(n\phi)d\phi$$

Equations (12) and (13) define the displacements of the tire treadband in terms of the forces acting on the treadband. Given the forces and moment (q_w, q_v, q_β) acting on the treadband, the tire treadband displacements (w, v, β) can be obtained by performing direct integration of equations (13). Alternatively, if the treadband displacements are known, then the forces and moment can be obtained by solving the integral equations.

5. GEOMETRIC AND KINEMATIC COMPATIBILITY IN ROLLING CONTACT WITHOUT SLIDING

Since the treadband is not directly in contact with the road, its displacements are not known in advance. The compatibility condition requires that the displacements of the treadband and tread rubber together must conform with the geometry and motion of flat road surface. Under a certain overall deflection d_0 of the tire, the geometric compatibility between the displacements of the treadband (w, v, β) and the normal and tangential deformations ($h, \gamma h$) of the tread rubber are considered in the next section [6].

Formulation of geometric relations on contact patch with no sliding

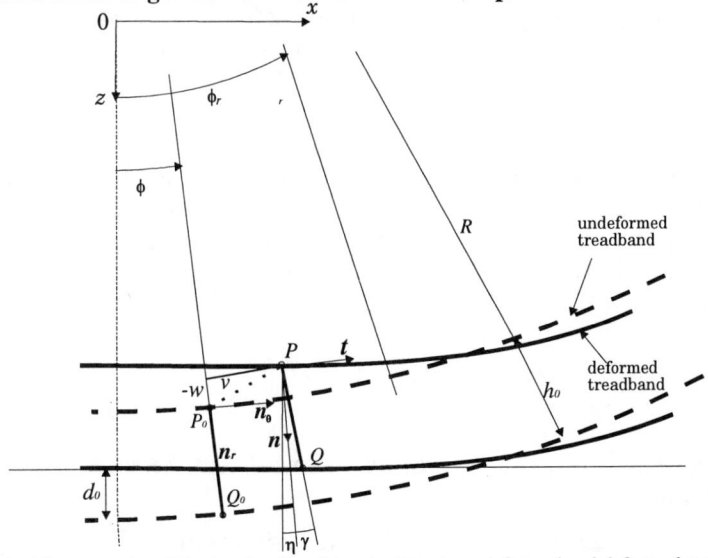

Fig.3: The geometry of the treadband and tread rubber in undeformed and deformed state

Fig.3 shows the geometry of the tire in the undeformed (thick dotted lines) and in the deformed (thick solid lines) states. Under normal loading, typical points P_0 and Q_0 respectively on the treadband and the tread surface in the initially undeformed state move to their new positions P and Q in the deformed state of the tire. The effect of change in curvature of the treadband in the deformed state is taken into account through the declination angle η between the normal to the ring and the vertical plumb line. The declination also contributes to the tangential displacement of points on the tread surface. The declination η and the rotation angle β of the treadband cross section at any given position ϕ are related by,

$$\eta = \phi + \beta \tag{14}$$

The position vector $\vec{r}_{OQ}$ becomes:

$$\vec{r}_{OQ} = \vec{r}_{OP} + \vec{r}_{PQ} = \left[(R+w)\vec{n}_r + v\vec{n}_\theta\right] + \left[h\vec{n} + \gamma h \vec{t}\right] = (x_Q \vec{i} + z_Q \vec{k}) \tag{15}$$

where

$$\begin{Bmatrix} \vec{n}_r \\ \vec{n}_\theta \end{Bmatrix} = \begin{bmatrix} \sin\phi & \cos\phi \\ \cos\phi & -\sin\phi \end{bmatrix} \begin{Bmatrix} \vec{i} \\ \vec{k} \end{Bmatrix}; \quad \begin{Bmatrix} \vec{n} \\ \vec{t} \end{Bmatrix} = \begin{bmatrix} \sin\eta & \cos\eta \\ \cos\eta & -\sin\eta \end{bmatrix} \begin{Bmatrix} \vec{i} \\ \vec{k} \end{Bmatrix} \tag{16a,b}$$

Here $(\bar{n}_r, \bar{n}_\theta)$ are unit vectors in the radial and tangential directions at points on the treadband axis and $(\bar{n}, \bar{t})$ are unit vectors in the normal and tangential directions of the treadband axis. Position vector $\bar{r}_{OQ}$ may be transformed in non-rotating Cartesian coordinates by unit vectors $(\bar{i}, \bar{k})$ with the help of the relation (16).

$$x_Q = (R+w)\sin\phi + v\cos\phi + h\sin\eta + \gamma h\cos\eta$$
$$z_Q = (R+w)\cos\phi - v\sin\phi + h\cos\eta - \gamma h\sin\eta \qquad (17a,b)$$

This coordinates (x_Q, z_Q) of points on the tread surface must satisfy the constraints imposed by the geometric compatibility and friction between the tread surface and the flat road. The mainly normal constraint due to the tire-flat road interface is:

$$z_Q = R_l \qquad (18)$$

where R_l is the height of the wheel rotation axis above the road.

The second contact constraint of free-rolling is the condition of rolling without sliding of all points on the tread surface inside the contact. These points travel at a uniform speed equal to the speed of translation V_t of the road surface. The linear speed and the angular velocity of the wheel of a free-rolling tire are related kinematically through the effective rolling radius R_e defined by:

$$V_t = R_e \Omega \qquad (19a)$$

Thus, the translation of point Q in the contact is

$$x_Q = R_e \phi \qquad (19b)$$

Combining equations (17),(18) and (19b), the normal h and the shear γh of the tread rubber deformation become:

$$h(\phi) = R_l \cos\eta + R_e \phi \sin\eta - (R+w)\cos(\phi-\eta) + v\sin(\phi-\eta)$$
$$\gamma h(\phi) = -R_l \sin\eta + R_e \phi \cos\eta - (R+w)\sin(\phi-\eta) - v\cos(\phi-\eta) \qquad (20a,b)$$

The unknown parameters R_e, R_l can be determined by specifying two boundary conditions, namely, that the normal and tangential tractions vanish at the front edge of the contact; At $\phi = \phi_f$, $h(\phi_f) = h_0$, $\gamma h(\phi_f) = 0$:

$$R_e = \left(h_0 \sin\eta(\phi_f) + (R+w(\phi_f))\sin\phi_f + v(\phi_f)\cos\phi_f \right) / \phi_f$$
$$R_l = h_0 \cos\eta(\phi_f) + (R+w(\phi_f))\cos\phi_f - v(\phi_f)\sin\phi_f \qquad (21a,b)$$

The forces acting on tread surface are

$$f_n(\phi) = k_{Et}(h(\phi) - h_0); \qquad f_t(\phi) = k_{Gt}\gamma h(\phi) \qquad (22a,b)$$

where $f_n(\phi)$ and $f_t(\phi)$ are the force intensities acting in the contact patch directed along the normal and tangent to the treadband. k_{Et} and k_{Gt} are the normal and shear stiffness of the tread rubber per unit length of the treadband. The coordinate transformation relating the tractions $(f_n, f_t), (\sigma, \tau)$ and (q_w, q_v, q_β) are:

$$\begin{Bmatrix} q_w(\phi) \\ q_v(\phi) \\ q_\beta(\phi) \end{Bmatrix} = \begin{bmatrix} \cos(\phi-\eta) & \sin(\phi-\eta) \\ -\sin(\phi-\eta) & \cos(\phi-\eta) \\ 0 & h \end{bmatrix} \begin{Bmatrix} f_n(\phi) \\ f_t(\phi) \end{Bmatrix}; \qquad \begin{Bmatrix} \sigma \\ \tau \end{Bmatrix} = \begin{bmatrix} \cos(\eta) & -\sin(\eta) \\ \sin(\eta) & \cos(\eta) \end{bmatrix} \begin{Bmatrix} f_n \\ f_t \end{Bmatrix} \qquad (23a,b)$$

where σ, τ are normal pressure and shear traction in the contact patch.

When we consider the case of the rotating tire, the windup rotation of the wheel relative to the ring through θ_r has to be taken into account. This is implemented by the coupling term in zeroth mode between wheel and ring motion. Introduction of θ_r in the compatibility constraint equation (17) is done simply by replacing ϕ and η with $\phi-\theta_r$ and $\eta-\theta_r$, respectively, so that the problem can be extended to the dynamic problem of the rotating tire.

6. INTEGRAL EQUATION OF THE CONTACT PROBLEM WITHOUT SLIDING

Essentially, the surface displacements of the tread rubber relative to the rigid wheel are prescribed by the boundary conditions in equation (20) of the rolling contact problem. However, these equations are non-linear and hence the modal decomposition technique cannot be used directly. The solution of the contact problem is obtained in two steps; In the first step, an approximate solution is found using linearized boundary condition. This linearization leads to a linear integral equation for the contact problem which is easily cast in a set of linear matrix equations in the unknown coefficients $\alpha_{n1}, \alpha_{n2}, \alpha_{n3}, \alpha_{n4}$ (see equation (26)). In the second step, the approximate (linearized) solution for normal and shear traction is used to calculate the treadband displacement. This step requires simply the numerical evaluation of an integral. The tread rubber deformation is calculated by demanding that it is compatible with the non-linear boundary conditions of the rolling contact. An improved estimate of the normal and shear tractions can be calculated easily. This iterative procedure using the method of successive substitution is repeated several times until the desired degree of the accuracy of the estimated solution is achieved. Following this two steps approach, the problem of convergence arising from the relatively small deformation of the stiff tread rubber has been overcome.

Solution using linearized boundary equations

Assuming that w, v, η are small compared to the free radius of the ring R, the non-linear boundary conditions (20a,b) and the transformation equation (23) are linearized by ignoring terms involving products of small displacements w, v and small angle η. The linearized force-deflection relations for the treadband read:

$$q_w(\phi) \doteq K_{Et}\left(\left(-h_0 + R_l + R_e\eta\phi\right)\cos\phi - (R+w)\cos^2\phi - \left(R\eta - \frac{1}{2}v\right)\sin 2\phi + (R_l - h_0)\eta\sin\phi\right)$$
$$- K_{Gt}\left((R_l\eta - R_e\phi)\sin\phi + (R+w)\sin^2\phi - \left(R\eta - \frac{1}{2}v\right)\sin 2\phi + R_e\eta\phi\cos\phi\right)$$
$$q_v(\phi) \doteq K_{Et}\left((h_0 - R_l - R_e\eta\phi)\sin\phi + \frac{(R+w)}{2}\sin 2\phi + (R\eta - v)\sin^2\phi + (R_l - h_0)\eta\cos\phi - R\eta\cos^2\phi\right)$$
$$+ K_{Gt}\left((R_e\phi - R_l\eta)\cos\phi - \frac{(R+w)}{2}\sin 2\phi + (R\eta - v)\cos^2\phi + R_e\eta\phi\sin\phi - R\eta\sin^2\phi\right)$$
$$q_\beta(\phi) \doteq K_{Gt}h_0\left(-R_l\eta + R_e\phi - (R+w)\sin\phi + (R\eta - v)\cos\phi\right) \quad (24a,b,c)$$

The implicit and coupled relation for the unknown coefficient α_{n1} in equation (12) using equation (24) becomes, for example:

$$\alpha_{n1} = \int_{\phi_f}^{\phi_r} \left[\begin{array}{l} K_{Et}\left(\begin{array}{l}(-R_l+h_0)\sin\phi - R_e\eta\phi\sin\phi + (R+w)\sin\phi\cos\phi \\ +(R\eta-v)\sin^2\phi + (R_l\eta-h_0\eta)\cos\phi - R\eta\cos^2\phi \end{array} \right) \\ +K_{Gt}\left(\begin{array}{l}-R_l\eta\cos\phi + R_e\phi\cos\phi - (R+w)\sin\phi\cos\phi \\ +(R\eta-v)\cos^2\phi + R_e\eta\phi\sin\phi - R\eta\sin^2\phi \end{array} \right) \\ +K_{Gt}h_0(1-n^2)(-R_l\eta + R_e\phi - (R+w)\sin\phi + (R\eta-v)\cos\phi)/R \end{array} \right] \sin(n\phi)d\phi \quad (25)$$

where v, w, R_l, R_e, η, $R_l\eta$ and $R_e\eta$ are in turn functions of ϕ_f and ϕ_r through the unknown coefficient $\alpha_{n1}, \alpha_{n2}, \alpha_{n3}, \alpha_{n4}$. Similar expressions can be obtained for α_{n2}, α_{n3} and α_{n4}. Expressing v, w, R_l, R_e, η using equations (12),(14) and (21) in terms of α_{n1}, α_{n2}, α_{n3}, α_{n4} and inserting these into equation (25) leads to a set of coupled equation in α_{n1}, α_{n2}, α_{n3}, α_{n4}. The expansion is truncated beyond the first N modes ($N \geq 30$) subject to an acceptable error bound of percentage in the tire deflection. Collecting terms containing unknown coefficients of α_{n1}, α_{n2}, α_{n3}, α_{n4} and other constants yields the following set of linear algebraic equations.

$$\alpha_{n1} = \sum_{i=1}^{N}[(a_{ni}^1)_1 \cdot \alpha_{i1} + (a_{ni}^1)_2 \cdot \alpha_{i2} + (a_{ni}^1)_3 \cdot \alpha_{i3} + (a_{ni}^1)_4 \cdot \alpha_{i4} - b_n^1] \quad (26a)$$

$$\alpha_{n2} = \sum_{i=1}^{N}[(a_{ni}^2)_1 \cdot \alpha_{i1} + (a_{ni}^2)_2 \cdot \alpha_{i2} + (a_{ni}^2)_3 \cdot \alpha_{i3} + (a_{ni}^2)_4 \cdot \alpha_{i4} - b_n^2] \quad (26b)$$

$$\alpha_{n3} = \sum_{i=1}^{N}[(a_{ni}^3)_1 \cdot \alpha_{i1} + (a_{ni}^3)_2 \cdot \alpha_{i2} + (a_{ni}^3)_3 \cdot \alpha_{i3} + (a_{ni}^3)_4 \cdot \alpha_{i4} - b_n^3] \quad (26c)$$

$$\alpha_{n4} = \sum_{i=1}^{N}[(a_{ni}^4)_1 \cdot \alpha_{i1} + (a_{ni}^4)_2 \cdot \alpha_{i2} + (a_{ni}^4)_3 \cdot \alpha_{i3} + (a_{ni}^4)_4 \cdot \alpha_{i4} - b_n^4] \quad (26d)$$

Equation (26) can be cast into a set of $4N$ linear equations which can be solved to give α_{n1}, α_{n2}, α_{n3} and α_{n4}. The treadband displacements and the contact forces can be determined for a given contact angle, ϕ_f and ϕ_r, which satisfy the boundary conditions of zero normal pressure at both the leading and trailing edges of the contact. We note that the boundary conditions at ϕ_f are automatically implemented at the same time by using the expression of R_e and R_l which prescribe conditions of zero traction at the leading edge of the contact. Once the front contact angle ϕ_f is found, the rear contact angle ϕ_r can be determined by iteration so as to meet the condition of zero normal pressure at the trailing edge of the contact.

Refined solution using the exact boundary conditions

Using the approximate solution for the tractions and the contact region (ϕ_f, ϕ_r) of the linearized problem, the exact solution is obtained by direct integration of equation (13) and successive substitution of this result into the exact boundary conditions of equation (20) and the coordinate transformation according to (23).

Consideration of Sliding Region

Sliding between the tire and the road is important not only as a source of wear but it also contributes to the rolling resistance. In the case of sliding, the kinematic constraint defined by equation (19) is no longer applicable in the sliding zone.

Assuming Coulomb friction to apply locally, the boundary condition on the shear traction is

$$\tau \leq \mu|\sigma|sgn(\tau) \quad \text{when } V_s = 0, \quad \text{and} \quad \tau = -\mu|\sigma|sgn(V_s) \quad \text{when } |V_s| > 0 \quad (27)$$

Comparing the tractive ratio τ/σ with any finite value of coefficient of friction μ, it turns out that sliding will start from the trailing edge of the contact even while the tire operates under free-rolling conditions.

The local sliding displacement x_s is defined by the relative distance between a point on the translating road surface and the corresponding point on the tread rubber surface as the point passes through the sliding zone.

$$x_s = (x_Q(\phi) - x_Q(\phi_s)) - R_e \cdot (\phi - \phi_s) \tag{28}$$

where ϕ_s is the angular position of beginning of sliding. The local sliding velocity V_s can be obtained by differentiating equation (28) with respect to time:

$$V_s = \left(\frac{\partial x_Q}{\partial \phi} - R_e\right)\Omega \tag{29}$$

The contribution of frictional power to the rolling resistance force in sliding zone:

$$P_s = \int_{\phi_s}^{\phi_r} \tau V_s dx_s \tag{30}$$

The contribution of hysteretic energy dissipation is given by:

$$P_h = F_x V_t \tag{31}$$

where F_x is the resultant frictional force.

7. QUASI-STATIC FREE-ROLLING TIRE

The numerical results for the case where the rolling speed is vanishingly small are given below. The solution in this section are obtained by assuming that there is no sliding in the contact patch.

Influence of normal load on the contact variables:

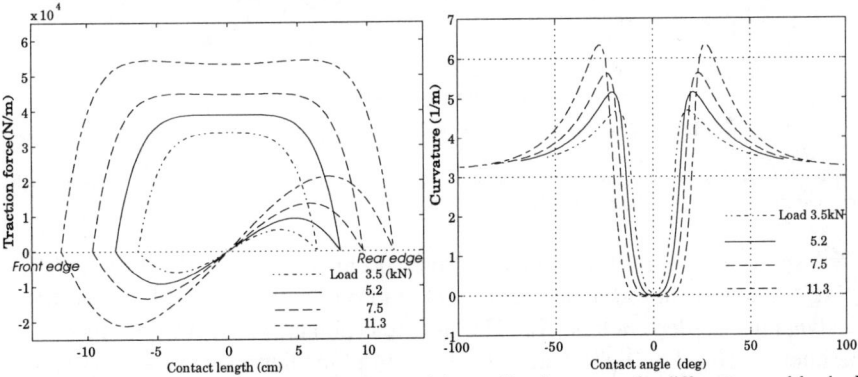

Fig.: 4.1, 4.2: The contact forces distributions and the treadband curvature for different normal loads; F_Z = 3.5, 5.2, 7.5, 11.3 (kN)

Fig.4.1 shows the extent of the contact and the distributions of normal and shear tractions for different normal loads. Both the contact length and the average level of the pressure increase with increasing load. At small vertical loads the normal pressure shows a convex shape, like a parabola; at medium loads the parabolic shape tends to become flat; at large loads the shape changes markedly exhibiting a dip in the central region. These trends are generally agreement with those reported in experimental studies [9]. At vanishingly small speeds the resultant rolling resistance friction force or rolling resistance moment tend to zero due to the negligible effect of damping at such low speed. Consequently the normal traction distribution and the position of the contact patch are symmetric with respect to the axis of the wheel. The anti symmetry of shear traction imply a zero rolling resistance force.

Fig.4.2 shows the change in treadband curvature under the same conditions. At large loads the curvature in the contact region has a small negative value. This means that the treadband tends to buckle inwards against the action of the inflation pressure which corresponds to the dip in the pressure distribution at large loads. The curvature peaks at some distance outside the region of contact and then decreases to approach its mean undeformed curvature ($1/R$) asymptotically at points away from the contact patch.

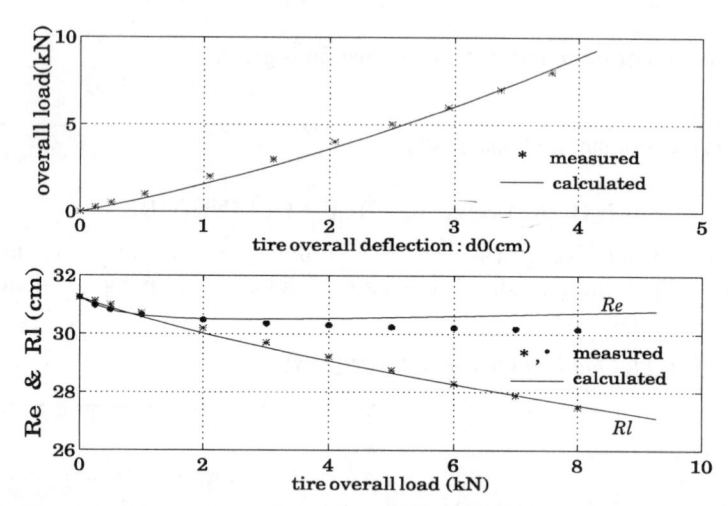

Fig.: 5.1: The relation between tire overall deflection and load according to the theory and experiment for the tire: 205/60R15;

Fig.: 5.2: The relation between tire overall load and effective rolling radius (marked 'o') and dynamic loaded radius (marked '*') under the assumption of without sliding.

Fig.5.1 shows the relation between the overall deflection d_0 and the overall load F_z. Fig.5.2 shows the influence of normal load on the effective rolling radius R_e and dynamic loaded radius R_l. The marked points represent the results of experiments [10]. For light loads, R_e is a little smaller than R_l as reported in experimental studies. At larger loads, the calculated R_e increases only slightly while the experimental results exhibit a slightly decreasing trend. This discrepancy observed at large deflections is probably due to the geometrical non-linearity associated with large displacements and rotations which are not taken into account in the present linear analysis of the tire deformation.

8. FREE-ROLLING AT HIGHER SPEEDS WITH FINITE FRICTION

The two different cases of free-rolling tires discussed in section 1 are analysed in this section. In addition the analysis addressees the free-rolling at high speeds and the influence of finite friction and damping in the tire.

Free-rolling of Driven and Driving wheels

	Driven Wheel	Driving Wheel
F_z (kN)	4.827	4.861
F_x (kN)	0.168	0
M_{rr} (Nm)	-48.35	-50.33
R_e (cm)	30.69	30.65

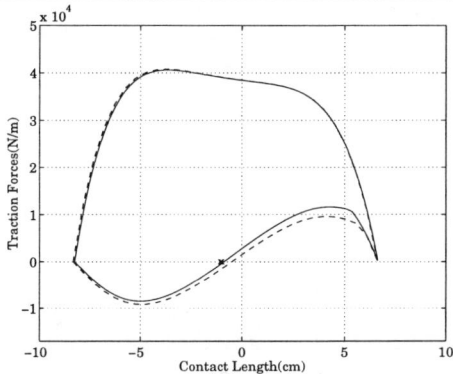

Fig.6: Free-rolling condition on the driven wheel (solid line) and driving wheel (dotted line) under d_0=2.5cm, Ω=150rad/sec and μ=0.65.

Table 2: Numerical results of the normal load, tangential force, rolling resistance moment and effective rolling radius, respectively, for the same overall deflection.

Fig.6 and Table 2 present the numerical results for the two free-rolling situations for the same overall deflection d_0=2.5cm and rotating speed 150rad/sec. Equation (12) indicates that damping tends to shift the maximum deflection towards the rear part of the contact patch. The total asymmetry of the resultant forces due to internal damping is caused both by the asymmetric distribution of normal and tangential tractions in the contact patch and asymmetric position of the contact patch with respect to the wheel axis. The resultant effect is the shifting of the centroid of normal pressure forward shown by 'x' $\approx$ -1.0cm. These asymmetric effects produce a rolling resistance moment M_{rr}. It is also observed that a region of sliding appears in the rear region of contact (roughly at 5.28cm). The location of the leading contact edge moves to -8.25cm while the rear contact edge is located at the coordinate to 6.62cm.

Occurrence of Sliding Zone and Contribution to Frictional Work

The traction forces distribution assuming no sliding ($\mu = \infty$) is shown by dotted line in Fig.7.1. Shear traction which applies for a finite friction coefficient (μ =0.65) according to the Coulomb law is shown by the solid line in the figure. The region of sliding in Fig.7.2 starts at the rear edge of the contact region and extends up to the point marked 'o'. The contributions to the total energy dissipation and the rolling resistance due to sliding and hysteresis are obtained using equations (30),

(31). At the rotating speed of 200rad/sec, the coefficient of rolling resistance has a value of 4.6% assuming an equivalent viscous damping coefficient of 450(Ns/m^2). The power dissipation P_s due to sliding is only 243 (Nm/s) whereas that due to hysteresis P_h is 13910(Nm/s). Clearly, the power loss caused by sliding is very small a mere 1.7% of the power loss due to the hysteretic cause.

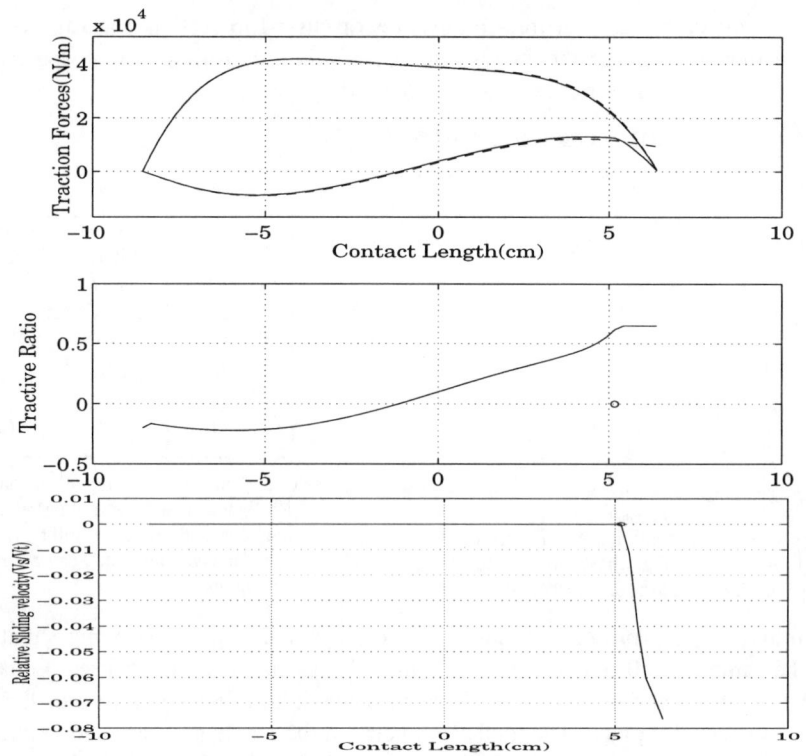

Fig.7.1. Traction forces of no sliding condition(dotted line) and of having finite friction coefficient μ =0.65 (solid line)

Fig.7.2. Plot of tractive ratios (τ / σ) along the contact patch when μ =0.65.

Fig.7.3. Plot of non-dimensional sliding velocity ratio (slide/roll) along the contact patch.

Fig.7.3 shows how the normalised local sliding velocity varies within the contact patch.

Influence of Damping

In the previous sections tire damping has been taken into account by including linear viscous damping elements in the viscoelastic foundation. However, the actual damping characteristics of a complex system such as the tire is very difficult to model precisely. This modelling aspect will be explored by considering three different models of damping. The suitability of the damping model may be judged by how well the predictions based on the model match experimental measurements of the rolling resistance and particularly its dependence on the speed of rolling.

1. A viscous damping with a constant rate which is independent of speed.
2. Structural damping represents the empirical model of material hysteresis defined by a damping rate which is inversely proportional to the angular velocity (frequency of deformation). The equivalent damping rate c_{eq} is [7]:

$$c_{eq} = \frac{\alpha}{\Omega} \qquad (32)$$

   where α is a constant independent of the frequency.
3. The third model considered is a Maxwell 4 element model described by Sakai [8] which consists of a parallel combination of Voigt and Maxwell spring-damper elements. This model can be reduced to the Voigt type by introducing an equivalent dynamic stiffness G' and an equivalent dynamic viscosity η':

$$G' = G_1 + G_2 \frac{\Omega^2 \tau_2^2}{1+\Omega^2 \tau_2^2}; \qquad \eta' = \eta_3 + \eta_2 \frac{1}{1+\Omega^2 \tau_2^2} \qquad (33)$$

   where G_1, η_3 and G_2, η_2 are respectively stiffness and viscosity parameters in Voigt and Maxwell elements proposed in [8]. The characteristic time τ_2 is known as the retardation time.

The present work concentrates on the effect of rolling speed and the nature of damping on the rolling resistance due to deformation hysteresis and frictional sliding.

Comparison of Modal Damping Ratios for Viscous and Structural Damping
Modal damping ratio is calculated for the cases of viscous and structural damping by using the modal parameters in equation (7.1) and critical damping as $2\sqrt{m_n k_n}$

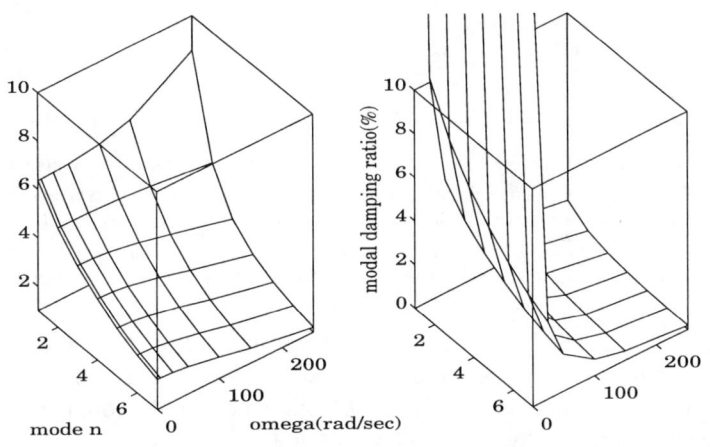

Fig.8.1: Variation of modal damping ratio with rolling speed for the case of constant rate viscous damping.

Fig.8.2: Variation of modal damping ratio with speed for the case of structural damping model.

Fig.8.1 and 8.2 show the influence of the damping model (viscous or structural) on the modal damping ratio. The damping ratio varies depending upon both the mode number and the rolling speed. At very low speed of rolling (Fig.8.1), the predicted trend showing a decrease of the modal damping ratio with increasing of mode number is in good agreement with experimental results [11]. The opposite trends of modal damping ratios with increasing speed of rolling according to the models of viscous and structural damping (especially for low mode numbers) explain the typical characteristics of rolling resistance as shown in Fig.9.4.

Contact Force Distribution and Rolling Resistance

Fig.9.1: constant viscous damping rate

Fig.9.2: structural damping

Fig.9.3: Maxwell 4 element type

(Fig.9.4): coefficient of RR for 3 damping models

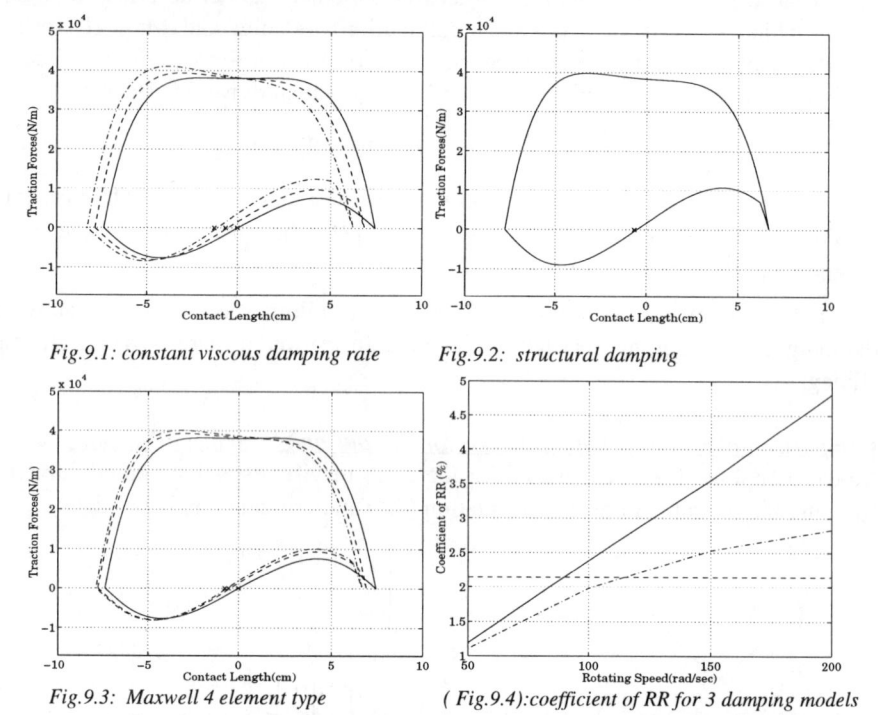

Fig.9.1,9.2,9.3. Distribution of traction forces at different speeds predicted by the different models of damping; 1(rad/sec):solid line, 100(rad/sec):dotted line, 200(rad/sec):dash-dotted line.

Fig.9.4: Rolling resistance F_x/F_z(%) predicted by three damping models; solid line(constant viscous damping), dotted line(structural damping), dash-dotted line(Maxwell 4 element type).

For the case of constant rate viscous damping, increasing the rolling speeds increases the asymmetry both in the distribution of forces and the position of the centroid of normal pressure as seen in Fig.9.1. As expected, the rolling resistance increases linearly as shown by the solid line in Fig.9.4.

The use of the structural damping model with a damping rate which varies inversely with the rolling speed results in identical distribution of tractions as seen in Fig.9.2. This leads to the constant value of the coefficient of rolling resistance independent of speed as shown by the dotted line in Fig.9.4.

If a 4 element Maxwell model is used to represent the damping, it turns out that the influence of rolling speed on the distribution of normal and shear tractions is less strong than that in the case of a constant rate viscous damping (see Fig.9.3). The rolling resistance increases with the speed, but with a decreasing rate as seen in Fig.9.4 by the dash-dotted line.

CONCLUSION

The rolling contact of tire is studied by considering the non-linear boundary conditions imposed by a flat road surface with finite friction on both the normal and the tangential deformation of the tire. The "ring on a viscoelastic foundation" model of tires has all the essential features to describe the rolling contact of tires.

The model of a ring on a viscoelastic foundation is used to study a range of problems of in-plane free-rolling of tires on a flat road.

1. The two different situations of free rolling of driven and driving wheels are analysed. It is found that the difference between these two situations with respect to the distribution of normal and tangential tractions is very small. So for all practical purpose the rolling resistance moment or couple in the two situations is almost identical.

2. It is shown that the rolling resistance on a flat and dry road with a finite $\mu=0.65$ is contributed mainly by the inelastic deformation (hysteresis) of the tire. The region of sliding which is restricted to a small part starting from trailing edge of the contact patch makes only a small contribution to the rolling resistance.

3. The analysis of the contact problem of free-rolling tire without any sliding in the contact patch (infinite μ) yields a unique definition of the effective rolling radius R_e. More generally, while rolling on roads with finite friction, R_e will exhibit a weak dependence on μ. The results of the analysis assuming no sliding show that R_e depends non-monotonically on the normal loads as seen in Fig.5.2.

4. Different models have been considered to represent the damping properties of rolling tires. One of the important criteria in assessing these models is the magnitude of the rolling resistance and the variation of the resistance with the speed of rolling. Each of the three models considered (constant-rate viscous damping, structural damping and Sakai model consisting of 4 spring and damper elements) show a characteristic variation of rolling resistance with speed. The structural damping model exhibits speed independent trend which is similar to that reported in the open literature. The reported trend showing a steep increase in rolling resistance at very high speeds remains to be investigated.

5. The analysis of the contact problem described in this paper may be extended to study tire behaviour in braking and traction. Another important extension

studied currently is the non-stationary rolling of tires on an uneven road surface.

REFERENCES

[1] Schuring,D.J, *The rolling loss of pneumatic tires*, Rubber Chemistry and Technology, 53(1):601-727, 1980
[2] Lippmann,S.A., Oblizajek,K.L. and Metters,J.J., *Sources of rolling resistance in radial ply tires*, SAE paper No. 780258, 1978.
[3] Gong,S., *A study of in-plane dynamics of tires.*, PhD thesis, Delft University of Technology, Faculty of Mechanical Engineering and Marine Technology, 1993.
[4] Soedel,W., and Prasad,M.G., *Calculation of natural frequencies and modes of tires in road contact utilising eigenvalues of the axisymmetric contact-free tire*, Journal of Sound and Vibration, 70(4):573-584, 1980.
[5] Huang,S.C., *The vibration of rolling tires in ground contact*, International Journal of Vehicle Design, 13(1):78-95, 1992.
[6] Klingbeil,W.W, and Witt,W.H., *Some consequences of Coulomb friction in modelling longitudinal traction*, Tire science and Technology, TSTCA, 18(1):13-65, 1990.
[7] Meirovitch, L., Elements of Vibration Analysis, McGraw-Hill Book Company, 1986
[8] Sakai,E.H, Tire Engineering *(in Japanese)*, Grand prix printing company, 1986
[9] Clark, S.K., Mechanics of Pneumatic Tires, National Bureau of Standards, Monograph 122, Washington, D.C., 1971
[10] Zegelaar,P.W.A., and Pacejka,H.B., *Dynamic tyre response to brake torque variations*, Vehicle System Dynamics, 2nd International Tyre Colloquium on Tyre Models for Vehicle Dynamic Analysis, Berlin, 1997
[11] Fejes,I. and Savkoor,A.R., *Modelling and identification of pneumatic tire*, 2nd International Conference on Motion and Vibration Control, Yokohama, Japan, 1994

APPENDIX:

Table1: The parameters of a ring model used in the numerical work

Description	Symbol	Values	Unit
sidewall radial stiffness	k_w	$6.30*10^5$	N/m^2
sidewall tangential stiffness.	k_v	$1.89*10^5$	N/m^2
equivalent viscous damping coefficient	c_{ea}	300~450	Ns/m^2
treadband bending stiffness.	EI	2.0	Nm^2
treadband mass density	ρA	3.15	Kg/m
treadband width	b	0.14	m
treadband mean radius	R	0.3	m
inflation pressure	p_0	$1.2*10^5$	N/m^2
tread normal stiffness.	k_{Et}	$47.7*10^6$	N/m
tread tangential stiffness.	k_{Gt}	$19.1*10^6$	N/m
tread thickness	h_0	0.0125	m

REPRESENTING TRUCK TIRE CHARACTERISTICS IN SIMULATIONS OF BRAKING AND BRAKING-IN-A-TURN MANEUVERS

P. FANCHER, J. BERNARD, C. CLOVER, and C. WINKLER

ABSTRACT

The tire characteristics and modeling concepts presented here are intended for use in studying the performance of trucks (including articulated vehicles) in braking and braking-in-a-turn maneuvers such as those included in recent versions of U.S. Federal Motor Vehicle Safety Standard (FMVSS) 121.

A semi-empirical tire model is used to provide example truck-tire characteristics. The semi-empirical model is based upon the theoretical concepts employed in brush type tire models. Tire dynamics are included using a relaxation length approach. These postulated tire concepts are employed in conjunction with measured or specified tire stiffnesses and tire-road frictional properties to predict longitudinal and lateral forces.

INTRODUCTION

The basic mechanical properties of the tire used in the model presented herein are cornering stiffness, longitudinal stiffness, lateral relaxation length, longitudinal relaxation length, carcass stiffness, pneumatic trail, maximum friction between tire and road surface at low sliding velocity, asymptotic minimum friction between tire and road surface at high sliding velocity, and a parameter that determines the shape of the friction function for the surface and tire combination. (See table 1.)

The model differs from previous brush type models in the treatment of the sliding region of the contact patch [1]. A measure of the sliding velocity is used to determine the friction limit for shear stress in the sliding region. Once the frictional properties are determined the point where sliding starts is calculated. Then the force components from the adhesion region and the sliding region are calculated using equations derived by integrating the shear stresses. When the contact patch is sliding, the force vector opposes the direction of sliding. The equations for steady-state tire force characteristics are given in appendix A.

The influences of tire dynamics are approximated by including relaxation effects in a simplified manner in which the steady-state deformation pattern of the tire develops over distance and time. The approach used is highly conceptual and insightful and follows material prepared by Bernard and Clover [2]. See appendix B.

The influences of varying brake torque and frictional conditions are examined using the tire model. Example analyses based on simulating a mobile tire tester that

Table 1. Tire Model Parameters

Name and symbol	Units	Case 1*	Case 2*
cornering stiffness, C_α	lbs/radian	43,200	43,200
longitudinal stiffness, C_s	lbs/unit slip	48,000	48,000
longitudinal relaxation length, σ_x	ft	0.5	0.5
lateral relaxation length, σ_y	ft	1.1	1.1
maximum friction parameter, μ_o		1.5	0.6
maximum friction parameter, μ_f		0.4	0.1
friction function shaping, V_f	ft/sec	27.5	41.0
pneumatic trail, x_p	in	1.5	1.5
carcass stiffness for aligning torque, C_y	lbs/in	4000	4000
tire radius, R	in	20	20

* Case 1: Representative values for a high friction road surface and F_z = 6000 lbs.
* Case 2: Representative values for a slippery wet road surface and F_z = 6000 lbs.
Note: 1 pound = 0.45 kilograms and 1 foot = 12 inches = 0.3 meters

operates with fixed slip angles and transient braking torque are presented to illustrate predictions of the effects of dynamic tire properties on tire test results for longitudinal and lateral forces.

The ability to obtain shear force data for truck tires is limited to a few tire testers, and, in most cases, it is not possible to obtain data that will distinguish between the virtues of various techniques for modeling the dynamics of truck tires. Consequently, there is a need to use conceptual reasoning to aid in selecting methods for representing tire shear force properties in a manner that is appropriate for the vehicle dynamics study to be undertaken using the model. This paper provides information and ideas that are intended to aid in the process of acquiring the understanding needed to make wise selections of tire models for vehicle dynamics studies.

EQUATIONS FOR STEADY STATE CHARACTERISTICS

The set of equations given in appendix A is based upon a semi-empirical tire model. Figure 1 illustrates the main elements of the computation of the steady-state longitudinal force at fixed slip conditions. There are a great number of simplifying assumptions incorporated into the model. A primary simplification results from assuming a uniform pressure distribution. This results in a fairly simple form for the model's equations. Since this is a semi-empirical model, the "end and edge effects" due to non-uniform pressure distributions, transitions between adhesion and sliding, etc. are accounted for by extracting stiffness and friction values to fit test data at various vertical loads. In other words, this is not meant to be a rigorous, detailed theory. It is intended to represent the basic phenomena in a relatively simple manner. The goal is a simplified model that uses a few parametric quantities to predict tire forces over a wide range of operating conditions.

The subject of "What is longitudinal slip?" is an issue for this (and all other) models when it comes to examining or using experimental data. In this work, tread deformation is computed with respect to the attachment points of the brush elements at the belt end of those elements. This means that the deformation pattern for longitudinal deflection in the adhesion region of the contact patch (collapsed along an equatorial line in the tire) is linearly related to s/(1 - s) where:

$$\text{slip, } s = (V - R\omega) / V \quad (1)$$

Although the next equation (equation 2) involves the need to avoid dividing by zero at locked wheel conditions ($\omega = 0$), it has the advantage that the deformation at a point at a distance x from the front of the contact patch (in the adhesion region) has a deformation $\delta(x)$ given by the following equation:

$$\delta(x) = x \, s_c \quad (2)$$

where

$$s_c = (V - R\omega) / R\omega \quad (3)$$

With regard to lateral slip and its tangent, one can propose a modified definition along the lines of equations

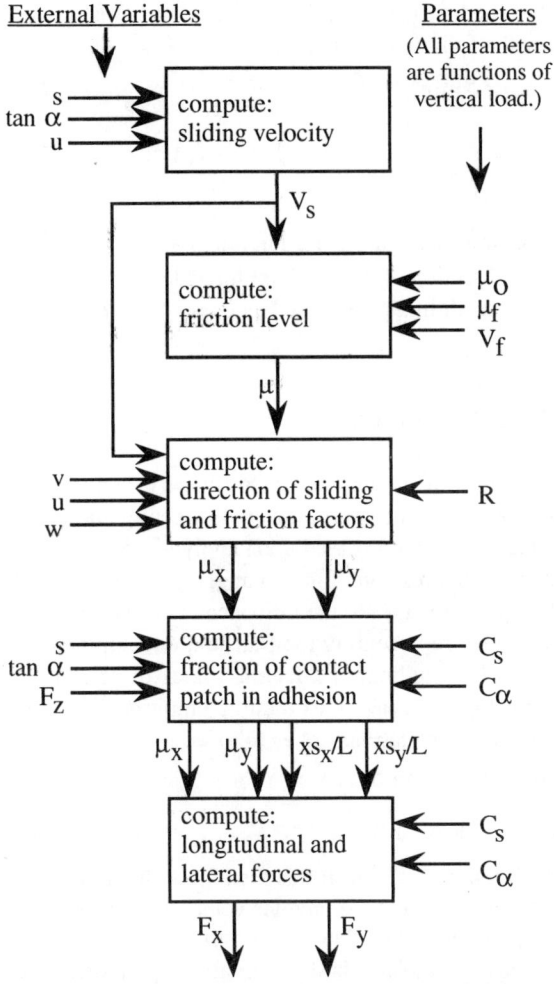

Figure 1. Computational diagram for the steady-state portion of the tire model

(2) and (3); viz.,

$$\tan \alpha_c = v / R\omega \quad (4)$$

The purpose of this consideration is to treat cases of combined slip in a manner that depends upon the rotational motion of the carcass/belt end of the brush elements. (For "free rolling" conditions $V = u = R\omega$ and equation (4) represents the conventional definition of lateral slip.)

With few additional caveats, the steady state equations given in appendix A are believed to be fairly representative of those presented in a number of brush type tire models (e.g., [3–5]). The primary difference involves the treatment of the friction between the brush elements and the road surface. In this case, friction (μ) is treated as a function of sliding velocity per the following equation:

$$\mu = \mu_{min} + (\mu_{max} - \mu_{min}) \exp(-|V_s| / V_f) \tag{5}$$

where the sliding velocity is

$$V_s = \{(u - R\omega)^2 + v^2\}^{0.5} \tag{6}$$

and V_f is a parameter that determines the shape of the friction function.

Furthermore, in combined slip, the direction of the resultant force depends upon an angle of friction, θ, such that, in pure sliding, the force opposes the direction of sliding. (See appendix A.)

TIRE DYNAMICS

The velocities involved in the steady-state model do not ordinarily include the influences of carcass motions or the transients involved with "rolling into" the steady-state condition. The following discussion is based upon applying the concepts in reference [2] to the development of expressions for slip angle and longitudinal slip, thereby providing a first order approximation to the influences of tire dynamics. Two first order differential equations for representing the transient development of slip and slip angle are created using longitudinal and lateral values of relaxation length parameters. The basis for this approach lies in examining empirical data and following the work of others. The resulting equations are as follows:

$$d/dt (\tan \alpha') = (v - u \tan \alpha') / \sigma_y \tag{7}$$

and

$$d/dt (s') = (u - R\omega - us') / \sigma_x \tag{8}$$

where v is the lateral velocity component at the wheel, u is the longitudinal velocity component in the wheel plane at the wheel, ω is the angular velocity of the hub, R is the effective radius of the tire, σ_y represents the lateral relaxation length, and σ_x represents the longitudinal relaxation length. (d/dt is the derivative operator.) (See appendix B.)

To include the influences of tire dynamics, the quantities $\tan \alpha'$ and s' are used in place of $\tan \alpha$ and s, respectively, as inputs to the steady-state section of the tire model.

EXAMPLE ANALYSES BASED UPON USING A MOBILE TIRE TESTER

In these analyses, the mobile tire tester is considered to be a device for presenting a truck tire to a test surface at a prescribed slip angle, vertical load, and velocity [6]. A friction brake is used to apply torque to the test wheel, thereby achieving various levels of longitudinal slip. Idealized equations representing the mobile tire tester are as follows:

$$I \, d\omega/dt = T_b - F_x R \tag{9}$$

TRUCK TIRE CHARACTERISTICS

$$u = V \cos \alpha \quad (10)$$

$$v = V \sin \alpha \quad (11)$$

$$\alpha = -\delta \quad (12)$$

where T_b is the braking torque (which is a given function of time); I is the moment of inertia of the tire, wheel, hub, brake rotor, etc.; V is velocity of the test vehicle; and δ is the steer angle of the test wheel.

Test results from a low-speed flat-bed tire tester indicate that the lateral relaxation length is about 1.1 ft (0.33 m) for a typical truck tire at rated load. (See figure 2.) (For fixed slip angle analyses, the lateral relaxation length (σ_y) has no influence on the results.)

Based upon test results from a mobile tire dynamometer, a longitudinal relaxation length of approximately 0.5 ft (0.15 m) will fit the lower frequency aspects of the test data shown in figure 3.

Figures 4 and 5 provide predictions of lateral force and longitudinal force during idealized mobile tire tester experiments. These results are based upon equations 7 through 12 and the steady state model given in appendix A.

To obtain the results shown in figure 4, the brake torque is increased approximately

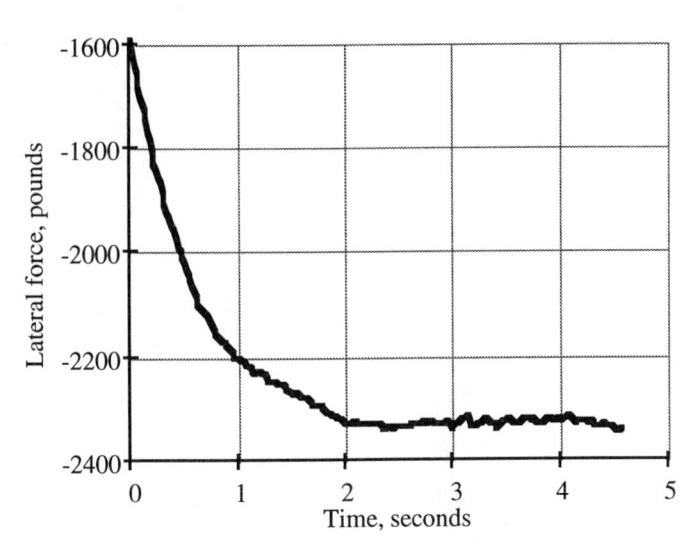

Figure 2. Lateral force transient, flat-bed tester, V = 2 ft/sec, α = 4 deg

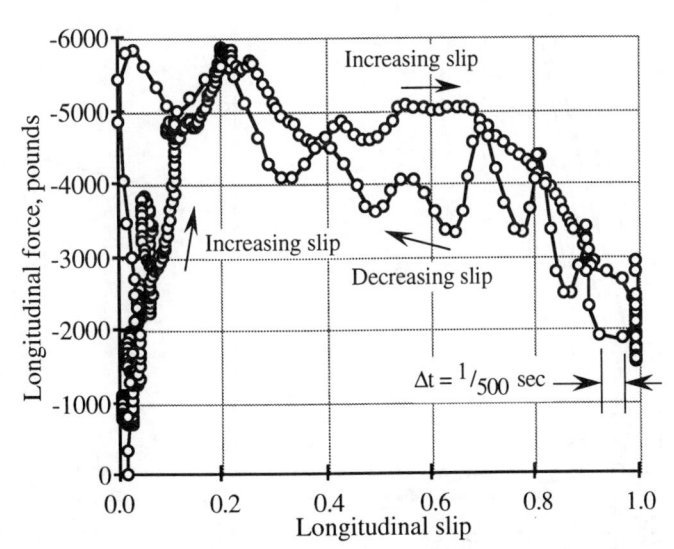

Figure 3. Longitudinal force during one complete braking cycle, mobile tester, V = 66 ft/sec, α = 0

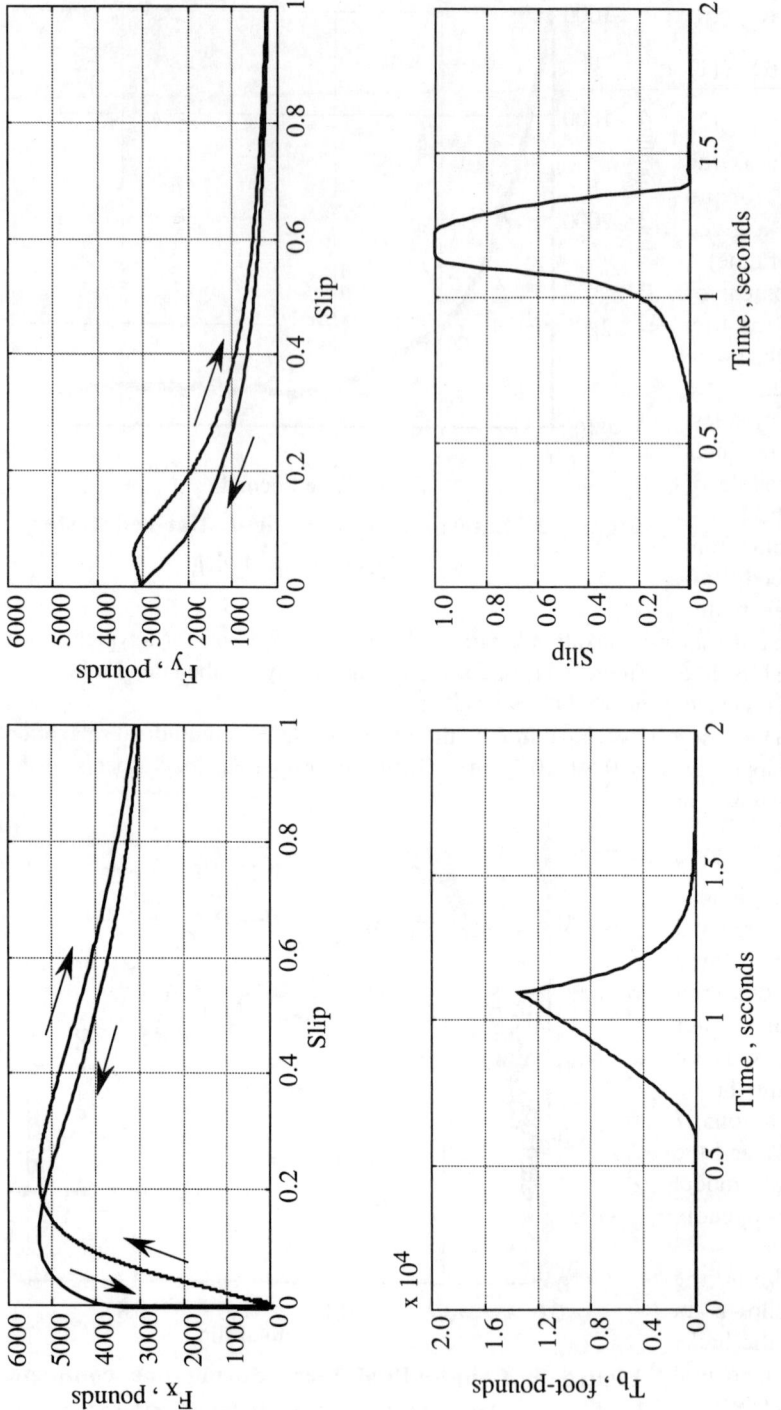

Figure 4. Mobile tire tester simulation with approximately linear increase in brake torque
Case 1 parameters, $V = 66$ ft/sec, $\alpha = 4$ deg, $\sigma_x = 0.5$ ft

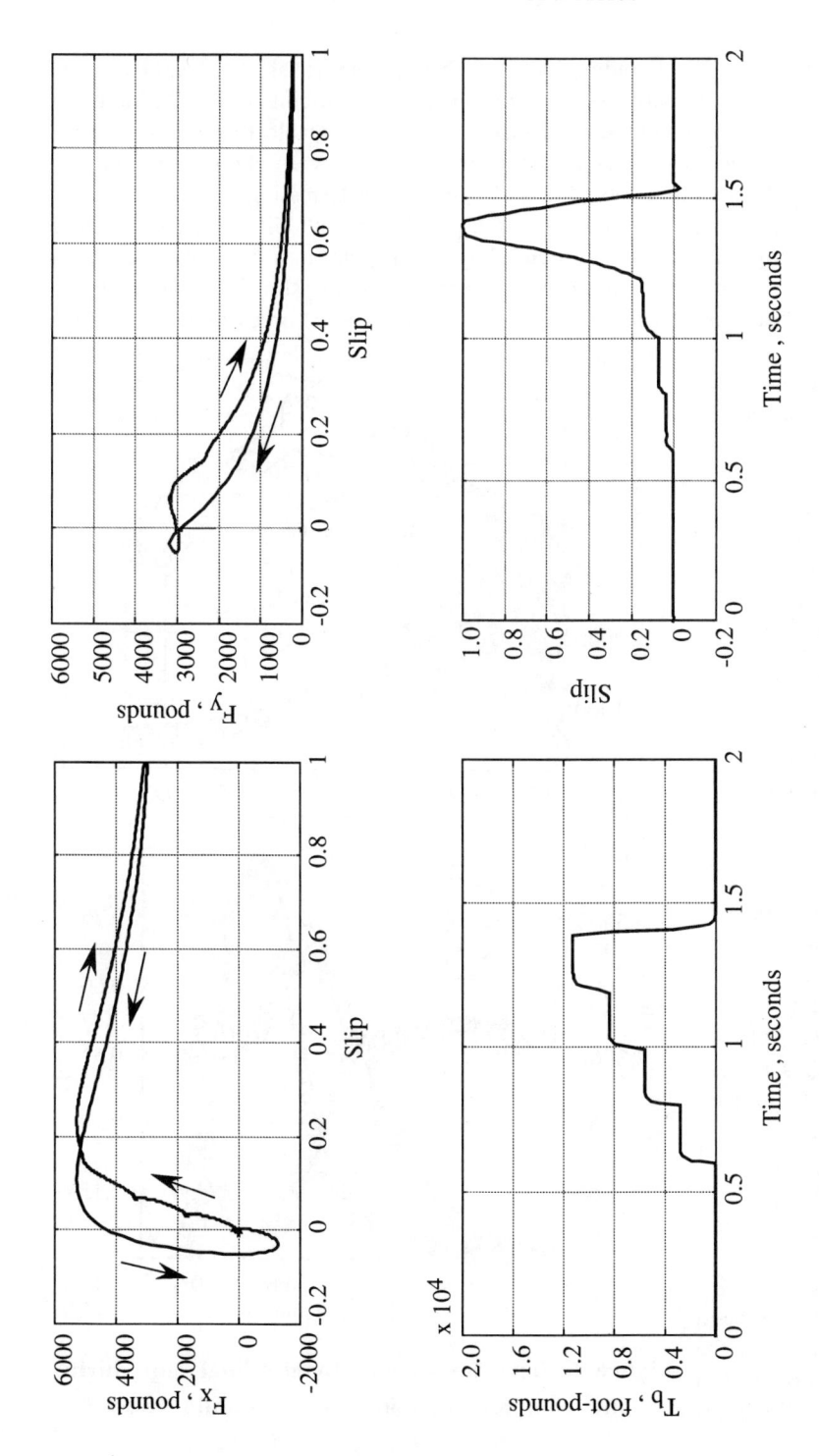

Figure 5. Mobile tire tester simulation with "stair-step" increase in brake torque Case 1 parameters, $V = 66$ ft/sec, $\alpha = 4$ deg, $\sigma_x = 0.5$ ft

linearly and then suddenly reduced to zero. This braking waveform causes the wheel to reach lockup momentarily and then to spin up to free rolling again. The values of the tire parameters used in these calculations are listed in table 1 as case 1. Clearly, the longitudinal and lateral force components predicted when slip is increasing are considerably different from those predicted when slip is decreasing rapidly.

The brake torque time history given in figure 5 is meant to illustrate the influence of steps in torque as might be introduced by a (poor) anti-lock braking system (or a jerky driver). The graphs of force components versus slip ($(u - R\omega)/u$) make interesting patterns due to the manner in which these forces develop as time increases.

Figure 6 shows raw data from an existing mobile tire tester. The data are gathered at a rate of 512 data points per second. The slip cycle goes from free rolling to close to wheel lock and quickly back to free rolling. The test was performed at a 5 degree slip angle at 66 ft/sec (19.8 m/sec). The data show the influence of tire dynamics on both the lateral as well as the longitudinal components of the tire force. As a function of slip, the lateral force is noticeably smaller when slip is decreasing than it is when slip is increasing.

Figure 4, although not meant to fit the tire data presented in figure 6, illustrates that the tire model

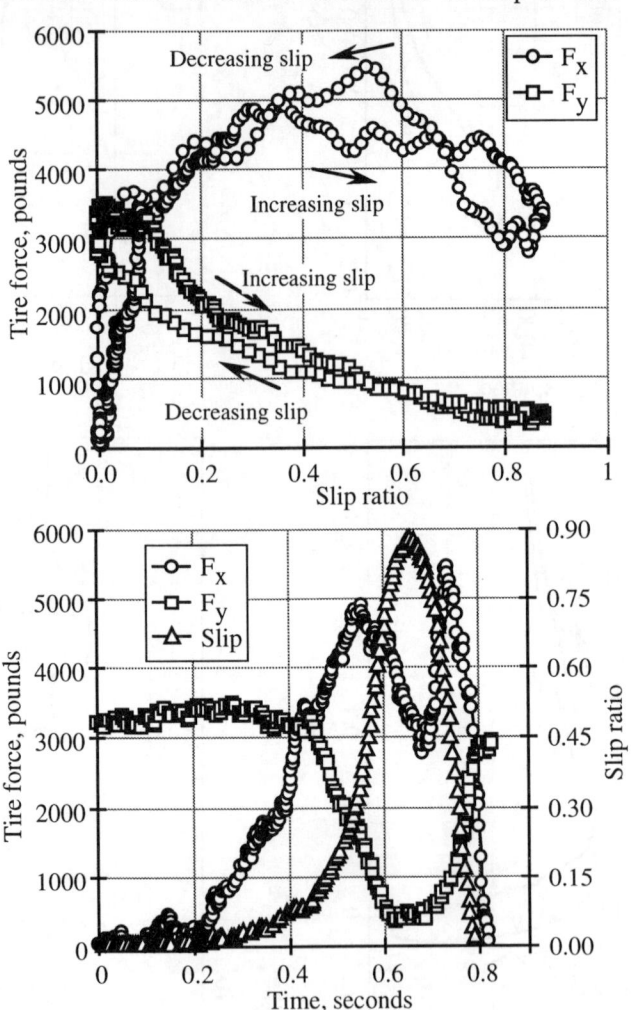

Figure 6. Tire forces and longitudinal slip during one complete braking cycle, mobile tester, V = 66 ft/sec, α = 5 deg

produces tire dynamic effects that are similar to those observed in the test data. If one ignores the high frequency variations in the test data, a comparison between figures 6 and 4 indicates that tire dynamics have been treated in a manner that is representative of the observed phenomena.

Figure 7 contains graphs based upon analytical results for an idealized mobile tire-tester. These results are intended to be representative of the types of force components that might be obtained during a braking in a turn maneuver performed on a slippery test surface. In this case the forward velocity, V, is presumed to decrease from 44 ft/sec (13.2 m/sec) and the slip angle is held constant at 4 degrees. Case 2 in table 3 gives the values of the tire parameters used in these calculations. Clearly there will be a large loss in side force when braking slip becomes large. The influences of tire dynamics increase as the velocity decreases.

CONCLUDING STATEMENTS

In seeking to understand the work presented here, it may be helpful to interpret the invariant tire-stiffness parameter suggested by Apetaur as being related to the relaxation length effect as observed from tire testing results [7]. Furthermore, the tire-stiffness parameter (relaxation length effect) might be viewed as representing carcass deformation properties in response to tread shear stresses as represented by a brush-type tire model. However, none of these matters have been resolved in this paper.

The choice made here for representing tire dynamics is only one amongst several possibilities suggested in the literature. Other possibilities that are being considered for representing tire dynamics include models with a torsional degrees of freedom between the tire's belt and the wheel's hub (e.g. [8]), using the rolling velocity (Rω) in place of the longitudinal velocity (u) in determining the time constant (e.g. [9]), adding a carcass elastic function (e.g. [7]), and arranging the tire dynamics to act only on the steady-state forces from the adhesion region so that the forces from the sliding region occur instantaneously without any need to roll into a new deformed state.

The choice of employing differential equations for transients in slip and slip angle has resulted here in a very simple set of equations. However, whether these equations are accurate enough is an important question. The range of application of this type of model is not well defined currently. Questions concerning sufficient fidelity or accuracy are difficult to answer. In the example shown here, the form of the results are as expected. The challenge now is to verify their accuracy, and hence their utility for predicting vehicle performance.

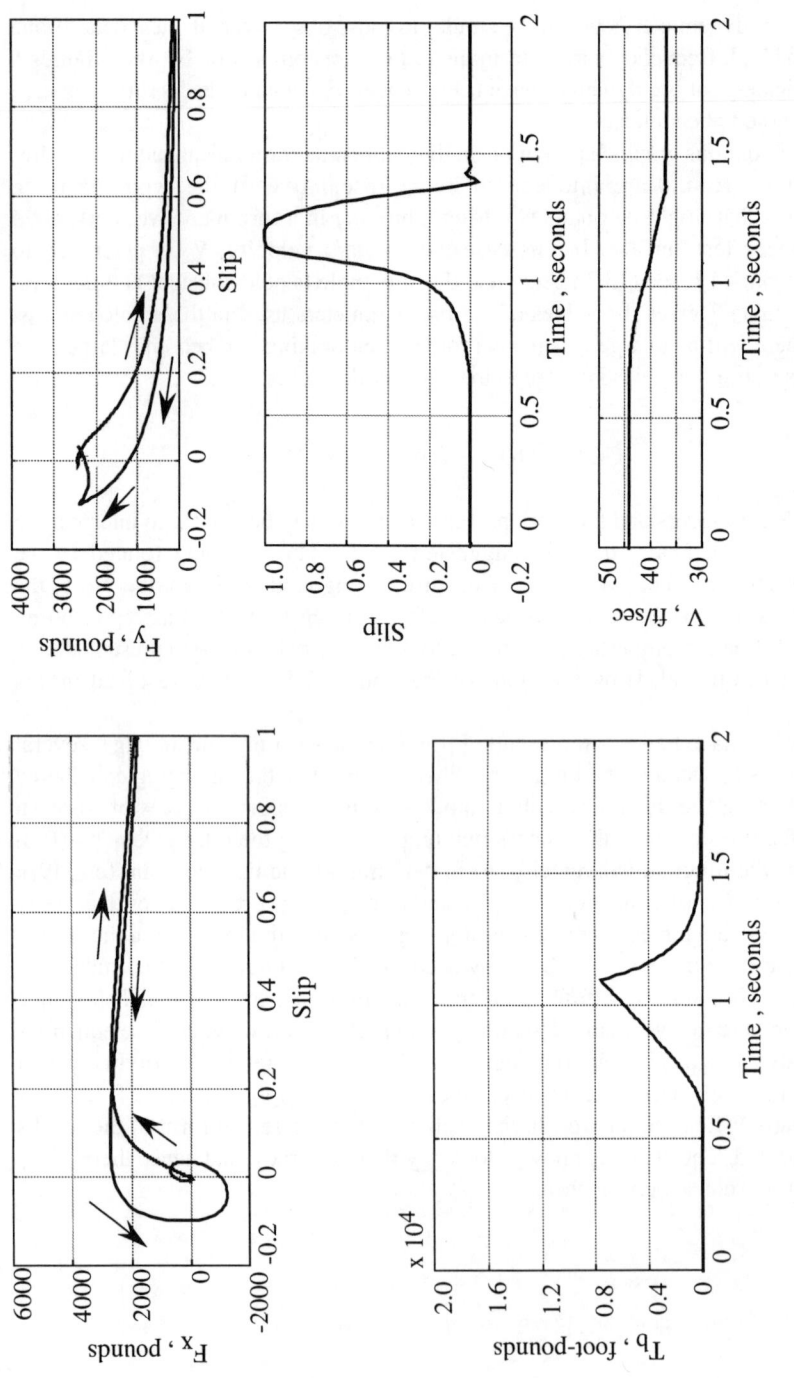

Figure 7. Mobile tire tester simulation with approximately linear increase in brake torque
Case 2 parameters, V = 44 ft/sec, α = 4 deg, σ_x = 0.5 ft

REFERENCES

1. Fancher, P.S. "Generic data for representing truck tire characteristics in simulations of braking and braking-in-a-turn maneuvers." University of Michigan Transportation Research Institute. Report number UMTRI-95-34. Ann Arbor, Michigan. September, 1995.
2. Bernard, J.E. and Clover, C.L. "Tire modeling for low-speed and high-speed calculations." SAE paper number 950311. Society of Automotive Engineers. Detroit. February, 1995.
3. Allen, W.R. et al. "Tire modeling requirements for vehicle dynamics simulation." SAE paper number 950312. Society of Automotive Engineers. Detroit. February, 1995.
4. Sakai, H. "Theoretical and experimental studies on the dynamic cornering properties of tyres." *International Journal of Vehicle Design*. Vol 2. No. 1–4. 1981.
5. Gim, G. and Nikravesh, P.E. "An analytical study of pneumatic tire dynamic properties: Part 1–3." *International Journal of Vehicle Design*. Vol. 11, No. 6. Vol. 12, No. 1–2. 1991.
6. Pottinger, M. "A combined cornering and braking test for heavy duty truck tires." *Road transport technology—4. Proceedings.* Fourth International Symposium on Heavy Vehicles Weights and Dimensions. University of Michigan Transportation Research Institute. Ann Arbor. 1995.
7. Apetaur, M. "Modelling of transient nonlinear tyre responses." *Proceedings* of the First International Colloquium on Tyre Models for Vehicle Dynamics Analysis. Delft, Netherlands. October, 1991. Pp. 116-128.
8. van Zanten, A., Ruf, W.D. and Lutz, A. "Measurement and simulation of transient tire forces." SAE paper number 890640. Society of Automotive Engineers. Detroit. February, 1989.
9. Jansen, S.T.H. et al. "Sensitivity of vehicle handling to combined slip tyre characteristics." *Proceedings* of the International Symposium on Advanced Vehicle Control. Aachen University of Technology. Aachen, Germany. June 1996.

APPENDIX A—TIRE EQUATIONS USED IN THE STEADY STATE MODEL

A. SLIDING VELOCITY

$$V_s = [(s)^2 + (\tan \alpha)^2]^{0.5} u \qquad (a1)$$

where V_s = sliding velocity, s = longitudinal slip, α = slip angle, and u = forward velocity component in the wheel plane.

B. FRICTION

$$\mu = \mu_f + (\mu_o - \mu_f)e^{-V_s/V_f} \qquad (a2)$$

where μ = frictional potential, μ_f = minimum friction at high sliding velocity, μ_o = maximum friction at zero sliding velocity, V_f = exponential velocity constant for "shaping" the mu versus s curve.

In general,
$$V_f = [V_s/(\ln((\mu_o - \mu_f)/(\mu - \mu_f)))] \quad (a3)$$

Example 1. For $\mu_o = 0.9$, $\mu_f = 0.4$, and $\mu = 0.5$ at 45 mph (66 ft/sec),

$$0.5 = 0.4 + (0.9 - 0.4) e^{-66/V_f}$$

or, $\quad V_f = 66/(\ln(0.5/0.1)) = 41$ ft/sec. $\quad$ for a "0.9 surface."

Note:
- For this example, 0.5 = the locked wheel (s =1) value when the tire is sliding at 66 ft/sec. In the next example, 0.25 = the locked wheel value.

Example 2. For $\mu_o = 0.5$, $\mu_f = 0.2$, and $\mu = 0.25$ at 45 mph (66 ft/sec),

$$0.25 = 0.2 + (0.5 - 0.2) e^{-66/V_f}$$

or, $\quad V_f = 66/(\ln(0.3/0.05)) = 36.8$ ft/sec $\quad$ for a "0.5 surface."

C. DIRECTION OF SLIDING AND FRICTION FACTORS FOR COMBINED SLIP

The angle of friction θ defines the direction of sliding such that:

$$\sin \theta = v/V_s \quad (a4)$$

and, $\quad \cos \theta = (u - R\omega)/V_s \quad (a5)$

The longitudinal friction factor is:

$$\mu_x = \mu \cos |\theta| \quad (a6)$$

The lateral friction factor is:

$$\mu_y = \mu \sin |\theta| \quad (a7)$$

Notes:
- Force components under total sliding oppose the direction of sliding. That is, θ defines the direction of sliding with respect to the wheel plane.
- The total friction is divided into lateral and longitudinal friction factors (capabilities). These factors determine the maximum amount of frictional force that can be generated in any direction.

D. FRACTION OF THE CONTACT PATCH IN ADHESION (COMBINED SLIP)

Longitudinally, for $s < 1$ and $s \neq 0$,

$$(xs_x/L)' = [(\mu_x) F_z (1 - s)] / [2 C_s |s|] \quad (a8)$$

where xs_x = the point in the contact where longitudinal sliding starts (and adhesion ends), L = the length of the contact patch, F_z = the vertical load, C_s = the longitudinal stiffness of the tire.

Note:
- For example, a typical tire might have:

$$C_S = 10 F_z - F_z^2/3000 \text{ lbs.} \tag{a9}$$

Laterally, for $\alpha \neq 0$,

$$(xs_y/L)' = [(\mu_y) F_z (1 - s)] / [2 C_\alpha | \tan \alpha |] \tag{a10}$$

where xs_y = the point in the contact where lateral sliding starts (and adhesion ends), L = the length of the contact patch, F_z = the vertical load, C_{alpha} = the lateral stiffness of the tire.

Note:
- For example, a typical tire might have:

$$C_\alpha = 0.9 \ C_S \text{ lbs.} \tag{a11}$$

If $(xs_x/L)' > 1$, $(xs_x/L) = 1$; otherwise, $(xs_x/L) = (xs_x/L)'$ (a12)

If $(xs_y/L)' > 1$, $(xs_y/L) = 1$; otherwise, $(xs_y/L) = (xs_y/L)'$ (a13)

Notes:
- If $(xs_x/L)' \geq 1$, the entire contact patch is in adhesion longitudinally.
- If $(xs_y/L)' \geq 1$, the entire contact patch is in adhesion laterally.
- The regions of adhesion can be different longitudinally and laterally. In the longitudinal adhesion region, C_S applies, and in the lateral adhesion region, C_α applies.

E. LONGITUDINAL AND LATERAL FORCES

$$F_x = -[C_S (xs_x/L)^2 (s/(1 - s)) + \text{sign}(s)(1 - (xs_x/L))(\mu_x) F_z] \tag{a14}$$

where F_x = the braking force. If $s = 0$, $F_x = 0$. If $s = 1$, $F_x = (\mu_x) F_z$.

$$F_y = -[C_\alpha (xs_y/L)^2 (\tan \alpha/(1 - s)) + \text{sign}(\alpha)(1 - (xs_y/L))(\mu_y) F_z] \tag{a15}$$

where F_y = the lateral force. If $\alpha > 0$, the lateral force is negative. If $\alpha < 0$, the lateral force is positive. If $s = 1$, $F_y = (\mu_y) F_z$.

Notes:
- Aligning torque AT may also be calculated using empirically obtained values for the pneumatic trail xp and the lateral deflection stiffness C_y for the tire: viz.,

$$AT = -xp\{F_{ya}[(4)(xs_y/L) - 3] + 3 F_{ys}(xs_y/L)\} + F_x F_y / C_y \tag{a16}$$

where $\qquad F_{ya} = -[C_\alpha (xs_y/L)^2 \tan \alpha/(1-s)]$

and $\qquad F_{ys} = -[(1 - (xs_y/L))(\mu_y) F_z] \text{sign}(\alpha)$

APPENDIX B—TIRE DYNAMICS

This model calculates a transient slip angle α' based on the lateral deformation of a hypothetical point P_0 at the longitudinal center of the contact patch. The distance σ_y, which is the so-called lateral relaxation length, indicates the longitudinal distance from P_0 to a point, P, along the carcass of the tire.

Figure B-1 presents a schematic diagram. The point P_0 is presumed to be following the road, thus its longitudinal velocity with respect to point P is u, and its lateral velocity with respect to the undeformed centerline of the contact patch is v.

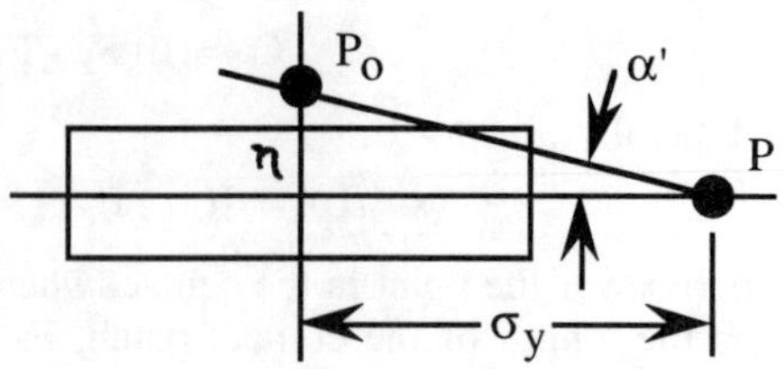

Figure B-1

Straightforward geometry yields

$$\tan(\alpha') = \eta/\sigma_y \tag{b1}$$

We now differentiate with respect to time by tracking the point currently at P_0. This yields

$$d(\tan\alpha')/dt = (v\,\sigma_y - u\,\eta)/\sigma_y^2 \tag{b2}$$

and after rearranging

$$d(\tan\alpha')/dt + u\tan(\alpha')/\sigma_y = v/\sigma_y \tag{b3}$$

Note that the steady solution is, as expected,

$$\tan(\alpha') = v/u \tag{b4}$$

Now consider the model for transient slip, s', which is based on the longitudinal deformation of a hypothetical point Q_0 currently at the longitudinal center of the contact patch. The distance σ_x, which is the so-called longitudinal relaxation length, indicates the longitudinal distance from Q_0 to a point Q along the carcass of the tire. The distance from point Q to the leading edge of the undeformed contact patch remains constant.

Figure B-2 presents a schematic diagram. The point Q_0 is presumed to be following the road, thus it's longitudinal velocity with respect to point Q is u. The corresponding point on the undeformed contact patch, currently located by ξ, has a velocity $R*\omega$ with respect to the point Q.

The classic definition of longitudinal slip yields

$$s' = (\sigma_x - \xi)/\sigma_x \tag{b5}$$

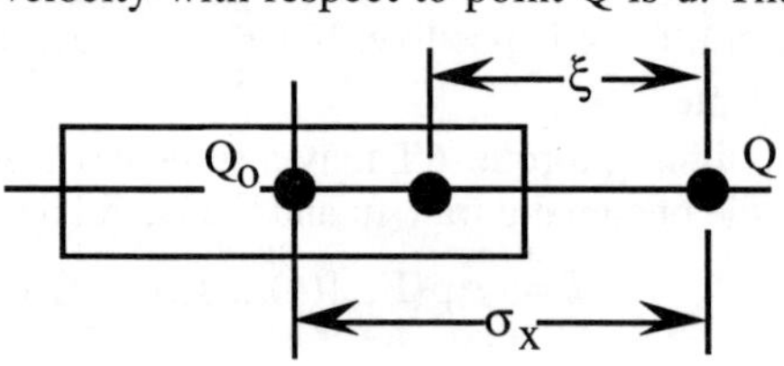

Figure B-2

We now differentiate with respect to time by tracking the two points currently ξ and σ_x respectively behind the point Q. This yields

$$ds'/dt = ((u - R\,\omega)\sigma_x - u(\sigma_x - \xi))/\sigma_x^2 \tag{b6}$$

This may be rearranged to get

$$ds'/dt + u\,s'/\sigma_x = (u - R\,\omega)/\sigma_x \tag{b7}$$

Note that the steady state solution is, as expected

$$s' = (u - R\,\omega)/u \tag{b8}$$

Finite element analysis and experimental analysis of natural frequencies and mode shapes for a non - rotating tyre

E. NEGRUS, G. ANGHELACHE and A. STANESCU

ABSTRACT

The paper presents the study of natural frequencies and mode shapes for a 175/70 R 13 tyre. The investigations are led in analytical and experimental directions. A finite element model is especially created for modal analysis and it takes account of tyre structure, constituent anisotropy and non-linear properties of materials. Test were effected in different spindle and patch boundary conditions and for two inflation pressure values.

INTRODUCTION

The natural frequencies and mode shapes of passenger car tyre play an important role in appearance of standing waves connected with the critical rotational speed, in noise generation due to rolling tyre, in the transmission of disturbances from tyre footprint into the vehicle. Also, the natural frequencies help to find some tyre characteristics like: treadband bending stiffness and carcass radial and tangential stiffness.

First, the tyre resonance study was experimentally effected and later the tyre model simulation was used for improving the knowledge in this field. The principal type of analytical model are:
- circular ring on an elastic foundation;
- finite element analysis.

This paper studies the radial natural frequencies and mode shapes for a 175/70 R 13 tyre.

FINITE ELEMENT MODEL

In analytical study of tyre resonance an original three dimensional finite element model of a 175/70 R 13 is used. This model is especially created for dynamic behaviour, it takes account of real tyre structure (rubber, textile and metallic cords), constituent anisotropy and non-linear properties of materials. It contains 210 shell elements of carcass cord, 90 shell elements of treadband metallic cords and 270 solid elements of rubber. For vibration analysis the model disregards the bead regions of tyre. Free-patch, fixed wheel are the first conditions imposed to the model.

The air pressure in the tyre leads to static preload in the tyre carcass, thus finite element model is first preloaded with the internal pressure of $2*10^5$ N/m^2 [4]. The deformed shape of tyre under inflation pressure is presented in figure 1. The finite element model works correctly, the obtained deformations being very closed to the experimental ones.

The second step is calculation of modal transient response [1] which consists of:
- extract model real eigenvalues and eigenvectors; the eigenvalues equation is:
$$[K - \omega^2 M] \{\Phi\} = 0$$
where K = stiffness matrix; M = mass matrix; $\{\Phi\}$= eigenvector; ω = circular frequency;
- display structure response.

DESCRIPTION OF EXPERIMENTAL WORK

Experimental natural frequencies and mode shapes were measured on the 175/70 R 13 tyre mounted on a 5.5 wide rim. Test were effected in the following conditions:
1. without degrees of freedom and free patch;

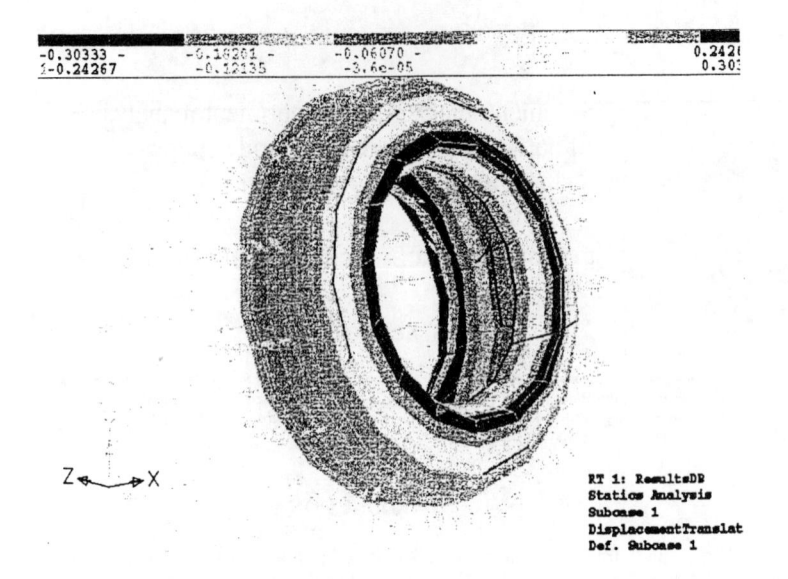

Figure 1 Finite element model of tyre: inflated shape

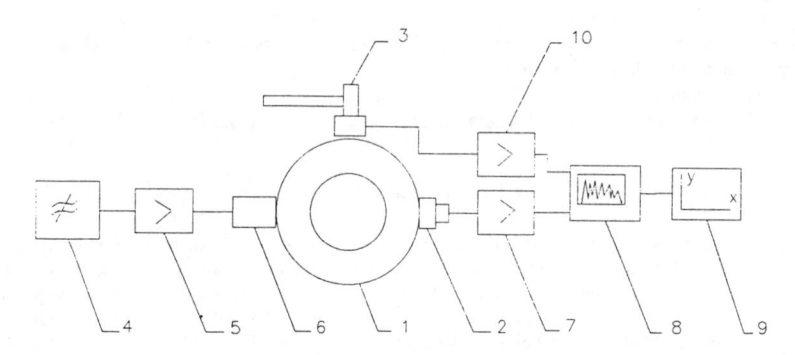

Figure 2 Experimental setup for tyre resonance study

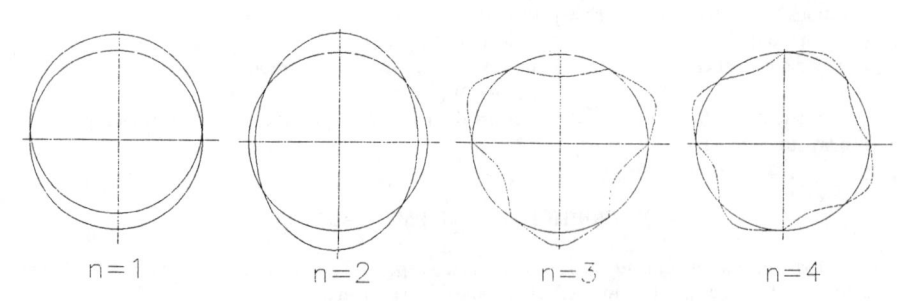

n=1 n=2 n=3 n=4

Figure 3 Circumferential mode shapes

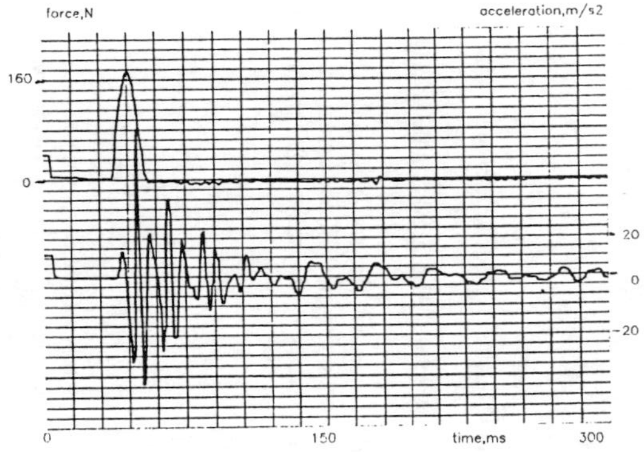

Figure 4 Time - history of input force (up) and treadband acceleration (down) 90° between measure points; $p=2*10^5$ N/m²

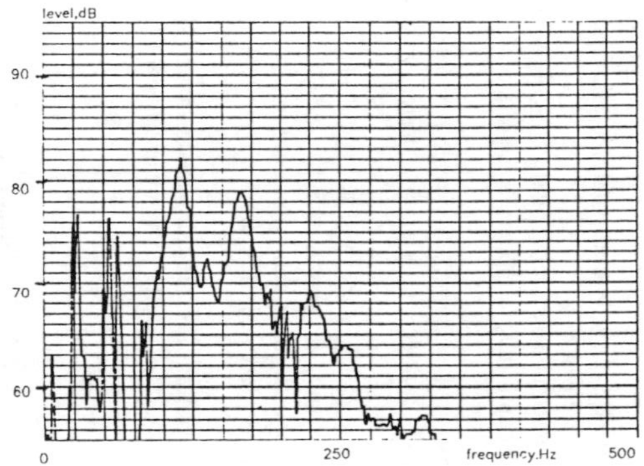

Figure 5 Spectrum of treadband radial acceleration in figure 4

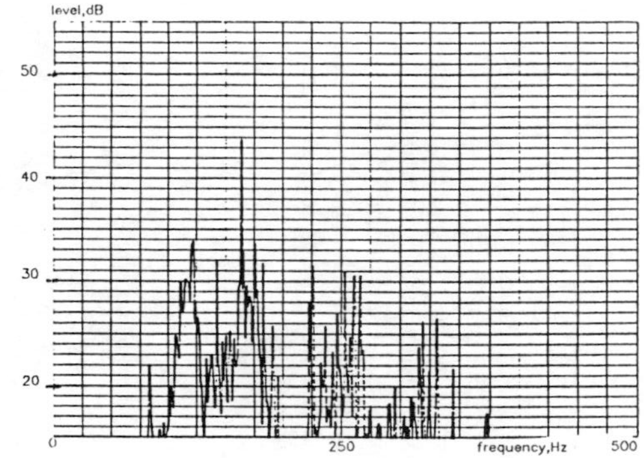

Figure 6 Measured frequency response: acceleration / force in figure 4

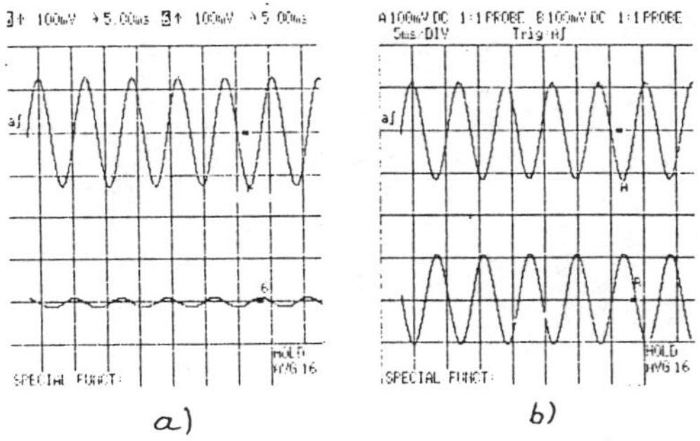

Figure 7 Time - history of treadband acceleration for the third mode shape measured in the following positions:
a) maximum (up) and nodal (down) points consecutively
b) maximum (up) and minimum (down) points consecutively

2. free wheel and free patch;
3. pinned wheel, fixed - patch and wheel loaded with vertical loads of 3000 N and 4000 N.

All types of tests were effected for two inflation pressure values: $1.5*10^5 \, N/m^2$ and $2*10^5 \, N/m^2$.

Two kinds of experiments were applied:

a) *obtaining resonance frequencies:* a hammer fitted with a load cell hits the middle of tyre treadband while an accelerometer measures the tyre's response. By using Fourier analysis techniques, the response function between the excitation and response points were obtained.

b) *obtaining the mode shapes:* treadband is excited with each of the natural frequencies provided to the tyre by a shaker and the circumferential acceleration (displacement) is measured along the treadband.

The experimental set-up for tyre resonance study is presented in figure 2. The main devices are: 1 - tyre; 2- accelerometer; 3 - hammer; 4 - signal generator; 5 - power amplifier; 6 - shaker; 7,10 - amplifier; 8 - real time analyser B&K; 9 - XY recorder.

Figure 3 shows the mode shapes 1 to 4 in the following conditions: wheel without degrees of freedom and free - patch, inflation pressure $2*10^5 \, N/m^2$. Similar results are presented in [2] and [3].

The damping behaviour of tyre structure and delay between input and output are outlined in figure 4. An example of treadband radial acceleration spectrum is presented in figure 5. Figure 6 shows the measured frequency response acceleration/force.

The treadband displacements were tested via the acceleration. For the third mode shape (143 Hz) we presented in figure 7 the vibration behaviour of three distinctive points: extreme and nodal (7a); extreme outside and extreme inside (7b).

The measured natural frequencies are listed in table 1.

Table 1 Natural frequencies of 175/70 R 13 tyre

n	free wheel and free - patch $p = 2*10^5 \, N/m^2$	without DOFs free patch $p = 2*10^5 \, N/m^2$	without DOFs free patch $p = 1.5*10^5 \, N/m^2$	fixed - patch vertical load $p = 2*10^5 \, N/m^2$; 3000 N
1	100	90	85	97
2	117	119	110	129
3	139	143	130	153
4	164	168	155	186
5	190	194	180	211
6	220	223	210	-

CONCLUSIONS

- Natural frequencies and mode shapes calculated by finite element model agree well with the experimental ones;
- The paper outlines the existence of damping behaviour tyre structure. Also it exists a delay between input excitation and output vibration of tyre structure;
- Increasing inflation pressure determines growing in natural frequencies for the same mode shape; a similar effect is produced by fixing the patch and loading tyre; contrary, with one exception, free - wheel and free - patch lead to decreasing in natural frequencies in the same inflation pressure conditions.

REFERENCES

1. Brauer, J.R., "What Every Engineer Should Know About Finite Element Analysis", Marcel Dekker.Inc., New York, USA, 1988.

2. Clark, S.K., "Mechanics of pneumatics tires", National Bureau of Standards, Washington, D.C., USA, 1981.

3. Gong, S., "A Study of In - Plane Dynamics of Tires", Delft University of Technology, Faculty of Mechanical Engineering and Marine Technology, Delft, Netherlands, 1993

4. Richards, T.R., Charek, L.T., and Scavuzzo, R.W., "The Effects of Spindle and Patch Boundary Conditions on Tire Vibration Modes", SAE Paper 860243, 1986.

Study on Real Time Estimation of Tire to Road Friction

S. YAMAZAKI*, O. FURUKAWA** and T. SUZUKI*

ABSTRACT

This paper presents an analytical approach for estimating the friction coefficient between a tire and the road. We analyzed the friction coefficient in real time, based on a brush type tire model with unlimited carcass stiffness. The validation experiments were conducted on wet and dry surfaces on an indoor drum type tire testing machine. The calculated friction coefficients agreed well with the measured values.

INTRODUCTION

Studies into safe vehicle operation have recently advanced; for example, the development of ABS(Anti-Locked Brake System), traction control technology, and yaw control technology. Safe automobile operation is controlled using high technology micro computers. This study to estimate the friction coefficient between a tire and the road in real time was conducted for a new development by W. R. Pasterkamp[1]. In this study, the friction coefficient between a tire and the road was estimated using cornering force and self-aligning torque during cornering; determining these measurements is very difficult.

This study was undertaken to facilitate advances in braking, driving force, and yaw control technologies. This paper presents an analytical approach to estimating the friction coefficient between a tire and the road in real time, based on a brush type tire model with unlimited carcass stiffness. The validation experiments were conducted on wet and dry surfaces on an indoor drum type tire testing machine. The calculated friction coefficients agree well with the measured values.

---

* Research Division 2, Japan Automobile Research Institute, 2530 Krima, Tsukuba,Ibaraki 305, Japan.
** Tochigi R & D Center, Honda R & D Co., LTD. 4630 Shimotakanezawa, Haga-Machi, Haga-gun, Tochigi 321-33, Japan.

ANALYSIS

The contact pressure p over the contact patch produced normal force; experiments have demonstrated that its average value over the contact patch width has a parabolic distribution in the circumference direction as

$$p = 4p_m \frac{\xi}{l}(1 - \frac{\xi}{l}) \qquad (1)$$

where the ξ-axis represents the longitudinal displacement of a tire, l is the contact length of the tire tread and p_{max} is the maximum value of the contact pressure distribution occurring at $\xi = l/2$, as shown in Fig.1. Normal force F_z is given by the following:

$$F_z = \frac{2}{3} p_m w l \qquad (2)$$

where w denotes the contact width.

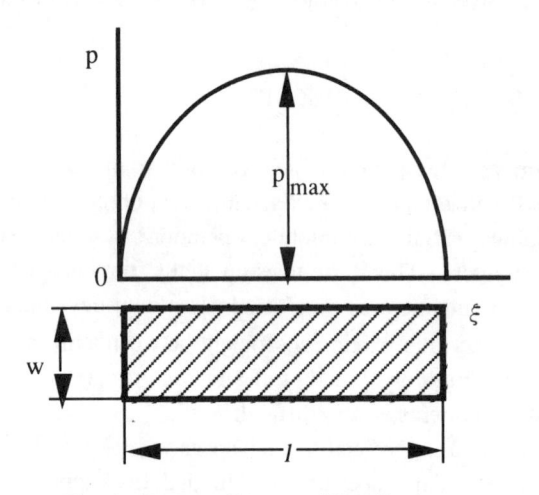

Fig. 1 Contact Pressure Distribution

The slip ratio S_s during braking is generally defined as

$$S_s = \frac{V_x - V_c}{V_x} \qquad (3)$$

where v_x is the vehicle velocity (on the test machine, the drum rotational velocity) and v_c is the velocity of the tire tread base. When we applied braking force to the vehicle, the drum velocity and the tire tread base velocity became different. As shown in Fig.2, during braking the traveling distance ξ of a point attached to the road surface from the front of the contact patch, in time Δt, can be expressed with

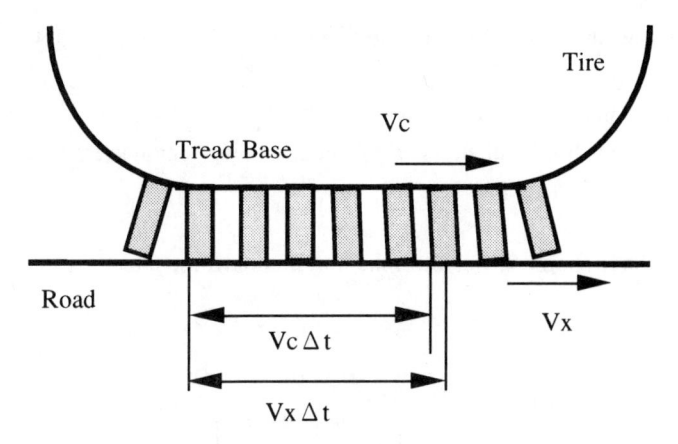

Fig. 2 Deformation of tread rubber

respect to the road coordinate system as

$$V_x \Delta t \tag{4}$$

During braking, the traveling distance ξ at the arbitrary point in the contact patch, in time Δt, can be expressed with respect to the road coordinate system as

$$V_c \Delta t \tag{5}$$

The relative longitudinal velocity ΔV_ξ can be written as

$$\Delta V_\xi = V_x - V_c \tag{6}$$

Displacement $\Delta \xi$ between the tire tread base and road surface is

$$\Delta \xi = \Delta V_\xi \Delta t \tag{7}$$

Eq.(7) can be expressed by using the slip ratio as

$$\Delta \xi = S_s \xi \tag{8}$$

As shown in Fig.3, when braking force is applied, an adhesion region and a sliding region occur in the contact patch. Where a is the adhesion length in the contact patch.

(1) Adhesion Region ($0 \leq \xi \leq a$)

The longitudinal elastic stress $\sigma_\xi^{(a)}$ was generated at the adhesion region in the contact patch and is given by

$$\sigma_\xi^{(a)} = k_x \Delta \xi = k_x S_s \xi \tag{9}$$

where k_x denotes the longitudinal stiffness rate per unit area of tread rubber.

(2) Sliding Region ($a < \xi \leq l$)

The longitudinal frictional stress $\sigma_\xi^{(s)}$ in the sliding region is expressed as

$$\sigma_\xi^{(s)} = \mu_x p \qquad (10)$$

where μ_x is the friction coefficient in the longitudinal direction.

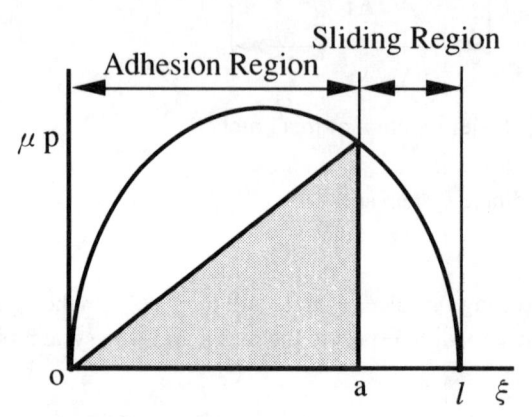

Fig. 3 Adhesion and sliding region in contact area

(3) Breakaway Point (at $\xi = a$)

The sliding rate S_n of the contact patch is written by Gim[1] as

$$S_n = 1 - n = 1 - \frac{a}{l} \qquad (11)$$

where n is the adhesion rate. To find the breakaway point, we locate where the longitudinal stresses of the adhesion and sliding regions become identical, i.e.

$$\sigma_\xi^{(a)} = \sigma_\xi^{(s)} \qquad (12)$$

Putting $\xi = a$ and substituting Eq.(11) into Eq.(12), we can obtain the equation

$$S_n = \frac{C_s S_s}{3\mu_x F_z} \qquad (13)$$

where C_s is the braking force per unit slip ratio when the slip ratio is very small. C_s refers to the braking stiffness and is given by the following:

$$C_s = \frac{1}{2} w k_x l^2 \qquad (14)$$

We can then obtain the relationship between the slip ratio S_s and the sliding rate as

$$S_s = \frac{3\mu_x F_z S_n}{C_s} \tag{15}$$

When a becomes 0, the tread contact patch is entirely in a sliding condition. When a=0, we can obtain $S_n=1$ from Eq. (12). Then the critical slip ratio S_{sc}, which occurs in the sliding condition of the contact patch, can be given from Eq.(15) as

$$S_{sc} = \frac{3\mu_x F_z}{C_s} \tag{16}$$

BRAKING FORCE

1) $S_s \leq S_{sc}$

The braking force can be written as

$$F_x = \int_0^l \sigma_\xi w d\xi$$
$$= C_s S_s l_n^2 + \mu_x F_z (1 - 3l_n^2 + 2l_n^3) \tag{17}$$

where $l_n = a/l$. From $C_s S_s = 3S_n \mu_x F_z$, we can obtain $\mu_x F_z = C_s S_s / 3S_n$. We can then obtain the final equation of the breaking force as

$$F_x = 3\mu_x F_z S_n \left(1 - S_n + \frac{S_n^2}{3}\right) \tag{18}$$

The friction coefficient is expressed as a function of the slip ratio by eliminating Sn from Eqs.(13) and (18).

$$9\mu_x^2 F_z^2 (F_x - C_s S_s) + 3\mu_x F_z C_s^2 S_s^2 - C_s^3 S_s^3 / 3 = 0 \tag{19}$$

The friction coefficient can be easily obtained by solving the above equation.

(2) $S_{sc} \leq S_s$

When the absolute value of the slip ratio is greater than the critical value, the adhesion length a becomes zero. There is no elastic deformation and stress, only frictional stress. The braking force may depend only on the frictional force as

$$F_x = \mu_x F_z \tag{20}$$

Therefore, the friction coefficient is

$$\mu_x = F_x / F_z \tag{21}$$

BRAKING TEST

(1) Test condition

A drum type tire test machine was used for our test. The drum diameter is 1.7 m. Table 1 shows the tire properties in dry surface conditions; Table 2 shows the tire properties in wet surface conditions. In wet conditions, a worn tire with a groove depth of 1.6 mm was used for the test. The inflation pressure was 190 kPa in both

Table 1 Tire properties (Dry Condition)

Tire:185/70R14(NEW), Inf.p.:190kPa

Load (kN)	2.94
Contact Width (mm)	103.0
Contact Length (mm)	97.5
Cornering Stiffness (kN/°)	0.705
Longitudinal Stiffness (kN/mm^3)	1.32E-04

Table 2 Tire properties (Wet Condition)

Tire:185/70R13(Groove;1.6mm), Inf.P.:190kPa

Load (kN)	3.0
Contact Width (mm)	150.0
Contact Length (mm)	120.0
Cornering Stiffness (kN/°)	1.07
Longitudinal Stiffness (kN/mm^3)	3.135E-04

test tires, and the drum speed was 30 km/h.

(2) Test procedure

The braking tests were conducted using the following factors:

a. The speeds of the test tire and the drum were the same.

b. The braking torque was applied slowly to the test tire. The drum speed was kept at 30m/h.

c. The longitudinal force and the slip ratio were measured in time.

COMPARISON OF CALCULATED AND EXPERIMENTAL RESULTS

(1) Dry surface

Figure 4 shows the experimental result of the relationship between the braking force and the slip ratio under dry surface conditions. Figure 5 shows a comparison between the calculated and experimental results. The sliding velocity is the relative

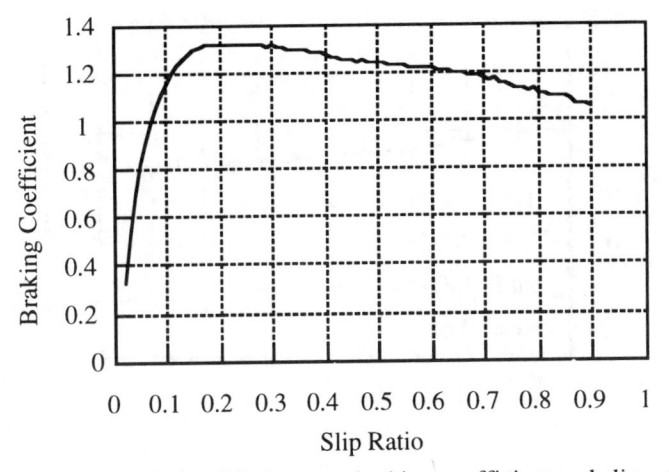

Fig. 4 The relationship between braking coefficient and slip ratio.

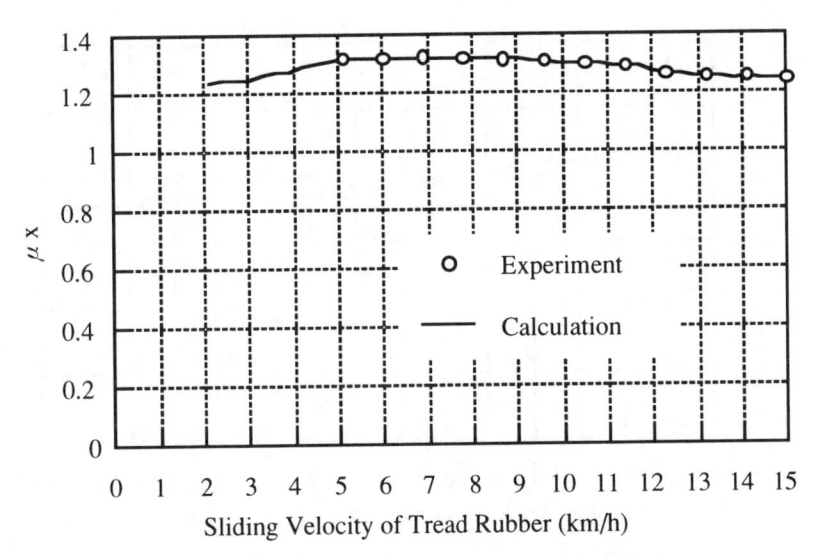

Fig. 5 Comparison of calculated and experimental results.

longitudinal velocity ΔV_ξ. When the test tire locked, the sliding velocity of the tire tread became equal to the drum velocity. We can confirm from Fig. 5 that the calculated friction coefficients agree well with the measured values.

(2) Wet surface

Figure 6 shows the experimental result of the relationship between the braking force and the slip ratio under wet surface conditions. Figure 7 shows a comparison

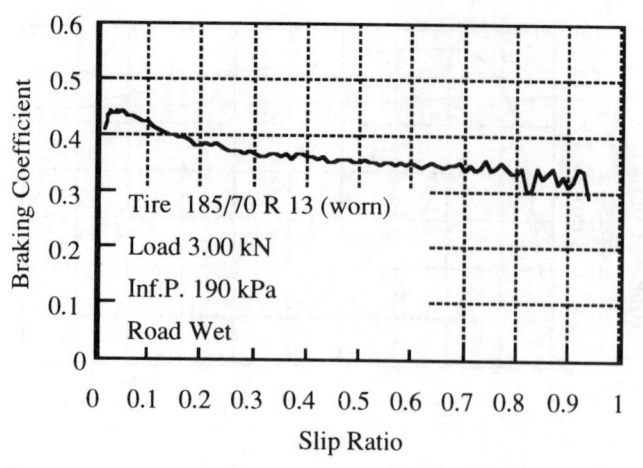

Fig. 6 The relationship between braking coefficient and slip ratio.

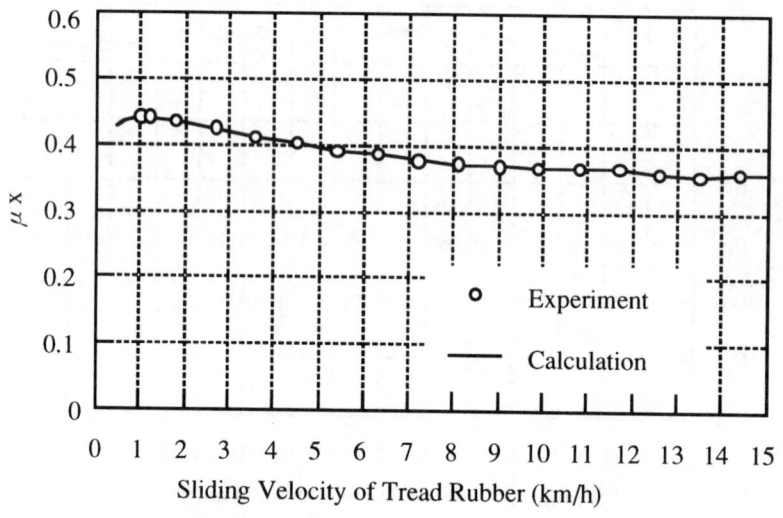

Fig. 7 Comparison of calculated and experimental results.

between the calculated and experimental results. We can confirm from Fig. 7 that the calculated friction coefficients agree well with the measured values. Figures 5 and 6 demonstrate that this estimation procedure yields the friction coefficient between a tire and the road in real time.

CONCLUSIONS

This paper described a procedure to estimate the friction coefficient between a tire and the road in real time which is easy to use and understand. The calculation method can be done quickly and can obtain the friction coefficient in a small slip ratio. Although braking force was used for the longitudinal force in this study, when this theory is applied to an actual vehicle, acceleration in a longitudinal direction can be used instead of braking force. This theory can be extended to driving conditions.

In future studies the usefulness of this theory will be confirmed by using an actual vehicle which is equipped with this estimation system.

REFERENCES

1) W.R.Pasterkamp and H.B.Pacejka, On Line Estimation Of Tire Characteristics For Vehicle Control, AVEC '94(1994.10), pp.521-526.
2) G.Gim, An Analytical Model Of Pneumatic tyres For Vehicle Dynamic Simulations, Part 2:Comprehensive Slip, International Journal of Vehicle Design, Vol.12, No.1, (1991) pp.19-39.

Magic Formula Tyre Model with Transient Properties

H. B. PACEJKA and I. J. M. BESSELINK

ABSTRACT

The tyre force and moment generating properties connected with the vehicle's horizontal motions are considered. Knowledge of tyre properties is necessary to properly design vehicle components and advanced control systems. For this purpose, mathematical models of the tyre are being used in vehicle simulation models. The steady-state empirical 'Magic Formula tyre model' is discussed. The aligning torque description is based on the concepts of pneumatic trail and residual torque. This facilitates its combined slip description. Following Michelin, weighting functions have been introduced to model the combined slip force generation. A full set of equations of the steady-state part of the model of the new version 'Delft Tyre 97' is presented. The non-steady state behaviour of the tyre is of importance in rapid transient maneuvres, when cornering on uneven roads and for the analysis of oscillatory braking and steering properties. A relatively simple model for longitudinal and lateral transient responses restricted to relatively low time and path frequencies is introduced.

INTRODUCTION

The widely used empirical tyre model, the newest version of which is presented here, is based on the so-called Magic Formula. The development of the model was started in the mid-eighties. In a cooperative effort TU-Delft and Volvo developed several versions (1987, 1989, 1991). In these models the combined slip situation was modelled from a physical view point. In 1993 Michelin introduced a purely empirical method using Magic Formula based functions to describe the tyre horizontal force generation at combined slip. In the previous version of 'Delft Tyre' this approach was adopted and the original description of the aligning torque has been altered to accommodate a relatively simple combined slip extension (cf. [11]). The model presented here is not restricted to small values of slip and the wheel may run backwards. A complete description of the steady-state response is given in the Appendix. The description of transient and oscillatory properties of the tyre-wheel combination has been improved. The dynamic behaviour may be covered by the model as long as the frequency of the wheel motion remains well below the first natural frequencies of the belt with respect to the rim (i.e. < ca. 15Hz) and if the wavelength of motion and road undulation is sufficiently large with respect to the tyre contact length (i.e. > ca. 1.5m).

STEADY-STATE MODEL FOR PURE SLIP

We refer to [3] for a detailed treatment of the pure slip part of this model (that is: at either lateral slip α or longitudinal slip κ). For the side force F_y and the fore and aft force F_x that part of the model remained unchanged. The formula reads:

$$y = D \sin[C \arctan\{Bx - E(Bx - \arctan Bx)\}] \quad (1)$$

with

$$Y(X) = y(x) + S_v \\ x = X + S_h \quad (2)$$

where

    Y: output variable F_x or F_y
    X: input variable α or κ

and

    B: stiffness factor
    C: shape factor
    D: peak value
    E: curvature factor
    S_H: horizontal shift
    S_V: vertical shift

The Magic Formula $y(x)$ typically produces a curve that passes through the origin $x=y=0$, reaches a maximum and subsequently tends to a horizontal asymptote. For given values of the coefficients B, C, D and E the curve shows an anti-symmetric shape with respect to the origin. To allow the curve to have an off-set with respect to the origin two shifts S_H and S_V have been introduced. A new set of coordinates $Y(X)$ arises as shown in Fig. 1.

The formula is capable of producing characteristics that closely match measured curves for the side force F_y and the fore and aft force F_x as functions of their respective wheel slip quantities: lateral slip α and longitudinal slip κ. Note that in this presentation α represents the tangent of the slip angle; at small lateral slip the difference is negligible.

Figure 1 (upper part) illustrates the meaning of some of the factors with the help of a typical side force characteristic. Obviously, coefficient D represents the peak value (with respect to the x-axis) and the product BCD corresponds to the slope at the origin ($x=y=0$). The shape factor C controls the limits of the range of the sine function appearing in the formula (1) and thereby determines the shape of the resulting curve. The factor B is left to determine the slope

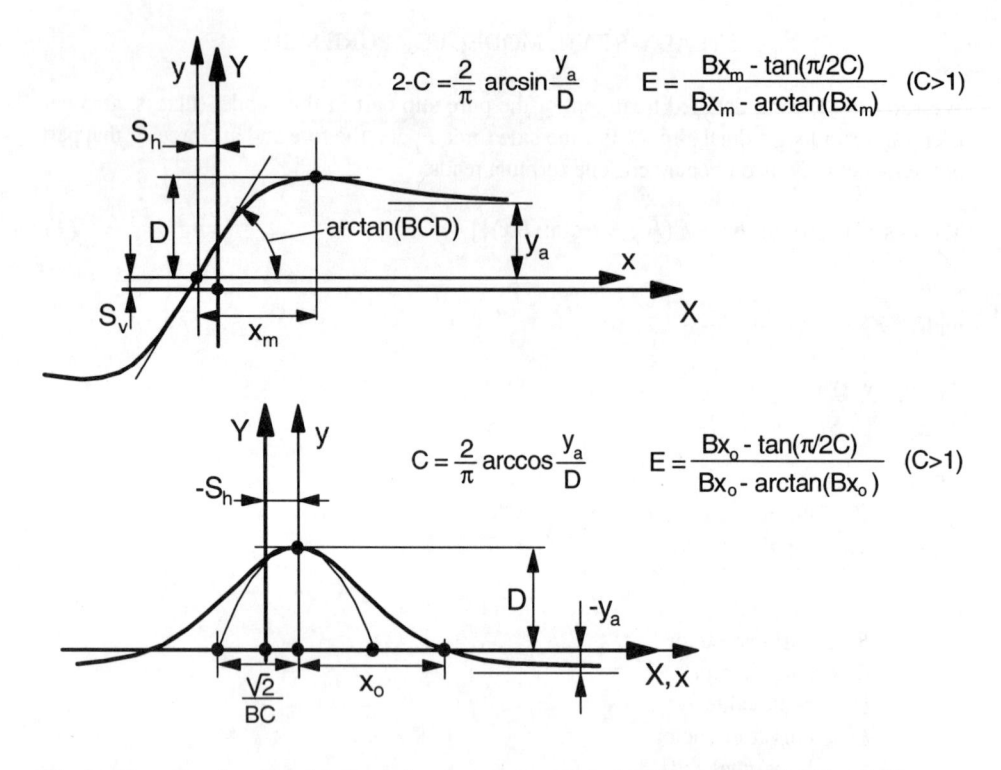

Fig. 1. Curves produced by the sine and cosine versions of the Magic Formula, Eqs.(1) and (6). Meaning of curve parameters have been indicated.

at the origin and is called the stiffness factor. The offsets S_H and S_V appear to occur when ply-steer and conicity effects and possibly the rolling resistance cause the F_y and F_x curves not to pass through the origin. Also, wheel camber will give rise to a considerable offset of the F_y vs α curves. Such a shift may be accompanied by a significant deviation from the pure anti-symmetric shape of the original curve. To accommodate such an asymmetry, the curvature factor E is made dependent of the sign of the abscissa (x).

$$E = E_o + \Delta E \cdot \text{sgn}(x) \tag{3}$$

Also the difference in shape that is expected to occur in the F_x vs κ characteristic between the driving and braking ranges can be taken care of.

The various factors are given functions of normal load and wheel camber angle. A number of parameters appears in these functions. A suitable regression technique is used to determine their values from measured data corresponding to the best fit (cf. [4]). One of the important functional relationships used is the load dependency of the cornering stiffness (or approximately: $BCD_y = \partial F_y/\partial \alpha$ at $\alpha + S_{Hy} = 0$).

$$BCD_y = K_y = p_{Ky1} \sin[2 \arctan(F_z/p_{Ky2})] \cdot (1 - p_{Ky3}|\gamma|) \quad (4)$$

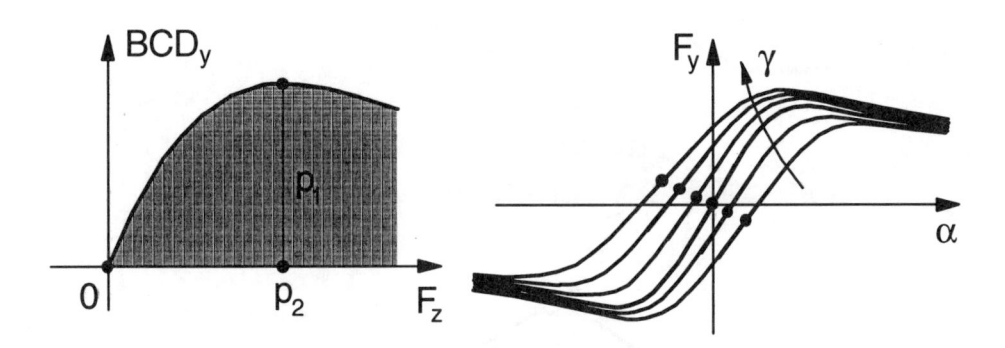

Fig. 2. Cornering stiffness versus vertical load and the influence of wheel camber, Eq. (4).

For zero camber, the cornering stiffness attains its maximum p_{Ky1} at $F_z = p_{Ky2}$. In Fig. 2 the basic relationship has been depicted. Apparently, for a cambered wheel the cornering stiffness is decreased.

The aligning torque M_z can now be obtained by multiplying the side force F_y with the pneumatic trail t and adding the usually small residual torque M_{zr} (Fig. 3). We have:

$$M_z = -t \cdot F_y + M_{zr} \quad (5)$$

with the pneumatic trail

$$t(\alpha_t) = D_t \cos[C_t \arctan\{B_t \alpha_t - E_t(B_t \alpha_t - \arctan(B_t \alpha_t))\}] \cdot c_\alpha \quad (6)$$

$$\alpha_t = \alpha + S_{Ht} \quad (7)$$

$$M_{zr}(\alpha_r) = D_r \cos[\arctan(B_r \alpha_r)] \cdot c_\alpha \quad (8)$$

$$\alpha_r = \alpha + S_{Hf} \quad (9)$$

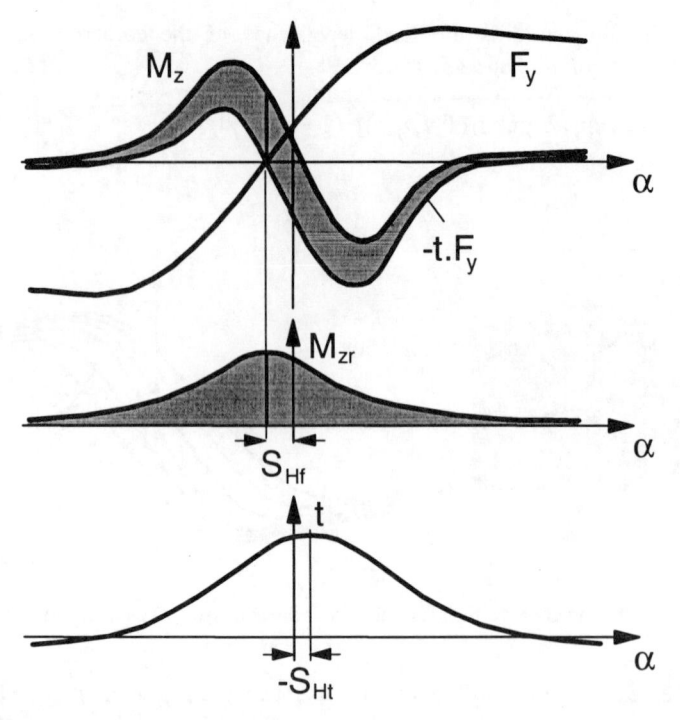

Fig. 3. Aligning torque characteristic produced by adding two parts: the product of side force and pneumatic trail and the residual torque, Eq. (7).

It is seen that both parts of the moment have been modelled using the Magic Formula, but instead of the sine function, the cosine function has been employed. In that way a hill-shaped curve is produced. In the Appendix more complete formulae are given including the consequence of running backwards.

In Fig. 1 (lower part) the basic properties of the cosine M.F. curve have been indicated (subscripts of factors deleted). Again, D is the peak value, C is a shape factor determining the level of the horizontal asymptote and now B influences the curvature at the peak (illustrated with the parabola). Factor E changes the shape at larger values of slip. The residual torque attains its maximum D_r at the slip angle where the side force becomes equal to zero. This is accomplished through the horizontal shift S_{Hf}. The peak of the pneumatic trail occurs at $\alpha = -S_{Ht}$. This formulation has proven to give excellent agreement with measured curves. The advantage with respect to the earlier versions where formula (3) is used for the aligning torque as well, is that we have now directly assessed the pneumatic trail which is needed (as in the previous version [3]) for handling the combined slip situation.

STEADY-STATE MODEL FOR COMBINED SLIP

In [3] the tyre's response to combined slip was modelled by using physically based formulae. A newer more efficient way is purely empiric. This method was developed by Michelin and published in [5]. It describes the effect of combined slip on both the lateral and the longitudinal forces. Weighting functions G are introduced which when multiplied with the original pure slip functions (that is Eq.(3)) produce the interactive effects of κ on F_y and of α on F_x. The weighting functions have a hill shape. They should take the value one in the special case of pure slip (κ or α equal to zero). When, for example, at a given slip angle α from zero increasing brake slip is introduced, the relevant weighting function for F_y may first show a slight increase in magnitude (becoming larger than one) but will soon reach its peak after which a continuous decrease follows. The cosine version of the Magic Formula is used to represent the hill shaped function:

$$G = D \cos[C \arctan(Bx)] \tag{10}$$

Here, G is the resulting weighting factor and x is either κ or α (possibly shifted). The coefficient D represents the peak value (slightly deviating from one if a horizontal shift of the hill occurs), C determines the level of the hill's base and B influences the sharpness of the hill. Apparently, the factor E was not needed to improve the fit. For the side force we obtain the following formulae:

$$F_y = G_{y\kappa} \cdot F_{yo} + S_{Vy\kappa} \tag{11}$$

$$G_{y\kappa} = \frac{\cos[C_{y\kappa}\arctan\{B_{y\kappa}(\kappa + S_{Hy\kappa})\}]}{\cos[C_{y\kappa}\arctan(B_{y\kappa}S_{Hy\kappa})]} \tag{12}$$

$$B_{y\kappa} = r_{By1}\cos[\arctan\{r_{By2}(\alpha - r_{By3})\}] \tag{13}$$

$$S_{Vy\kappa} = D_{Vy\kappa}\sin[r_{Vy5}\arctan(r_{Vy6}\kappa)] \tag{14}$$

In Eq.(11) F_{yo} denotes the side force at pure side slip obtained from Eq.(1); $S_{Vy\kappa}$ is the vertical shift sometimes referred to as the κ-induced ply-steer. This function varies with longitudinal slip κ as indicated in Eq.(14). Its maximum $D_{Vy\kappa}$ decreases with increasing magnitude of the slip angle (cf. App.). The factor $B_{y\kappa}$ influences the sharpness of the hill shaped weighting function (12). As indicated, the hill becomes more shallow (wider) at larger slip angles (then $B_{y\kappa}$ decreases according to Eq.(13)). The other coefficients r appearing in the formulae are treated as constant parameters. The combined slip relations for F_x are similar. However, a vertical shift was not needed to be included. In Fig. 4 a three-dimensional graph is shown

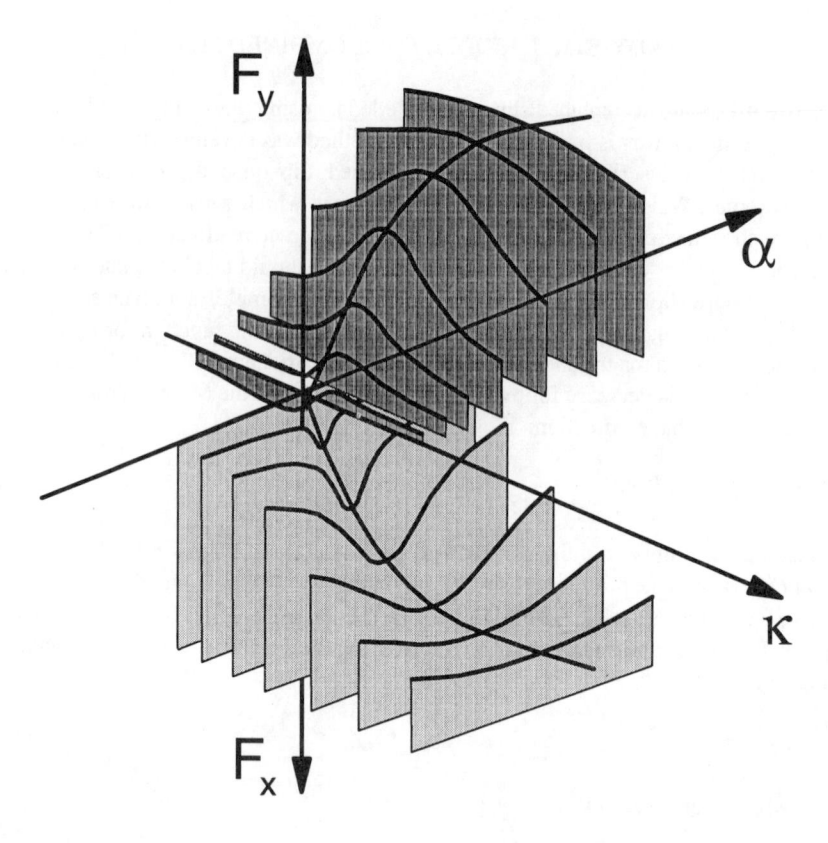

Fig. 4. Combined slip lateral and longitudinal force 3-D diagram showing the interaction of both slip components.

indicating the variation of F_x and F_y with both α and κ.

Regarding the aligning torque, physical insight has been employed to model the situation at combined slip. The arguments α_t and α_r (including a shift) appearing in the functions for pneumatic trail and residual torque are replaced by equivalent lateral slip quantities incorporating the effect of κ on the composite slip level. Besides, an extra term is included to account for the fact that an additional moment arm s arises for F_x as a result of camber γ and lateral tyre deflections through F_y (similar to the model description of [3]). The latter fact may give rise to a sign change of the aligning torque in the range of braking.

$$M_z = -t(\alpha_{t,eq}) \cdot F_y + M_{zr}(\alpha_{r,eq}) + s(F_y, \gamma) \cdot F_x \tag{15}$$

$$\alpha_{t,eq} = \sqrt{\alpha_t^2 + \left(\frac{K_x}{K_y}\right)^2 \kappa^2} \cdot \text{sgn}(\alpha_t) \tag{16}$$

and similar for $\alpha_{r,eq}$.

As mentioned before, the complete set of steady-state formulae has been listed in the Appendix. Parameters p, q, r and s of the model are non-dimensional quantities. In addition, user scaling factors λ have been introduced. With that tool the effect of changing friction coefficient, cornering stiffness, camber stiffness etc. can be quickly investigated in a qualitative way without having the need to implement a completely new tyre data set. Scaling is done in such a way that realistic relationships are maintained. For instance, when changing the cornering stiffness and the friction coefficient in lateral direction (through λ_{Ky} and $\lambda_{\mu y}$), the abscissa of the pneumatic trail characteristic is changed in a way similar to that of the side force characteristic.

MODEL FOR TRANSIENT RESPONSE

In previous publications, the concept of the relaxation length was employed to account for the compliance of the carcass with respect to the rim that is responsible for the lag in the response to lateral and longitudinal slip (cf. Refs. [6,7]). For small values of slip this way of approach is adequate. At larger slip levels, the lag diminishes according to experimental evidence. Also to avoid carcass deflections becoming too large, relaxation lengths may be introduced that decrease with increasing contact line deformation gradients [7]. The computations, however, become cumbersome due to possible instabilities and the situation at combined slip may become complex.

Therefore, another way to attack the problem is introduced. A contact patch is defined that can deflect horizontally with respect to the lower part of the rim. Only translations are allowed relative to the wheel plane. A mass point is attached to the moving contact patch to enable the calculation of the forces F_x^* and F_y^* and the moment M_z^* acting between road and tyre as a response to the slip velocity V_s^* of the mass point and to the camber angle. The effect of turn slip has not been taken into account.

It is assumed that this response to contact patch slip motions is instantaneous because a deflection is not needed to be developed here. In [8] this approach was adopted to model the longitudinal and circumferential dynamics of the tyre wheel combination in an attempt to study the effect of road undulations on anti-lock brake control. It may be noted that the carcass compliance together with the slip model of the contact patch automatically takes care of the wheel load dependent lag and also of the decrease in lag at increased (combined) slip. Figure 5 shows the model configuration in top and side view.

In addition, the contribution of three important gyroscopic effects related with belt distorsions are taken into account: the gyroscopic couple M_{zgyr} due to the rate of change of the lateral tyre deflection $\dot{v}_c$, the slip angle induced by this couple (through belt yaw distorsion) and the effect of the wheel yaw rate induced gyroscopic couple about the x-axis that acts on the tyre deflection v_c through the internal force F_{ygyr}. Moreover, the non-lagging part of the camber force is modelled to act directly on the wheel rim. Consequently, this part $F_{y\gamma NL}$ is first subtracted from the ground force F_y^*. The non-lagging part (i.e. the instantaneous response to pure camber variations) is a fraction of the total camber force. Sometimes negative fractions have been observed to occur [10].

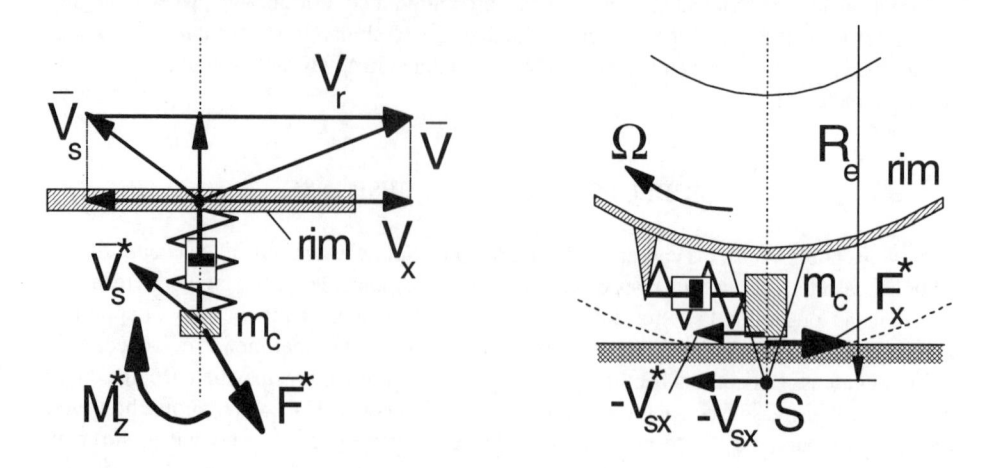

Fig. 5. Tyre model in top and side view, showing deflected contact patch with mass m_c.

The equations for the relative longitudinal and lateral motions u_c and v_c of the contact patch mass m_c with simplified acceleration terms read:

$$m_c \ddot{u}_c + k_{Fx}\dot{u}_c + c_{Fx}u_c = F_x^* \qquad (17)$$

$$m_c \ddot{v}_c + k_{Fy}\dot{v}_c + c_{Fy}v_c = F_y^* - F_{y\gamma NL} + F_{ygyr} \qquad (18)$$

Stiffnesses of the lower part of the carcass as measured on the standing tyre are introduced as well as damping coefficients. Damping may be needed to avoid self-excited vibrations when operating in ranges of slip where the characteristic slopes down with increasing slip. The slip velocity vector of the contact patch reads:

$$\bar{V}_s^* = \bar{V}_s + \begin{pmatrix} \dot{u}_c \\ \dot{v}_c - V_x \psi_{gyr} \end{pmatrix} \qquad (19)$$

and the slip vector of the contact patch:

$$\begin{pmatrix} \kappa^* \\ \alpha^* \end{pmatrix} = -\frac{\bar{V}_s^*}{|V_x|} \tag{20}$$

where $|V_x|$ denotes the absolute value of the forward component of the velocity of the wheel centre (or better: of the contact centre or point of intersection). The slip speed V_s of the wheel rim is defined as the horizontal velocity of slip point S that is attached to the rim a distance R_e (the effective rolling radius) below the wheel centre (Fig. 5).

The horizontal forces $F_{x,y}^*$ and the moment M_z^* may now be calculated as steady-state responses (cf. Appendix) to the slip of the contact patch. After that, the forces $F_{x,y}$ and the moment M_z of interest (which are supposed to act on the assumedly rigid tyre wheel disc) are calculated as follows:

$$F_x = F_x^* - m_c \ddot{u}_c \tag{21}$$

$$F_y = F_y^* - m_c \ddot{v}_c \tag{22}$$

$$M_z = M_z^* + M_{zgyr} \tag{23}$$

The following expressions hold for the various quantities introduced.

$$F_{y\gamma NL} = F_z(p_{Ty8} + p_{Ty9} df_z) \cdot \gamma \tag{24}$$

with df_z the normalized load increase (cf. App. Eq.(30)).

$$M_{zgyr} = -c_{zgyr} R_o m_t \Omega \dot{v}_c \tag{25}$$

with R_o the wheel radius, m_t the tyre mass, Ω the wheel speed of revolution.

$$F_{ygyr} = c_{ygyr} R_o m_t \Omega \frac{d\psi}{dt} \tag{26}$$

with ψ the wheel yaw angle.

$$\psi_{gyr} = c_{\psi gyr} \frac{M_{zgyr}}{R_o^2 c_{Fy}} \tag{27}$$

For a steel belted radial car tyre the non-dimensional coefficients c_{zgyr}, c_{ygyr} and $c_{\psi gyr}$ may all be estimated to be approximately equal to 0.5. The lateral stiffness c_{Fy} may be obtained by using its relation with the relaxation length σ_α through the cornering stiffness $C_{F\alpha}$ ($=K_y$).

$$\sigma_\alpha = \frac{C_{F\alpha}}{c_{Fy}} \tag{28}$$

Various types of motions have been simulated using the model. Limited space does not allow us to present results and discuss the performance of the model.

REFERENCES

1. H.B. Pacejka and R.S. Sharp, 'Shear force development by pneumatic tyres in steady state conditions: A review of modelling aspects'. Veh.Sys.Dyn. 20 (1991), pp. 121-176.
2. W.R. Pasterkamp and H.B. Pacejka, 'On-line estimation of tyre characteristics for vehicle control'. AVEC '94, Japan, Oct. 1994.
3. H.B. Pacejka and E. Bakker, 'The Magic Formula tyre model'. Proceedings 1st Tyre Colloquium, Delft, Oct.1991, Suppl. to Veh.Sys.Dyn., Vol. 21, 1993.
4. J. van Oosten and E. Bakker, 'Determination of Magic Formula tyre model parameters'. Proceedings 1st Tyre Colloquium, Delft, Oct. 1991, Suppl. to Veh.Sys.Dyn., Vol. 21, 1993.
5. P. Bayle, J.F. Forissier and S. Lafon, 'A new tyre model for vehicle dynamics simulations'. Automotive Technology International '93, pp. 193-198.
6. H.B. Pacejka, 'In-plane and out-of-plane dynamics of pneumatic tyres'. Veh.Sys.Dyn. 10 (1981), pp. 221-251.
7. H.B. Pacejka and T. Takahashi, 'Pure slip characteristics of tyres on flat and on undulated road surfaces'. AVEC '92, Yokohama, Sept. 1992.
8. P. van der Jagt, H.B. Pacejka and A.R. Savkoor, 'Influence of tyre and suspension dynamics on the braking performance of an anti-lock system on uneven roads'. EAEC Conf. '89, C382/047, IMechE 1989.
9. P.W.A. Zegelaar and H.B. Pacejka, 'In-plane dynamics of tyres on uneven roads', Proc. 14th IAVSD Symposium, Ann Arbor, Aug. 1995, Suppl. to Veh.Sys.Dyn., 1996.
10. A. Higuchi and H.B. Pacejka, 'The relaxation length concept at large wheel slip and camber'. This Proceedings.
11. H.B. Pacejka, 'The Tyre as a Vehicle Component'. Proc. XXVI FISITA Congress, 1996 (CD-Rom).

APPENDIX Delft Tyre 97 (steady-state part)

Superscript * refers to velocities and forces connected with the contact patch mass. Disregard this if only steady-state responses are required. Non-dimensional model parameters p, q, r and s have been introduced. For the user's convenience a set of scaling factors λ is available to examine the influence of changing a number of important overall parameters. The default value of these factors is set equal to one. We define in addition:

- V magnitude of speed of wheel (contact centre) (≥ 0),
- V_x forward component of speed of wheel (contact centre),
- V_r ($=R_e\Omega=V_x-V_{sx}$) forward speed of rolling,
- R_o unloaded tyre radius,
- R_e effective rolling radius,
- Ω wheel speed of revolution
- ρ tyre radial deflection,
- F_{zo} nominal load (≥ 0),
- F'_{zo} adapted nominal load

$$F'_{zo} = \lambda_{F_{zo}} F_{zo} \tag{29}$$

df_z the normalized vertical load increment

$$df_z = \frac{F_z - F'_{zo}}{F'_{zo}} \tag{30}$$

$|V'_x|$ downward limited absolute forward wheel speed

$$|V'_x| = \max(|V_x|, \epsilon_x), \quad (\epsilon_x = 0.01) \tag{31}$$

κ^* longitudinal slip of contact patch (protected against singularity)

$$\kappa^* = -\frac{V^*_{sx}}{|V'_x|} \tag{32}$$

α^* lateral slip of contact patch (protected against singularity)

$$\alpha^* = -\frac{V^*_{sy}}{|V'_x|} \tag{33}$$

V' downward limited speed V

$$V' = \max(V, \epsilon_x), \qquad (\epsilon_x = 0.01) \tag{34}$$

c_α = cosine of slip angle (protected against singularity)

$$c_\alpha = \frac{V_x}{V'} \tag{35}$$

Longitudinal Force (pure longitudinal slip)

$$F_{xo}^* = D_x \sin[C_x \arctan\{B_x \kappa_x - E_x(B_x \kappa_x - \arctan(B_x \kappa_x))\}] + S_{Vx} \tag{36}$$

$$\kappa_x = \kappa^* + S_{Hx} \tag{37}$$

$$C_x = p_{Cx1} \cdot \lambda_{Cx} \qquad (>0) \tag{38}$$

$$D_x = \mu_x \cdot F_z \tag{39}$$

$$\mu_x = (p_{Dx1} + p_{Dx2} df_z) \cdot \lambda_{\mu x} \qquad (>0) \tag{40}$$

$$E_x = (p_{Ex1} + p_{Ex2} df_z + p_{Ex3} df_z^2) \cdot \{1 - p_{Ex4} \text{sgn}(\kappa_x)\} \cdot \lambda_{Ex} \qquad (\leq 1) \tag{41}$$

$$K_x = F_z \cdot (p_{Kx1} + p_{Kx2} df_z) \cdot \exp(-p_{Kx3} df_z) \cdot \lambda_{Kx} \quad (= B_x C_x D_x = \frac{\partial F_{xo}}{\partial \kappa_x} \text{ at } \kappa_x = 0) \tag{42}$$

$$B_x = K_x / (C_x D_x) \tag{43}$$

$$S_{Hx} = (p_{Hx1} + p_{Hx2} df_z) \cdot \lambda_{Hx} \tag{44}$$

$$S_{Vx} = F_z \cdot (p_{Vx1} + p_{Vx2} df_z) \cdot \lambda_{Vx} \cdot \lambda_{\mu x} \tag{45}$$

Lateral Force (pure side slip)

$$F_{yo}^* = D_y \sin[C_y \arctan\{B_y \alpha_y - E_y(B_y \alpha_y - \arctan(B_y \alpha_y))\}] + S_{Vy} \tag{46}$$

$$\alpha_y = \alpha^* + S_{Hy} \tag{47}$$

$$\gamma_y = \gamma \cdot \lambda_{\gamma y} \tag{48}$$

$$C_y = p_{Cy1} \cdot \lambda_{Cy} \qquad (>0) \tag{49}$$

$$D_y = \mu_y \cdot F_z \tag{50}$$

$$\mu_y = (p_{Dy1} + p_{Dy2} df_z) \cdot (1 - p_{Dy3} \gamma_y^2) \cdot \lambda_{\mu y} \qquad (>0) \tag{51}$$

$$E_y = (p_{Ey1} + p_{Ey2} df_z) \cdot \{1 - (p_{Ey3} + p_{Ey4} \gamma_y) \text{sgn}(\alpha_y)\} \cdot \lambda_{Ey} \qquad (\leq 1) \tag{52}$$

$$K_y = p_{Ky1} F_{zo} \sin[2 \arctan\{F_z/(p_{Ky2} F_{zo} \lambda_{Fzo})\}] \cdot (1 - p_{Ky3}|\gamma_y|) \cdot \lambda_{Fzo} \cdot \lambda_{Ky}$$
$$(= B_y C_y D_y = \frac{\partial F_{yo}}{\partial \alpha_y} \text{ at } \alpha_y = 0) \tag{53}$$

$$B_y = K_y/(C_y D_y) \tag{54}$$

$$S_{Hy} = (p_{Hy1} + p_{Hy2} df_z + p_{Hy3} \gamma_y) \cdot \lambda_{Hy} \tag{55}$$

$$S_{Vy} = F_z \cdot \{p_{Vy1} + p_{Vy2} df_z + (p_{Vy3} + p_{Vy4} df_z) \gamma_y\} \cdot \lambda_{Vy} \cdot \lambda_{\mu y} \tag{56}$$

Aligning Torque (pure side slip)

$$M_{zo}^* = -t \cdot F_{yo}^* + M_{zr} \tag{57}$$

$$t(\alpha_t) = D_t \cos[C_t \arctan\{B_t \alpha_t - E_t(B_t \alpha_t - \arctan(B_t \alpha_t))\}] \cdot c_\alpha \tag{58}$$

$$\alpha_t = \alpha^* + S_{Ht} \tag{59}$$

$$M_{zr}(\alpha_r) = D_r \cos[\arctan(B_r \alpha_r)] \cdot c_\alpha \tag{60}$$

$$\alpha_r = \alpha^* + S_{Hf} \qquad (= \alpha_f) \tag{61}$$

$$S_{Hf} = S_{Hy} + S_{Vy}/K_y \tag{62}$$

$$\gamma_z = \gamma \cdot \lambda_{\gamma z} \tag{63}$$

$$B_t = (q_{Bz1} + q_{Bz2} df_z + q_{Bz3} df_z^2) \cdot (1 + q_{Bz4} \gamma_z + q_{Bz5}|\gamma_z|) \cdot \lambda_{Ky}/\lambda_{\mu y} \qquad (>0) \tag{64}$$

$$C_t = q_{Cz1} \qquad (>0) \tag{65}$$

$$D_t = F_z \cdot (q_{Dz1} + q_{Dz2} df_z) \cdot (1 + q_{Dz3} \gamma_z + q_{Dz4} \gamma_z^2) \cdot (R_o/F_{zo}) \cdot \lambda_t \tag{66}$$

$$E_t = (q_{Ez1} + q_{Ez2} df_z + q_{Ez3} df_z^2) \cdot \{1 + (q_{Ez4} + q_{Ez5} \gamma_z) \arctan(B_t C_t \alpha_t)\} \qquad (\leq 1) \tag{67}$$

$$S_{Ht} = q_{Hz1} + q_{Hz2} df_z + (q_{Hz3} + q_{Hz4} df_z) \gamma_z \tag{68}$$

$$B_r = q_{Bz9} \lambda_{Ky}/\lambda_{\mu y} + q_{Bz10} B_y C_y \tag{69}$$

$$D_r = F_z \cdot \{q_{Dz6} + q_{Dz7} df_z + (q_{Dz8} + q_{Dz9} df_z) \gamma_z\} \cdot R_o \cdot \lambda_{Mr} \cdot \lambda_{\mu y} \tag{70}$$

Longitudinal Force (combined) (based on [5])

$$F_x^* = D_{x\alpha} \cos[C_{x\alpha} \arctan\{B_{x\alpha}(\alpha^* + S_{Hx\alpha})\}] \tag{71}$$
$$B_{x\alpha} = r_{Bx1} \cos[\arctan(r_{Bx2} \kappa^*)] \cdot \lambda_{x\alpha} \quad (>0) \tag{72}$$
$$C_{x\alpha} = r_{Cx1} \quad (>0) \tag{73}$$
$$D_{x\alpha} = F_{xo}/\cos[C_{x\alpha} \arctan(B_{x\alpha} S_{Hx\alpha})] \tag{74}$$
$$S_{Hx\alpha} = r_{Hx1} \tag{75}$$

Lateral Force (combined) (based on [5])

$$F_y^* = D_{y\kappa} \cos[C_{y\kappa} \arctan\{B_{y\kappa}(\kappa^* + S_{Hy\kappa})\}] + S_{Vy\kappa} \tag{76}$$
$$B_{y\kappa} = r_{By1} \cos[\arctan\{r_{By2}(\alpha^* - r_{By3})\}] \cdot \lambda_{y\kappa} \quad (>0) \tag{77}$$
$$C_{y\kappa} = r_{Cy1} \quad (>0) \tag{78}$$
$$D_{y\kappa} = F_{yo}/\cos[C_{y\kappa} \arctan(B_{y\kappa} S_{Hy\kappa})] \tag{79}$$
$$S_{Hy\kappa} = r_{Hy1} \quad (<0.1 \text{ (drive-slip)}) \tag{80}$$
$$S_{Vy\kappa} = D_{Vy\kappa} \sin[r_{Vy5} \arctan(r_{Vy6} \kappa^*)] \cdot \lambda_{Vy\kappa} \tag{81}$$
$$D_{Vy\kappa} = \mu_y F_z \cdot (r_{Vy1} + r_{Vy2} df_z + r_{Vy3} \gamma) \cdot \cos[\arctan(r_{Vy4} \alpha^*)] \tag{82}$$

Normal Load

$$F_z = \max(p_{z1} \rho \cdot \lambda_{Cz} + p_{z2} \dot{\rho} \cdot \lambda_{Kz}, \ 0) \quad (\geq 0) \tag{83}$$

Overturning Couple

$$M_x = (q_{x1} F_y + q_{x2} \gamma) \cdot F_z \cdot \lambda_{Mx} \tag{84}$$

Rolling Resistance Torque

$$M_y = \frac{2}{\pi}(q_{y1} + q_{y2}F_x) \cdot F_z \cdot \arctan V_r \cdot \lambda_{My} \tag{85}$$

Aligning Torque (combined)

$$M_z^* = -t \cdot F_y' + M_{zr} + s \cdot F_x \tag{86}$$

$$t = t(\alpha_{t,eq}) = D_t \cos[C_t \arctan\{B_t \alpha_{t,eq} - E_t(B_t \alpha_{t,eq} - \arctan(B_t \alpha_{t,eq}))\}] \cdot c_\alpha \tag{87}$$

$$F_y' = F_y^* - S_{Vy\kappa} \tag{88}$$

$$M_{zr} = M_{zr}(\alpha_{r,eq}) = D_r \cos[\arctan(B_r \alpha_{r,eq})] \cdot c_\alpha \tag{89}$$

$$s = \{s_{sz1} + s_{sz2}(F_y/F_{zo}) + (s_{sz3} + s_{sz4}df_z)\gamma\} \cdot R_o \cdot \lambda_s \tag{90}$$

$$\alpha_{t,eq} = \sqrt{\alpha_t^2 + \left(\frac{K_x}{K_y}\right)^2 \kappa^{*2}} \cdot \text{sgn}(\alpha_t) \tag{91}$$

$$\alpha_{r,eq} = \sqrt{\alpha_r^2 + \left(\frac{K_x}{K_y}\right)^2 \kappa^{*2}} \cdot \text{sgn}(\alpha_r) \tag{92}$$

Note that in the formulation for the pure slip aligning torque the second term of the expression (69) for B_r has been recently introduced as a replacement of the first term; the suggested value for $q_{Bz10} = 1.5$ with $q_{Bz9} = 0$.

SOME MODELS AND METHODS OF PNEUMATIC TIRE MECHANICS

BELKIN A.E. *, BUKHIN B.L. **, MUKHIN O.N. **, NARSKAYA N.L. *

(*Moscow State Technical University,

**Tire Research Institute, Moscow)

SUMMARY

Pneumatic tire mechanics is divided into two separate sections, namely external tire mechanics, dealing with the effect of tires on motor-car dynamics, and internal tire mechanics, concerned with the stress and thermal state of tires and their service life depending on loading, structure design and material properties. Connecting links between these two sections are the output characteristics of tires which determine the operational qualities of a car. Since the authors are specialized in the field of internal tire mechanics this paper, for the most part, is concerned with the last section. It shall present a survey of the models and methods, worked out by Russian experts.

1. EXTERNAL TIRE MECHANICS

External tire mechanics is considered elastic wheel mechanics. It started developing much earlier than internal tire mechanics, paving ways for the calculation of tire characteristics by means of ring or shell-type models. Owing to this fact, the elastic wheel mechanics was based either on experiment or on very simple models.

In the USSR it was academician *E.I.Chudakov* who began investigating elastic wheel mechanics and his work was continued by his followers [1]. Nowadays, research of this kind is carried out by many educational and research institutes of the automobile industry. As a rule, book publications on the theory of the motor car contain sections dedicated to elastic wheel mechanics [2].

Chudakov and his school have made a detailed analysis of all kinds of static loading of the tire and have collected a great amount of experimental data [3]. They have classified types of tire stationary rolling, and studied stiffness, shock absorption and road grip properties of tires, rolling resistance, rolling of multi-wheel systems [4] and other characteristics.

At present, a great number of institutions are investigating the output characteristics of tires. Here are the main profiles of some of them.

The *Tire Research Institute* in Moscow is studying a complex of characteristics determining car motion security on dry and wet roads [5,6]. The tests are being carried out on various test stands, among them is one with an internal rolling surface. Systematic investigations of wear, noise and rolling resistance are being undertaken. A system of tire output characteristics determining the operational features of a car is given in [7].

The *Siberian Auto-Road Institute* is examining, by means of test stands of their own design, the influence of tire characteristics on tire motion stability and car dirigibility, including the characteristics of transient processes, non-stationary cornering and vibrations [8,9,10].

At *Volgograd State Technical University* a systematic study of the thermal state and the wear of tires is being conducted [11].

Different models are being applied to describe the tire as an element of the dynamic system "road-tire-automobile-driver" [12]. Vertical vibrations are described by Kelvin's model, in which a spring and a damper are connected in parallel. Unfortunately, the dependence of the model properties on rolling speed and vibration frequency had not been examined and was not taken into account in calculations. In the statistic modelling of car motion on a road with a random profile the contact zone of variable length has been simulated by a distributed system of Kelvin's elements with nonlinear springs [13].

The investigation of lateral motions primarily uses the cornering model which takes into account the interrelation between the cornering coefficient and loading as well as other factors [14]. The effect of longitudinal forces is described in terms of the sliding concept and the rolling radius depends on the wheel torque and tractive force.

In 1945, academician *M.V.Keldish* started developing tire models for nonstationary rolling modes with the aim of preventing the shimmy of aircraft wheels [15]. The rolling wheel is considered a non-holonomic system. During motion, the part of the tire running surface which is in contact with the ground surface is immovable (grip zone) and its midline coincides with the wheel rolling trajectory. The conditions of coincidence are the kinematic bonds (according to Keldish's theory they are the coordinate and the tangent in the contact centre). The further development of the theory and its application for the calculation of motion stability are described in the books [16,17,18,19].

Motion on a deformable surface (i.e. on plyable ground) is a special problem. The analysis system developed at the Moscow Tire Research Institute [20] is based on a combination of a ring-tire model with a ground assumed as an elasto-plastic medium.

2. INTERNAL TIRE MECHANICS

Research in internal tire mechanics began in the late 20's to the early 30's; it was initiated by research groups at some large tire manufacturing companies (*Goodyear, Dunlop,* etc.). In the Soviet Union, studies of this kind were first conducted at some departments of the combine *"Krasniy Treugolnik"* and later at the *Moscow Tire Research Institute*, established in 1941.

With regard to the companies' research teams one of the most important advances of the industrial groups was the development of the "equilibrium configuration theory" based on the netlike shell model and intended for calculating diagonal tires. The results of industrial research were either never published at all or were published with great delay. Similar investigations were frequently conducted simultaneously by different companies. Thus, the equilibrum configuration theory of a pneumatic tire was worked out independently by *J.F.Purdy* [21] (USA), *J.Hanus* [22] (France), *R.Hadekel* [23] (G.B.), *R.S.Rivlin* [24] (USA), *W.Hofferberth* [25] (Germany). All these works remained unknown to Soviet experts until the 60's or even 70's. Thus the equilibrum configuration theory was developed independently of foreign researchers by *A.A.Lapin* [26] and *V.L.Biderman* [27] in 1946-52.

In the 50-60's, schools doing research on pneumatic tire mechanics were set up at universities. Since then the high scientific potential of universities has been utilized for solving the various complex problems of tire mechanics. Existing achievements in the field of composite mechanics, shell theory, friction and wear theory, visco-elasticity, dynamics of non-holonomic systems, etc. have also been applied. The most famous scientific schools were headed by *S.K.Clark* (USA), *F.Boehm* (Germany), *T.Akasaka* (Japan). In the USSR it was *V.L.Biderman*. Russian tire experts have always carefully followed (and still follow) the available papers published by foreign schools and all their results. It is worth mentioning here that *F.Boehm*'s results published in the "Ingenieur-Archiv" [28] as well as his doctor thesis [29] are well known in Russia and have made a great impact on the calculation and theoretical analysis of radial tires here. It must be admitted that some new ideas suggested in F.Boehm's doctor thesis, especially concerning the rolling tire, require further development.

Let us consider in detail the two most commonly used tire models - the ring on an elastic base (Fig.1) and the pneumatic rotation shell (Fig.2).

The model of the ring on an elastic base was initially used for calculating the critical speed of tire rolling [30]. The bending stiffness of the ring was not taken into account. A formula for the critical speed of rolling has been deduced and its application for diagonal tires was quite successful.

In [31] the contact problem of the tire pressed on a rigid plane has been solved on the basis of a ring model with bending stiffnes. The tread ply was missing in this model. The characteristic feature of the solution with such a model is the presence of concentrated contact forces along the edges of the contact zone. The calculation results and experimental data were not compared, but it can be expected that the contact length and the model deflection are substantially less than the experimental values for the appropriate tire.

For solving the contact problem a ring model with a supplementary outer elastic ply, which simulates the properties of tread, seems appears to be more appropriate. Such a model was first applied by *E.Fiala* [32] for calculating cornering characteristics.

The contact problem of pressing the tire on a rigid plane was solved on the basis of Fiala's model by *O.N.Mukhin* [33]. In [34] the solution was generalized for the case of pressing the tire on the surface of a test drum.

The model of a ring with an outside elastic layer is suitable for analizing radial tires. When the tire is calculated with the help of the above one-dimensional model, only results reflecting the interdependence between the average contact length and the vertical load on the wheel agree quite well with experiments (Fig.3). The calculated loading characteristic is more rigorous than the measured one. In order to improve the correspondence with experiment the problem of straightening the running part of a radial tire in meridional direction in the centre of the contact surface has been solved [35]. Characteristic correlations between calculated values and experimental data can be seen in Figs. 4 and 5.

The ring model with a tread layer was also used for calculating the lateral, angular and torsional stiffnesses of a radial tire [33], as well as the characteristics of the stationary cornering of a wheel fitted with a radial tire [36].

The solutions of the above problems have been combined to form a complete computer program package for automated control calculation [37,38]. The program facilitates the calculation and analysis of the design stage of various tire structures for a series of the most important parameters, such as the inner forces in the carcass and belt caused by internal pressure and local pressing, tire deflections as a result of wheel loading, contact pressure distribution, lateral and angular stiffnesses, cornering characteristics, intensity of frictional effects in contact for a given wheel rolling mode with cornering.

Thus, despite the obvious simplicity of the one-dimensional models it is possible to obtain a number of important tire characteristics with the help of such models. However, one-dimensional models can not provide a profound investigation of the stress state in tire elements. For this kind of investigation the more informative two-dimensional models of a pneumatic rotation shell are employed. The shell structure and its elasticity properties are defined depending on the actual tire design.

For diagonal tires a special model in the form of a momentless netlike shell has been developed. The very first problem solved on the basis of this model was the internal pressure analysis, which was termed the "equilibrium configuration theory" [27]. As a result, some equations were derived for describing the profile of an inflated tire, methods of numerical solution were worked out and formulae for estimating the forces in cord threads were deduced. The calculation was based on the assumption of the non-extensibility of the cord threads and of the low forces in the rubber.

The work [27] in the USSR gave a start to a separate line of inquiry into tire mechanics, which involves the application of netlike shell models. Algorithms and programs [39] were elaborated for calculating the equilibrium configuration of an inflated tire, which enables researchers to nomograph the tire profile [40]. The nomographs have found wide application in tire designing.

The next problem to be solved was the calculation of tires loaded with centrifugal forces. An approximate treatment of this problem was conducted by

Ritz's variational method [41], cord was taken as inextensibile and cord and rubber deformation energy was neglected in calculating the shell potential. With the help of computers, the numerical solution of the problem was obtained in [42] exactly by means of a step-by-step method of consecutive loading. Hence, the problem of calculating the axisymmetrical deformation of diagonal tires was completely resolved.

It is much more difficult to analyse the non-axisymmetrical loading of a tire. Some researchers [43,44] succeeded in deriving equations describing netlike shell strain at arbitrary loading. In one case [43] they considered displacements (arbitrary in magnitude) of the shell made of inextensibile threads, while in the other [44] displacements were considered small, but thread extensibility was taken into account.

Thus, the works [43,44] have laid the foundation for the theory of analyzing non-axisymmetrical deformations of the diagonal tire by the netlike shell model. Since this theory is of immense complexity, very few analytical solutions of practical interest for tire professionals can be found. It is worth mentioning one analytical research devoted to defining the critical rolling speed of a tire [45].

Some works [46,47,48] are devoted to numerical methods of investigation the tire by the netlike shell model. The author of [46] has solved the practically important problems concerning the effect of local loads on a pneumatic tire. These problems have been solved by expanding the solution in trigonometric series and by further numerical integration of ordinary differential equations for every term of the series. In the papers [47,48], also dealing with the local load effect analysis of the tire, the solution of equilibrium equations for netlike shell was received by the finite-difference method. Comparing the two ways of analysing, one should emphasize the reliability and high accuracy of the method using the trigonometric series, on the one hand, and universality of the finite-difference method, which solves nonlinear problems, and among them the contact problem, on the other hand.

In the paper [49], the diagonal tire model used is an orthotropic momentless rotational shell with an elastic outer layer simulating the tread. The contact problem of tire pressed on a rigid plane is solved in a geometrically non-linear statement. Comparing the analytical results with experimental data the author came to the conclusion that strain fields and loading characteristics for truck tires can be precisely analysed with the help of this model. As regards motor-car tires, the analysis gave an undervaluation of the radial stiffness. This can evidently be explained by the fact that the motor-car tire runs at a lower internal pressure, and an essential part in providing its load-carrying capacity is based on the bending stiffness of the carcass, which is not taken into account in the momentless theory.

Further improvement of the designing model of radial tire is connected with the application of the moment theory of multi-layered shells. This problem is discussed in detail in the book by *E.I.Grigolyuk* and *G.M.Kulikov* [50] which contains a thorough investigation of the axisymmetrical state of a tire in the state of inflating. The authors devote special attention to the elaboration of a geometrically nonlinear theory of multi-layered shells on the basis of Timoshenko's hypothesis and the "broken normal" hypothesis using Reissner's mixed variational principle. The elaborated theories have been applied by the authors to analyse diagonal and radial tires. The analysis shows that the application of Timoshenko's kinematic hypothesis to the whole package of rubber-cord layers of the tire can lead to considerable errors in analysing the stress state in the belt edge area. To avoid this disadvantage the authors use Timoshenko's hypothesis for every rubber-cord layer separately, i.e. they prefer the "broken normal" hypothesis.

More complex problems of the non-axisymmetrical loading of diagonal tires have also been considered within the limits of a multi-layered shell model [51]. Here the method of local variations is used to solve the problem of tire-road contact. The essence of this method is as follows. The total energy functional of a tire pressed on the road is expressed in terms of displacements of nodal points. Then, by successively varying the displacements of each point, values at which the energy is minimal can be found. The successive displacements of the nodal points make it possible to observe the contact of the tire with any surface. This method is not difficult to implement, however, the process takes too long because of the fact that it is necessary to accomplish multiple variations of thousands of displacements.

With regard to the application of multi-layered shell models with generalized variant of Timoshenko's kinematic hypothesis, it should be admitted that there is a certain limitation in its implementation due to the schematic and nondefinite division into layers. The stresses, obtained as a result of the analysis, quite obviously depend on the division into layers. Naturally, such a model can be effectively used for multi-layered large-size truck tires because it takes proper account of the nonuniform stress distribution across the tire thickness and allows a number of adjoining plies to be combined into one, thus reducing the total number of layers in the model. This is important for tire analysis, in particular for trucks, whose carcass has dozens of plies.

Now let us turn to the shell models of radial tires. The basic difference in the structure of radial tires does not allow the same models which had been worked out for diagonal tires analysis are to be used. This made the elaboration of new models necessary.

One of the first working shell models was suggested by *Boehm* in [52], where the shell is considered to consist of two membrane layers, which model carcass and belt, respectively. The distribution of forces between the carcass and belt layers is defined by a function which was termed the "girding function". This function is selected and defined in varying values across the width of the belt layer, after which the appropriate tire configuration can be determined. It should be emphasized that the introduced concept of a girding coefficient has proven itself quite useful and was introduced to the practice of tire designing.

This model was further developed by *Biderman* in [53], where the running part of the tire was taken as a three-layer orthotropic shell with two bearing membrane layers, modelling the carcass and the belt, and with a rubber interlayer separating them, which worked only in respect to transverse shearing. The side wall of the tire was modelled as a one-layer momentless shell. Due to its relative simplicity and its informativeness and precision the model proved quite suitable for analysing radial tires. In [53] it was employed for analysing the stress state caused by internal pressure.

Later on, this approach to the modelling of radial tires was developed in [54,55,56], where the behaviour of tires under non-axisymmetrical load corresponding to the local pressure in contact with the road was investigated. In [54] the analysis was carried out on the basis of the Lagrange variational principle in a geometrically non-linear statement by a method of local variations. The linearized theory of small deformations in shells, preloaded by internal pressure, is used in work [55]. The problem of stresses in the tire is solved in ordinary trigonometrical series, by using numerical integration for every harmonic of the stress state. The results thus obtained need to be revised in the contact spot area where displacements are already not small. Paper [56] describes the algorythm of a multi-level analysis of the tire according to the nonlinear theory. In this case, the finite element method is used to revise the stress state in the contact zone. The previously obtained solution in trigonometrical series [55] is considered here as an initial approximation in order to set up an iterational process of revision.

The belt of some radial tires has protecting plies which can be placed, according to the design, both over the working layers, i.e. on the tread side, or under the working layers, i.e. on the carcass side. These plies are intended to either protect the belt ply from damage or to take load from the rubber near the belt edges. A method for calculating of the tires with protecting plies at the belt has been elaborated in work [57].

At present, analyses of tire stress states under operating load are generally being performed in static statements. However, the nature of tire deformation at high rolling speeds differs considerably from the strain in a motionless or slowly rolling tire. Dynamic effects make themselves felt at much lower than critical speeds. Work [58] proposes a solution to the dynamic contact problem for the radial tire as a three-layer visco-elastic shell during stationary and frictionless rolling. As an example, Fig.6 shows the level lines of the carcass deflections of a tire which is rolling at 200 km/h. It reveals how waves emerge on the side wall of the tire which indicates existance of belt waves.

Around the late 60's researchers began to apply the method of finite elements to analyse pneumatic tires. This enabled them to expand and make considerably more complex the class of models used for the tire as a whole and for its parts as well. Due to the FEM it became possible to treat the tire as a three-dimensional body. In the beginning scientists mainly used already existing universal program packages, such as NASTRAN, MARC, ADINA, NONSAP, SUPERB and others. By means of these powerful programs researchers could turn to analysing tires under internal pressure and contact load on the basis of both shell models and three-dimensional models. However, in [59,60] it is pointed out that there is only a rather approximate correspondence between calculation and experimental data. For instance, the survey paper [59] stresses that despite a good agreement between numerical data and measurements as regards deflection and the shape of inflated and local pressed profiles, the discrepancy in the data on stress-strain state reach 70%. The major reason for such considerable discrepancies is that in tire designing it is necessary to take into account a series of special features in the behaviour of a rubber-cord structure and this, naturally, limits the scope for the application of universal FEM packages. A short discussion of the FEM application for calculating of pneumatic tires can be found in the survey papers of *R.A.Ridha* [59] and *H.Rothert* et al. [61].

A demand arose for elaborating new special FEM programs or supplementing the existing program packages with new elements intended for calculating pneumatic tires and other rubber-cord structures. A substantial number of works have been conducted along this line. For example, papers [62,63,64,65] describe the use of finite elements for modelling the behaviour of rubber and rubber-cord composites.

The recent work [66] by *Faria L.O., Oden J.T.*, et al. will be mentioned as an example of the latest developments in the stress-strain analysis of radial tires. The finite element model of tires suggested by the authors was employed to solve contact problems with friction which arise in the case of cornering as a result of wheel plane turning. The three-dimensional finite elements used in the model can take into account large strains and the incompressibility of rubber.

3 . CONCLUSION

In conclusion, we can point out the trend to work out more complex three-dimensional models, which can take into consideration many peculiarities of the structure in the operational state. It can be expected that, in future, the FEM may become the basic method for tire analysis. Nevertheless, it is our opinion, for today, the three-dimensional tire model needs extreme powerful computers and too great accuracy of the input data to be used in everyday designing practice. Therefore, less complicate models do not loose their topicality and will still play their part in tire designing practice. The one and two-dimensional tire models are also being constantly improved and provide even more reliable results. At today's stage of development the methods of tire analysis on the basis of the above methods are characterised by a trend of creating object-oriented program packages [37,38,67], which find wide application in the design offices of tire plants.

REFERENCES

1. Чудаков Е.И. Избранные труды, т.1. Теория автомобиля. М.,АН СССР, 1961. 463 с. (Chudakov E.I., Selected works. Vol.1, Moscow, USSR. Academy of Sciences, 1961, p. 463).

2. Литвинов А.С., Фаробин Я.Е. Автомобиль: Теория эксплуатационных свойств. М.,"Машиностроение",1989. 240 с. (Litvinov A.S., Farobin Ya.E., Automobile: Theory of operational characteristics. Moscow, "Mashinostroenie", 1989, p. 240).

3. Кнороз В.И. и др. Работа автомобильной шины. М.,"Транспорт", 1976. 238 с. (Knoroz V.I. et al., Behaviour of the automobile tire. Moscow, "Trasport", 1976, p. 238).

4. Петрушов В.А. и др. Сопротивление качению автомобилей и автопоездов. М., "Машиностроение", 1975. 225 с. (Petrushov V.A. et al., Rolling resistance of automobile and truck-trains. Moscow, "Mashinostroenie", 1975, p. 225).

5. Калинковский В.С. и др. Метод определения характеристик безопасности шин.//5-й симпозиум "Проблемы шин и резинокордных композитов". 1993, с.70-81. (Kalinkovski V.S. et al., A method of estimating characteristics of tire security. // 5th Symposium "Problems of tires and rubber-cord composites", Moscow, 1993, pp. 70-81).

6. Калинковский В.С. и др. Стабилизирующие моменты шин при уводе на мокрой асфальто-бетонной поверхности. // 6-й симпозиум "Проблемы шин и резинокордных композитов". 1995, с.105-108. (Kalinkovski V.S. et al., Tire stabilizing moments in the case of cornering on a wet asphalt-concrete surface // 6th Symposium "Problems of tires and rubber-cord composites", Moscow, 1995, pp. 105-108).

7. Бухин Б.Л. Выходные характеристики пневматических шин. ЦНИИТЭнефтехим, М., 1978. 83 с. (Bukhin B.L., Output characteristics of pneumatic tires. CNIITEneftekhim, Moscow, 1978 p. 83).

8. Каня В.А., Дик А.Б. Определение стабилизирующего момента шины. // 5-й симпозиум "Проблемы шин и резинокордных композитов". 1993, с.82-88 (Kanya V.A, Dik A.B., Determination of the stabilizing moments of a tire// 5th Symposium "Problems of tires and rubber-cord composites", Moscow, 1993, pp. 82-88).

9. Капралов С.С. и др. Экспериментальное исследование влияния характеристик увода шин на управляемость автомобиля.//5-й симпозиум "Проблемы шин и резинокордных композитов". 1993, с.89-96 (Kapralov S.S. et al., Experimental investigation of tire cornering effects on motor-car dirigibility // 5th Symposium "Problems of tires and rubber-cord composites", Moscow, 1993, pp. 89-96).

10. Капралов С.С. и др. Метод определения текущих значений оценочных параметров шин.// 6-й симпозиум "Проблемы шин и резинокордных композитов". 1995, с.109-116 (Kapralov S.S. et al., Method of determining the current values of tire estimating parameters // 6th Symposium "Problems of tires and rubber-cord composites", Moscow, 1995, pp. 109-116).

11. Новопольский В.И. и др. Прогнозирование температуры протектора в пятне контакта автомобильной шины. // 3-й симпозиум "Проблемы шин и резинокордных композитов". 1991, с.70-78 (Novopolski V.I. et al., Predicting the temperature of protector in the contact area of a motor-car tire // 3rd Symposium "Problems of tires and rubber-cord composites", Moscow, 1991, pp. 70-78).

12. Хачатуров А.А. и др. Динамика системы дорога-шина-автомобиль-водитель. М., "Машиностроение", 1976. 535 с. (Khachaturov A.A. et al. , Dynamics of the road-tire-car-driver system. Moscow, "Mashinostroenie", 1976, p. 535).

13. Ненахов А.Б., Бухин Б.Л. Динамическая нагруженность шин при вертикальных колебаниях. //Исследование механики пневматической шины. М.:ЦНИИТЭнефтехим, 1988, с.77-95 (Nenakhov A.B., BukhinB.L., Dynamic loading of the tire at vertical vibrations // A study of pneumatic tire mechanics. CNIITEneftekhim, Moscow, 1988, pp. 77-95).

14. Бухин Б.Л. Введение в механику пневматических шин. М., "Химия", 1988. 224 с. (Bukhin B.L., An introduction into pneumatic tire mechanics. Moscow, "Khimia", 1988, p. 224).

15. Келдыш М.В. Шимми переднего колеса трехколесного шасси. М., ЦАГИ. вып. 564, 1945. 33 с. (Keldish M.V. Shimmy of the front wheel of a three-wheel chassis. Moscow, CAGI. Iss. 564, 1945, p. 33).

16. Неймарк Ю.И., Фуфаев Н.А. Динамика неголономных систем. М., "Наука", 1967. 520 с. (Neymark Yu.I., Fufaev N.A., Dynamics of non-holonomic systems. Moscow, "Nauka", 1967, p. 520).

17. Тураев Х.Т. и др. Теория движения систем с качением. Ташкент, "Фан", 1987 (Turaev H.T. et al., Theory of motion for rolling systems. Tashkent, "Fan", 1987).

18. Тураев Х.Т. Моделирование и исследование динамики колесных транспортных машин с деформируемыми шинами. Ташкент, "Фан", 1995 (Turaev H.T., Modelling and investigation of the dynamics of transport wheel motor-vehicles with deformable tires. Tashkent, "Fan",1995).

19. Левин М.А., Фуфаев Н.А. Теория качения деформируемого колеса. М., "Наука". 1989 (Levin M.A., Fufaev N.A., Rolling theory of deformable wheel. Moscow, "Nauka", 1989) .

20. Ле Хонг Хань и др. Метод расчета характеристик пневматических шин, определяющих проходимость автомобиля. // Межд. конф. по каучуку и резине, Москва, 1984, В-33 (Le Hong Han et al., A method of analysing pneumatic tire characteristics determining the cross-country mobility of a car // International Rubber Conference, Moscow, 1984, B-33).

21. Purdy J.F., Day R.B., // Goodyear Research Laboratory Report. 1928.

22. Hanus J., L'enveloppe mince.// Internal publication of Soc. Anon. Pneumatique. Dunlop. 1946.

23. Hadekel R., The Mechanical Characteristics of Pneumatic Tyres.// S.+T. Memo 10/52 T.P.A. 1952.

24. Rivlin R.S., The deformation of a membrane formed by inextensible cords.// Archive for Rational Mech. and Analysis, vol. 2, 1959, pp. 447-476.

25. Hofferberth W., Zur Festigkeit des Luftreifens. // Kautschuk Gummi, N 9, 1956, pp. 225-231.

26. Лапин А.А. Резино-кордовые оболочки как упругие и силовые элементы машин. // Расчеты упругих элементов машин и приборов, МВТУ, N 16, Машгиз, 1952 (Lapin A.A., Rubber-cord shells as elastic and load-bearing machine elements. // Analyses of elastic components of machine elements and instruments, MVTU, No. 16, Mashgiz, 1952).

27. Бидерман В.Л. Расчет формы профиля и напряжений в элементах пневматической шины, нагруженной внутренним давлением.// Труды НИИШП, сб. 3, М.: Госхимиздат, 1957, с. 16-51 (Biderman V.L., Analysis of the profile shape and stresses in parts of a pneumatic tire loaded with internal pressure// NIIShP Transactions Iss. 3, Moscow. Goskhimizdat, 1957, pp. 16-51).

28. Boehm F., Mechanik des Gurtelreifens.// Ingenieur-Archiv, v. 35, No. 2, 1966, s. 82-101.

29. Boehm F., Zur Mechanik des Luftreifens. // Dr. techn.- Dissertation, TH Stuttgart 1966.

30. Бидерман В.Л. Критическая скорость качения пневматической шины //Расчеты на прочность. Вып.7. М.: Машгиз, 1961, с. 324-349 (BidermanV.L., The critical speed of rolling pneumatic tire.// Strength Analyses Iss. 7. Moscow, Mashgiz, 1961, pp. 324-349).

31. Бидерман В.Л., Бухин Б.Л. К расчету шин с меридиональным расположением нитей корда в каркасе. // Расчеты на прочность. Вып.9. М.: Машгиз, 1963, с.34-47 (Biderman V.L., Bukhin B.L. , On the analysis of tires with a meridional arrangement of cord threads in the carcass. // Strength Analyses. Iss. 9. Moscow, Mashgiz, 1963, pp. 34-47).

32. Fiala E., Seitenkraefte am rollenden Luftreifen. // VDI - Zeitschrift, 1954. Vol. 96, pp. 973-979.

33. Мухин О.Н. Расчет жесткостных характеристик автомобильных шин типа Р. // Расчеты на прочность. Вып.15. М.: "Машиностроение", 1971, с.58-87 (Mukhin O.N., Analysis of stiffness characteristics of radial car tires. // Strength Analyses. Iss. 15. Moscow, "Mashinostroenie", 1971, pp. 58-87).

34. Мухин О.Н. Расчет механических характеристик меридиональной шины, обжатой на барабан. // Механика пневматических шин. М.: НИИШП, 1976, с.136-147 (Mukhin O.N. , Analysis of the mechanical characterictics of a meridian tire pressed on a drum.// Mechanics of pneumatic tires. Moscow, NIIShP, 1976, pp.136-147).

35. Мухин О.Н. Расчет прогиба радиальной шины с учетом меридиональной кривизны беговой дорожки. // Механика пневматических шин как основа рационального конструирования и прогнозирования эксплуатационных свойств. М.: НИИШП, 1974, с.12-25 (Mukhin O.N., The radial tire deflection analysis accounting the meridian curvature of the running track. // Mechanics of pneumatic tires as a basis for rational designing and predicting operational characteristics. Moscow, NIIShP, 1974, pp.12-25).

36. Истирание резин. // Бродский Г.И. и др. М.: "Химия", 1975, с.143-164 (Abrasion of Rubber. // Brodski G.I. et al., Moscow, "Khimia", 1975, pp.143-164).

37. Mukhin O.N., Inner pressure and radial loading analysis of radial pneumatic tyres. P.1. Inner pressure. // Prostor expo. TRI. Moscow, 1992, No. 4, pp.15-74.

38. Mukhin O.N., Inner pressure and radial loading analysis of radial pneumatic tyres. P.2. Radial loading. // Prostor expo. TRI. Moscow, 1994, No.1, pp.39-116.

39. Бидерман В.Л., Бухин Б.Л., Николаев И.К. Расчет равновесной конфигурации резинокордной оболочки вращения на ЭВМ. // Каучук и резина, 1966, № 5, с.33-35 (Biderman V.L., Bukhin B.L., Nickolaev J.K., A computer analysis of the equilibrium configuration of a rubber-cord shell of rotation.// Kautchuk i rezina, 1966, No. 5, pp.33-35).

40. Бидерман В.Л. и др. Атлас номограмм равновесных конфигураций пневматических шин. М.: "Химия", 1967, 36 с. (Biderman V.L. et al., A nomograph atlas of equilibrium configurations of pneumatic tires. Moscow, "Khimia", 1967, p. 36).

41. Бухин Б.Л. Расчет напряжений и деформаций в пневматических шинах при их вращении. // Расчеты на прочность. Вып.6, М.: Машгиз, 1960, с.56-65 (Bukhin B.L., Analysis of stresses and strains in pneumatic tires in rotation. // Strength Analyses. Iss. 6. Moscow, Mashgiz, 1960, pp. 56-65).

42. Бухин Б.Л., Гильдман И.М., Каплинский Э.М. Симметричная деформация безмоментной сетчатой оболочки вращения. // Каучук и резина, 1969, № 11, с.36-39 (Bukhin B.L., Gildman J.M., Kaplinski E.M., Symmetrical deformation of a momentless netlike shell of rotation.// Kautchuk i rezina, 1969, No. 11, pp.36-39).

43. Бидерман В.Л. Дифференциальные уравнения деформации резинокордных оболочек вращения. // Расчеты на прочность в машиностроении. Труды МВТУ, № 89, М.: Машгиз, 1958, с.119-146 (Biderman V.L., Differential equations for deformation of rubber-cord shells of rotation. // Strength analyses in mechanical engineering. MVTU Transactions. No.89, Moscow, Mashgiz, 1958, pp.119-146).

44. Бидерман В.Л., Бухин Б.Л. Уравнения равновесия безмоментной сетчатой оболочки.//Изв.АН СССР. МТТ, 1966, № 1, с.81-89 (Biderman V.L., Bukhin B.L., The equilibrium equations of a momentless netlike shell. // Izv. AN USSR, MTT, 1966, No. 1, pp.81-89).

45. Бидерман В.Л., Бухин Б.Л. Расчет критической скорости качения пневматической шины. // Изв. АН СССР, ОТН, Механика и машиностроение, 1961, № 1, с.52-57 (Biderman V.L., Bukhin B.L. , Analysis of the critical speed of a rolling pneumatic tire.// Izv. AN USSR, OTN. Mechanics and mechanical engineering, 1961, No. 1, pp.52-57).

46. Бухин Б.Л. Теория безмоментных сетчатых оболочек вращения и ее приложение к расчету пневматических шин. Докт. диссертация, МВТУ, 1972, 309 с. (Bukhin B.L , The theory of momentless netlike shells of rotation and its application for analysing pneumatic tires. Doctor degree thesis, MVTU, 1972, p. 309).

47. Дьяконов Е.Г., Николаев И.К. О решении уравнений равновесия сетчатых оболочек методом сеток. // Механика пневматических шин как основа рационального конструирования и прогнозирования эксплуатационных свойств. М.: НИИШП, 1974, с.75-91 (Diakonov E.G., Nicolaev I.K., On solving equilibrium equations of netlike shells by the mesh-analysis method.// Mechanics of pneumatic tires as a basis for rational designing and predicting operational characteristics. Moscow, NIIShP, 1974, pp.75-91).

48. Дьяконов Е.Г., Николаев И.К. О численном решении некоторых нелинейных краевых задач теории сетчатых оболочек.//Труды III Всесоюзной конференции "Численные методы решения задач теории упругости и пластичности", т.1, Новосибирск, 1974, с.85-101 (Diakonov E.G., Nikolaev I.K., On numerical solutions of some nonlinear boundary problems in netlike shell theory.// Transactions of the III All-Union Conference "Numerical methods of solving problems of elasticity and plasticity theory. Vol. 1, Novosibirsk, 1974, pp. 85-101).

49. Николаев И.К. Математическая модель диагональной шины. //Механика пневматических шин. М.: НИИШП, 1976, с. 5-36 (Nikolaev I.K., A mathematical model of diagonal tire. // Mechanics of pneumatic tires. Moscow, NIIShP, 1976, pp. 5-36).

50. Григолюк Э.И., Куликов Г.М. Многослойные армированные оболочки. Расчет пневматических шин. М.: "Машиностроение", 1988, 287 с. (Grigolyuk E.J., Kulikov G.M., Multi-layered reinforced shells. Calculation of pneumatic tires. Moscow, "Mashinostroenie'', 1988, p.287).

51.Кваша Э.Н., Плеханов А.В., Прусаков А.П. Неклассический вариант моментной теории пневматических шин. // Международная конференция по каучуку и резине. Секция В. М., 1984, В 50. (Kvasha E.N., Plekhanov. A.V., Prusakov A.P., A non-classical version of the moment theory of pneumatic tires. // International Rubber Conference. Section B. Moscow, 1984, B 50).

52. Boehm F. Zur Statik und Dynamik des Guertelreifens. // ATZ, 69, No. 8, 1967, s. 255-261.

53. Бидерман В.Л., Левковская Э.Я. Расчет напряжений и деформаций, вызываемых давлением, в шинах типа "Р". // Изв. вузов. Машинострое-ние, 1969, № 3, с.107-112 (Biderman V.L., Levkovskaya E.Ya., The analysis of stresses and strains caused by pressure in radial tires.// Izv. Vuzov. Mashinostroenie. 1969, No. 3, pp.107-112).

54. Фотинич О.В. Численный метод исследования неосесимметричных деформаций пневматических шин. Канд. диссертация. М., МВТУ, 1972, 205 с. (Fotinitch O.V., A numerical method of investigating non-axisymmetrical tires deformations. Candidate degree thesis. Moscow, MVTU, 1972, p.205) .

55. Белкин А.Е. Расчет шин радиальной конструкции как трехслойных ортотропных оболочек вращения.//Расчеты на прочность. Вып.30. М.: "Машиностроение", 1989, с.40-47 (Belkin A.E., Analysis of radial tires as three-layered orthotropic shells of rotation. // Strength Analyses. Iss. 30. Moscow, "Mashinostroenie'', 1989, pp. 40-47).

56. Белкин А.Е., Чернецов А.А. Методика расчета напряженно-деформированного состояния легковых радиальных шин по нелинейной теории трехслойных оболочек. // Вестник МГТУ. Серия "Машиностроение". М.: МГТУ, 1993, № 2, с.113-125. (Belkin A.E., Tchernetsov A.A., Strategy of analysing the stress-strain state of radial car tire by the non-linear theory of three-layered shells.// Vestnik MGTU. Ser."Mashinostroenie'', Moscow, 1993, No. 2, pp.113-125).

57. Белкин А.Е., Уляшкин А.В. Анализ напряженно-деформированного состояния резинокордных оболочек вращения с учетом эффектов перекрестного армирования. // Изв. вузов. Машиностроение, 1991, № 10-12, с. 37-42. (Belkin A.A., Uliashkin A.V., Analysis of the stress-strain state in rubber-cord shells of rotation considering the cross-wise reinforcement effect.// Izv. Vuzov. Machinostroenie , 1991, No. 10-12, pp. 37-42).

58. Белкин А.Е., Нарская Н.Л. Динамический контакт шины как вязкоупругой оболочки с опорной поверхностью при стационарном качении. // Каучук и резина, 1996, № 1, с.23-26. (Belkin A.E., Narskaya N.L., Dynamical contact of the tire as a visco-elastic shell with a bearing

surface in the state of stationary rolling. // Kautchuk i rezina, 1996, No. 1, pp.23-26).

59. Ridha R.A., Computation of stresses, strains and deformations of tires. // Rubber chemistry and technology. 1980, No. 4, pp. 849-902.

60. Shoemaker P., Tire engineering by finite element modelling. // SAE 840065, 1984, p.12.

61. Rothert H., Gebbeken N., Jagusch J., Kaliske M., Recent developments in the numerical tire analysis. // International Rubber Conference. Moscow, 1994. Vol.1, pp. 246-252.

62. Cescotto S., Fonder G., A finite element approach for large strains of nearly incompressible rubber-like materials. // Int. J. Solids Structures. 1978, No. 15, pp. 589-605.

63. Ishihara K., Development of three-dimensional membrane element for the finite element analysis of tires. // Tire science and technology, 1992, No. 1, pp. 23-36.

64. Simo J., Taylor R., Quasi-incompressible finite elasticity in principal stretches. // Computer methods in applied mechanics and engineering. 1991, No. 85, pp. 273-310.

65. Feng K., Statische Berechnung des Guertelreifens unter besonderer Beruecksichtigung der kordverstaerkten Lagen. Diss. TU Berlin 1955, Fortschr.-Ber. VDI Reihe 12, Nr.258, VDI- Verlag, Duesseldorf 1995, 150 S.

66. Faria L.O., Oden J.T.,Yavari B.,Tworrydlo W.W., Bass J.M., Becker E.B., Tire modelling by finite elements. // Tire science and technology. 1992, No. 1, pp. 33-56.

67. Белкин А.Е., Беликов А.Ю., Нарская Н.Л., Уляшкин А.В Элементы автоматизированного проектирования и расчет напряженного состояния радиальных шин. // Каучук и резина, 1993, № 2, с.11-14. (Belkin A.E., Belikov A.Yu.,Narskaya N.L., Ulyashkin A.V., Elements of automated design and analysis of the stress-strain state of radial tires. // Kautchuk i rezina. 1993, No.2, pp. 11-14).

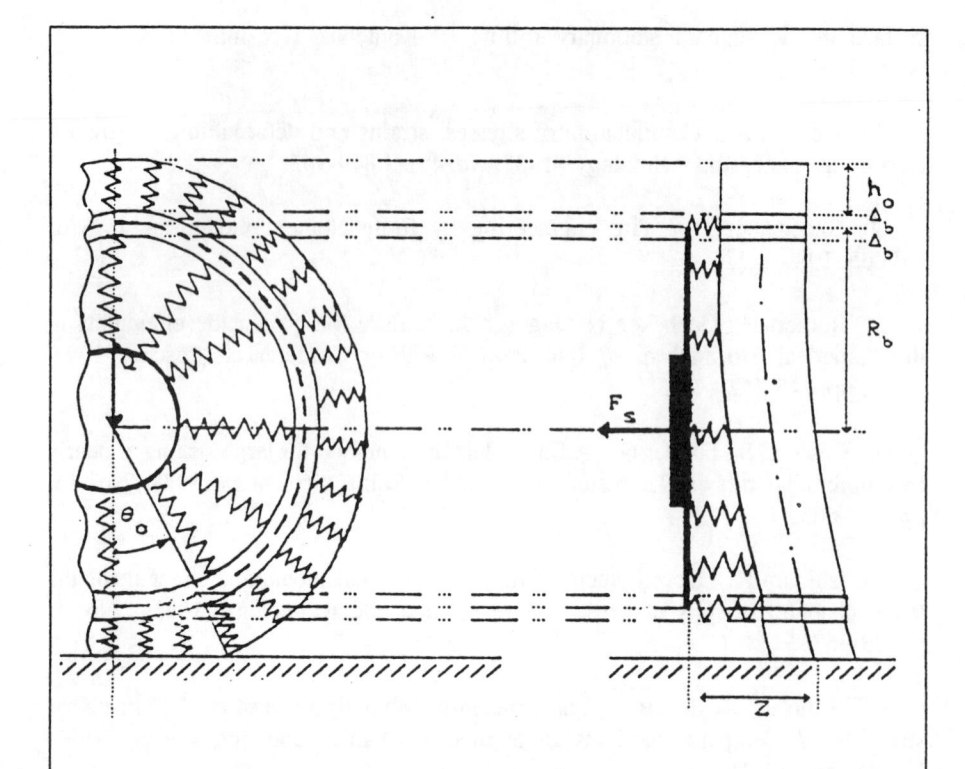

Fig.1 Model ring with elastic base and external elastic layer imitating tread - analogous to the E. Fiala model.

Q - radial load,

F_s - lateral load,

R_b - radius of a working ring,

Δ_b - half of thickness of a working ring,

h_o - tread thickness,

Z - lateral displacement.

PNEUMATIC TIRE MECHANICS 267

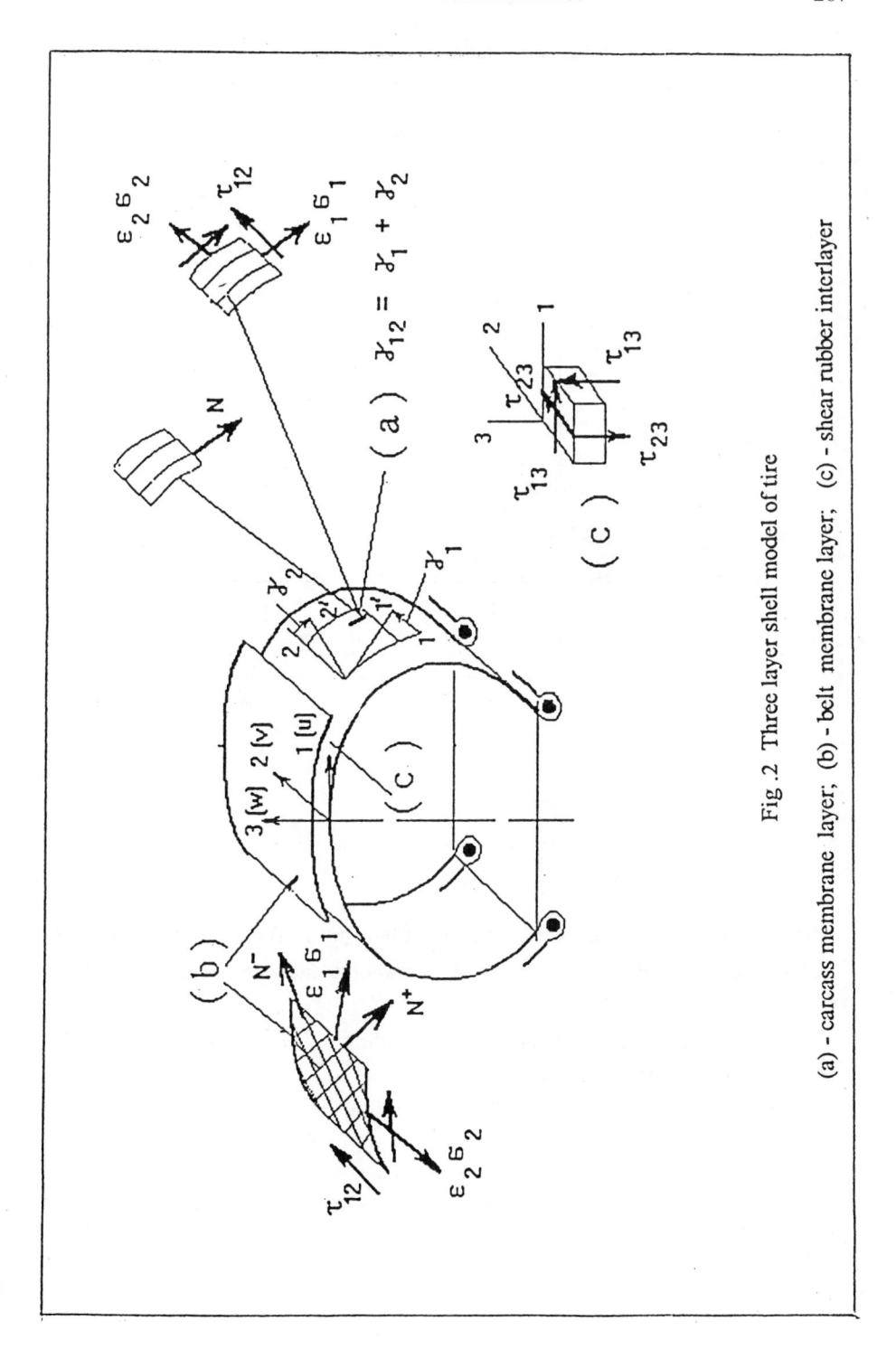

Fig. 2 Three layer shell model of tire

(a) - carcass membrane layer; (b) - belt membrane layer; (c) - shear rubber interlayer

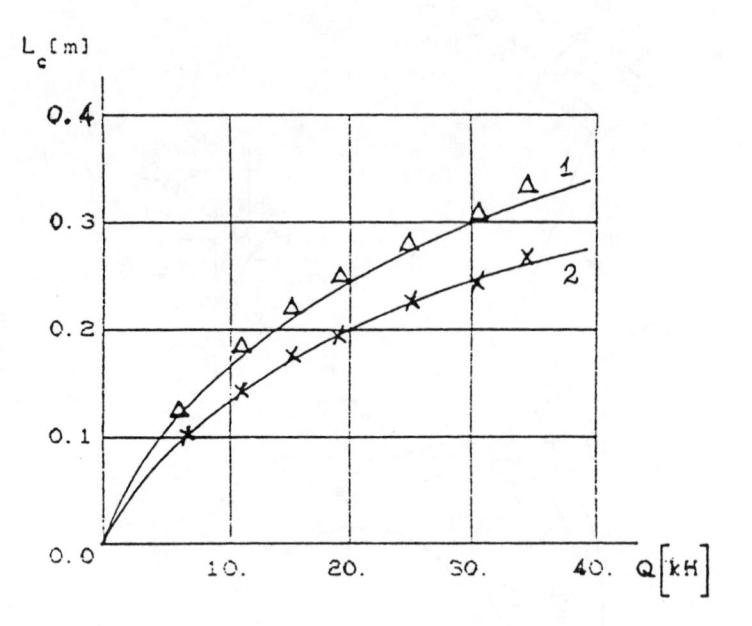

Fig. 3 Dependence of averaged contact length L_c on normal load Q on a wheel of 260-508R tire when loaded on plane surface and test durm with standard radius R_d = 0.796 m.

Solid lines - calculation curves for tire loading on plane surface (1) and external surface of a test durm (2).

Δ - experimental dots for loading on surface,

$\times$ - experimental dots for loading on a test durm.

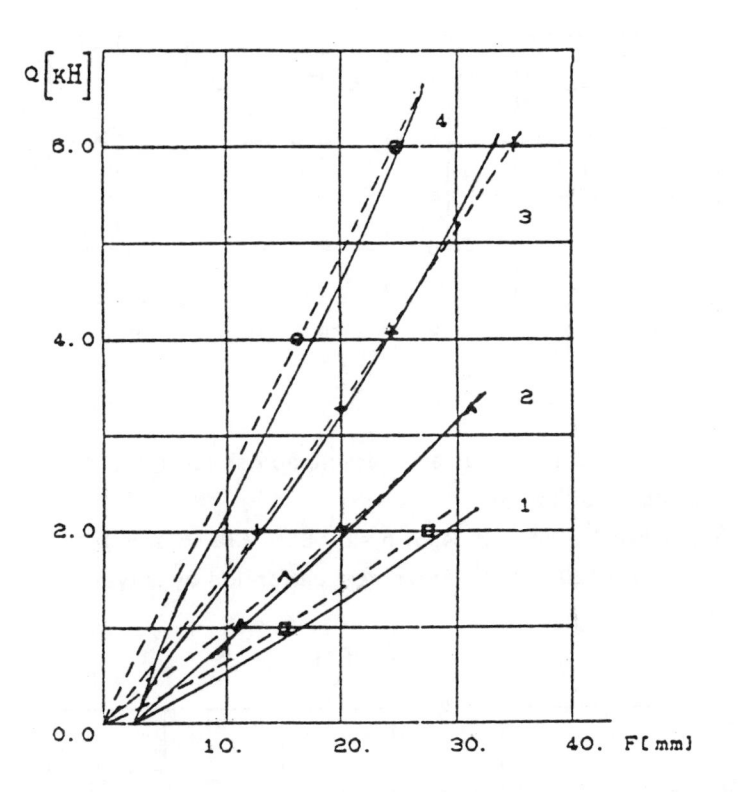

Fig. 4 Load-deformation characteristics under various inner pressures for passenger tire 175/70R13 mod. ИН-251 with single ply carcass 22B and 2-plies belt s.c. 4Л22.

Q - vertical load   F - deflection of a tire
1 - p = 0.05 MPa, 2 - p = 0.1 MPa, 3 - p = 0.2 MPa, 4 - p = 0.3 MPa
    ▫ - experiment with p = 0.05 MPa,
    ────── - calculation curves
    ʌ - experiment with p = 0.1 MPa,
    ✗ - experiment with p = 0.2 MPa,
    ○ - experiment with p = 0.3 MPa.
(experimental data obtained by Melnikov V.Ya.)

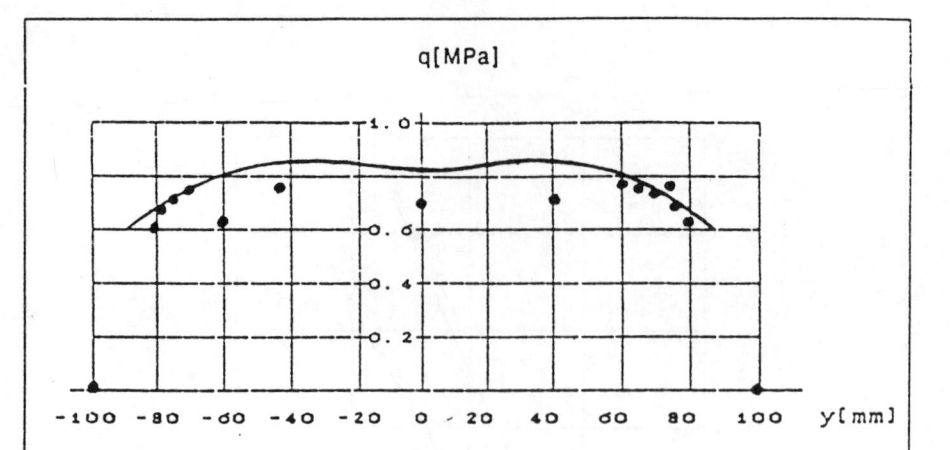

Fig. 5 Distribution of contact pressure across footprint in the middle of contact

Tire 260-508R mod. ИН-142Б, p = 0.5886 MPa (6 atm.),
Q = 20210 N (2060 kgf) (experiment performed by Mayer A. Z.)

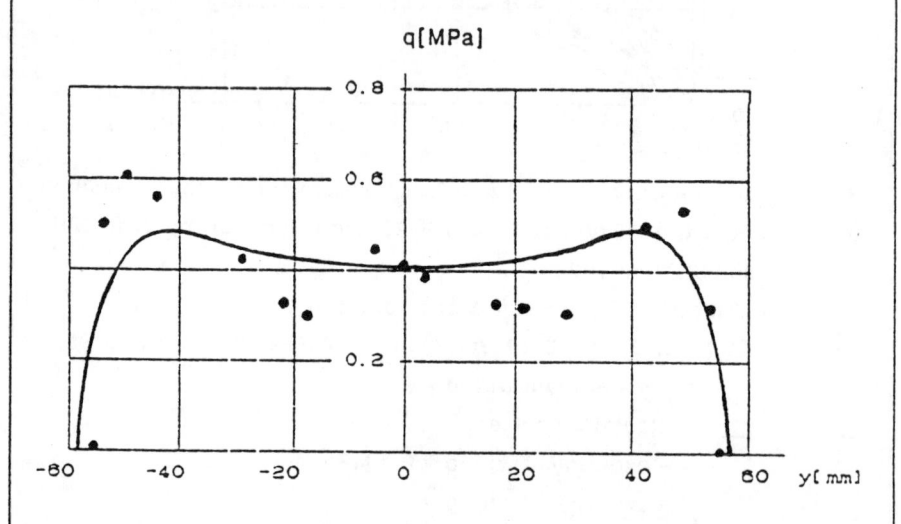

Tire 175/70R13 mod. ИН-251 with single-ply carcass 22В
and 2-plies belt s.c. 4Л22. p = 0.1962 MPa (2 atm.),
Q = 3973 N (405 kgf) (Experiment performed by Komarov S. N.).
■ - experimental values; solid lines - calculating curves.

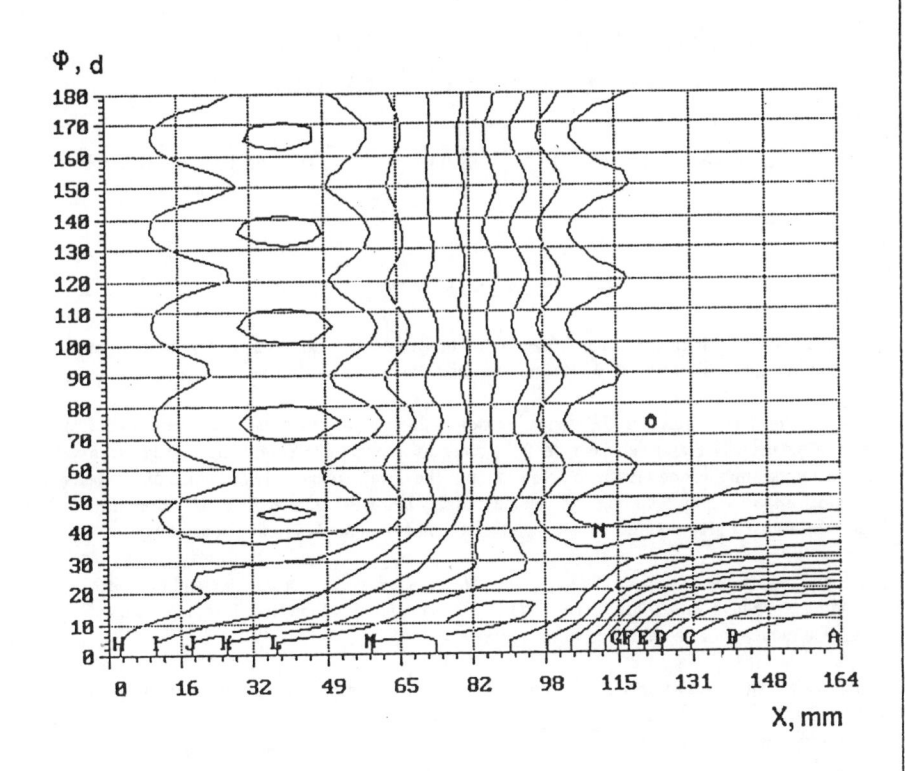

Fig. 6 Tire side wall waves caused by belt waves at rolling speed 200 km/h

x - meridianal coordinate measured from bead to top,
φ - angle of tire rotation measured from contact exit to contact entrance,
A,B,... - level lines for function of carcass normal displacement W, mm
("+" corresponds to outside direction)
A: -20; B: -19; C: -17; D: -15; E: -13; F: -11; G: -9;
H: -7; I: -6; J: -4; K: -2; L: -1; M: +1; N: +2; O: +4

TYDEX Workshop: Standardisation of Data Exchange in Tyre Testing and Tyre Modelling

J.J.M. VAN OOSTEN, H.-J. UNRAU, A. RIEDEL and E. BAKKER

ABSTRACT

As a result of the '1st International Colloquium on Tyre Models for Vehicle Dynamics Analysis' in 1991, the international 'TYDEX Workshop' working group was established.

This workshop concentrated on the standardisation of the exchange of tyre measurement data and the interface between tyre and vehicle models in order to improve the communication between vehicle manufacturers, suppliers and research organisations. The development and knowledge of tyre behaviour is of great importance to both tyre and vehicle industry and will be intensified. Therefore the TYDEX Workshop showed great interest from all parties to come to some kind of standardisation.

In the two expert groups - one focused on *Tyre Measurements - Tyre Modelling* and the other on *Tyre Modelling - Vehicle Modelling* - the TYDEX-Format and the Standard Tyre Interface have been developed, which will be explained in this paper.

Furthermore a short overview of the European TIME project aiming at a standard tyre testing procedure will be given, which is reliable and consistent with realistic driving conditions. Elaborating standard testing procedures is one of the important consequences from the TYDEX Workshop.

1. INTRODUCTION

In Vehicle Dynamics studies the description of the interaction forces between the tyres and road is one of the most important items. In order to reduce development time and costs, computer simulation of vehicle dynamics becomes more and more important in an early phase of the development process of new vehicles or improvement of existing vehicles. Requirements to the accuracy of the description of the interaction forces between tyre and road have increased considerably, causing a need for more accurate (often also more complex) tyre models. Likewise more accurate and complex testing is required to enable the desired Vehicle Dynamics analysis in simulation environments.

In general vehicle manufacturers do not have the objective to specialize on tyre modelling and tyre testing, but like to have access to appropriate tyre test data and tyre models in order to fulfil their needs in vehicle modelling. Tyre manufacturers, universities and supplementary research institutes support vehicle manufacturers with knowledge on tyre modelling and supply their customers with tyre test data. The communication between receiving companies and suppliers appears to be difficult and very time consuming.

As a result of the "1st International Colloquium on Tyre Models for Vehicle Dynamics Analysis" in 1991, the international TYDEX Workshop was established aiming at the improvement of the TYre Data and modelling EXchange between automotive companies and suppliers.

This presentation will explain the background, objectives and results of the TYDEX Workshop and will give an overview of the European TIME project, which objectives are in line with the TYDEX ideas.

2. OBJECTIVES AND APPROACH

With the general objective in mind, which is the improvement of the communication between companies and suppliers in the field of tyre testing and tyre modelling, the TYDEX Workshop defined two expert groups with the following goals:
- The first expert group (*Tyre Measurements - Tyre Modelling*) has as its main goal to specify an interface between tyre measurements and tyre models. The interface could be described as a definition of a method or format to describe tyre measurement data in such a way that it contains all necessary items to fit tyre models to the underlying data. The format shall also allow for a description of the test conditions and for a transmission of tyre model parameters.
- The second expert group (*Tyre Modelling - Vehicle Modelling*) has as its main goal to specify a general interface between tyre models and vehicle models. With the aid of this interface it is possible to exchange a tyre model by another one without changing the vehicle model.

Defining an interface between tyre measurements and tyre models requires information on the type of desired vehicle simulations and their objectives, in order to know the application area of the tyre models. Table 1 shows a list of typical driving manoeuvres [1] and their reflection on the requirements on the tyre models in terms of bandwidth for in-plane and out-of-plane behaviour. This example illustrates that different tyre models are required with different level of complexity applicable for the different driving manoeuvres, which confirms the need for a standard interface between vehicle models and tyre models.

Tab. 1. Possible driving manoeuvres and their requirements on the tyre model.

driving manoeuvres	road surface	F_z	F_x	F_y	M_z
Standstill					
hydropuls	even	< 30			
steering	even				stat.
starting	even	< 5	< 5		
Straight line					
constant speed	even	stat.	stat.		
	uneven rough	< 30	< 30		
accelerating/braking	even	< 5	< 5		
	uneven rough	< 30	< 30		
jolting	even		< 5		
shimmy	even			< 30	< 30
longitudinal ruts	even	stat.		stat.	stat.
cross-wind section	even	stat.		< 5	< 5
µ-step	even	stat.	< 5	< 5	< 5
Curve driving					
step steer input	even	< 5		< 5	< 5
steady state circle	even	stat.	stat.	stat.	stat.
	uneven rough	< 30		< 30	< 30
power off/accel/braking	even	< 5	< 5	< 5	< 5
	uneven	< 30	< 30	< 30	< 30
sinusoidal steer input	even	< 5	< 5	< 5	< 5
µ-step	even	stat.	< 5	< 5	< 5

F_z Vertical wheel load
F_x Longitudinal force
F_y Lateral force
M_z Self aligning torque
stat.   static/stationary

From theoretical point of view the borderline between the vehicle model and tyre model should not exclude any tyre model, but for a practical interface one must limit the number of in- and output variables. The TYDEX Workshop tried to find an optimum between practical use with a maximum of application possibilities. Furthermore the information about the road required by the tyre model depends on the tyre model. A general purpose interface between tyre model and vehicle model implies also a definition of the interface between tyre model and road model.

In Section 3.2 the proposed solution for a standard interface, developed by the TYDEX Workshops, is explained.

The tyre parameters used in tyre models should be derived from tyre measurement data. Tyre test procedures depend largely on the type of application of the tyre model within the vehicle model. Not only differences in tyre test procedures but also in tyre test devices can cause differences in tyre test results. Even for a well known tyre property like cornering stiffness rather large inconsistencies can exist. In Figure 1 [2] results are shown of a survey by a group of European vehicle manufacturers, tyre suppliers and research institutes on the comparability of test devices with respect to cornering stiffness.

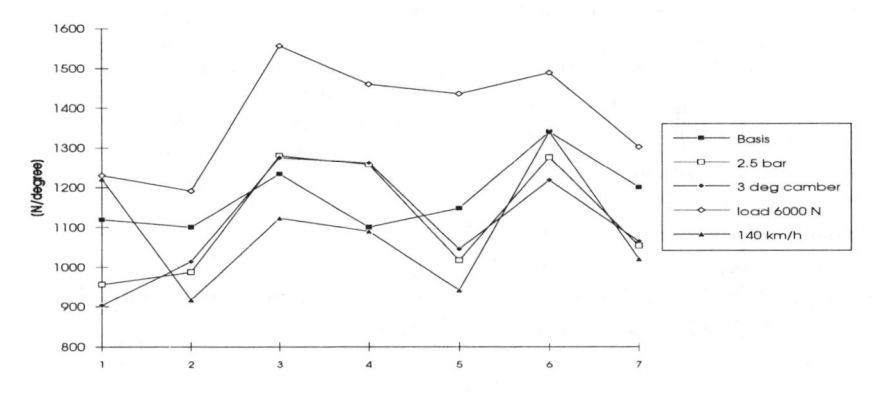

Fig. 1. Comparison of cornering stiffness determined with 7 different test devices. Basis: 60 km/h, 2.0 bar, no camber, 4 kN vertical load.

Although the measurement sequence was prescribed in detail, the differences (between maximum and minimum measured values for each configuration) vary from 21% up to 46%!

At present a general tyre test procedure resulting in tyre test data valid for any tyre model is a theoretical dream. The first step towards standardization is the definition of a standard file format, in which tyre test data can be presented and transferred between companies. A first proposal for a standardized format by the TYDEX Workshop will be discussed in Section 3.1.

As a result of the discussions within TYDEX about guidelines for tyre procedures aiming at tyre test results, which are consistent with realistic driving conditions, the European project TIME was started. In Chapter 4 the objectives and a short overview of the TIME activities are explained.

The TYDEX Workshop is supported by a number of vehicle manufacturers (cars and trucks), tyre manufacturers, other suppliers and research institutes. Representatives of these companies and institutes formed and participated in the expert groups. Table 2 lists the companies who support the Standard Tyre Interface (STI) and the TYDEX-Format [3]. At present the Standard Tyre Interface [4] is already available in multi-body simulation packages like ADAMS, DADS, MADYMO, FADIS and Mesaverde. Examples of supporting the TYDEX-Format are DELFT-TYRE, IPG-TIRE, the TIME project and the SWIFT project.

Tab. 2. Companies supporting the TYDEX Workshops standards.

BMW AG	Germany
Robert Bosch GmbH	Germany
Centro Ricerche FIAT	Italy
Continental AG	Germany
Daimler-Benz	Germany
FHT Esslingen	Germany
Ford Werke AG	Germany
Goodyear S.A.	Luxembourg
IPG GmbH	Germany
Mercedes Benz AG	Germany
Michelin	France
NedCar	The Netherlands
Porsche AG	Germany
PSA	France
Steyr-Daimler-Puch	Austria
Volvo Car Corp.	Sweden
Volvo Truck Corp.	Sweden
TNO	The Netherlands
Toyota (TCL)	Japan
Univ. of Berlin	Germany
Univ. of Delft	The Netherlands
Univ. of Karlsruhe	Germany
Univ. of Vienna	Austria
Volkswagen AG	Germany

3. RESULTS OF THE TYDEX WORKSHOPS

3.1 Standardized Interface between Tyre Measurements and Tyre Models

As already mentioned the TYDEX-Format was developed and unified to make the tyre measurement data exchange easier. The structure of the data files, abbreviations for the parameters, axis systems and definitions are fixed by the TYDEX-Format Manual.

The Structure of the Data Files

The TYDEX-Format was developed for the filing of tyre measurement data, but filing of tyre model data is also possible. An example of the file structure can be seen in Figure 2. The data file is divided in different sections having special keywords as a header.

The data file may consist of up to 12 keywords:

**HEADER	**MODELPARAMETERS
**COMMENTS	**MODELCOEFFICIENTS
**CONSTANTS	**MODELCHANNELS
**MEASURCHANNELS	**MODELOUTPUT
**MEASURDATA	**MODELEND
**MODELDEFINITION	**END

In connection with these keywords essentially the following has to be considered:
- All keywords start in column 1 with 2 asterisks.
- The keywords **HEADER and **END must be put in every file. Other keywords are optional.
- If sections concerning the tyre model are used, the keywords **MODELDEFINITION and **MODELEND must appear.

The meaning of the different sections beginning with above mentioned keywords is explained in the following (see Figure 2).

The first section is marked with the keyword **HEADER. This section contains information to identify the measurement or model data. This is suitable for the file management. The section consists of only 5 parameters. Columns 1 to 8 give an abbreviation for the parameters which are shown in columns 11 to 39. The abbreviations are fixed in the TYDEX-Format Manual and must not be changed as they clearly identify the parameter. From column 51 on information on the parameters follows.

The next section marked by the keyword **COMMENTS contains information which cannot be put into other blocks or which is additional user information. Lines after this keyword have free format and an unlimited number of comment lines is allowed. This section ends before the next keyword starting with two asterisks (**).

In the section **CONSTANTS measurement data, model data and character information are given which are constant for the whole measurement or for the whole model data set given in the respective file. Usually, all information about tyre, rim and test conditions is given in this section. Numerical values given in the

```
Column
1         11                                    41        51          61       71     80
**HEADER
RELEASE   Release of TYDEX-Format                         1.3
MEASID    Measurement ID                                  05039ABC
SUPPLIER  Data Supplier                                   MICHELIN
DATE      Date                                            20/02/97
CLCKTIME  Clocktime                                       09:50
**COMMENTS
This section can be used to put in any comment. The format is free. Blank lines can be
used.
**CONSTANTS
NOMWIDTH  Nominal Section Width of Tyre        mm         185
ASPRATIO  Nominal Aspect Ratio                 %          70
TYSTRUCT  Tyre Structure                                  radial
RIMDIAME  Nominal Rim Diameter                 inch       13
INFLPRES  Inflation Pressure                   bar        2.5
INCLANGL  Inclination Angle                    deg        -3
AMBITEMP  Ambient Temperature                  deg C      25
**MEASURCHANNELS
MEASNUMB  Measurement Point No.
RUNTIME   Running Time                         s          0.01
FZH       Vertical Force                       kN         0.001
SLIPANGL  Slip Angle                           deg
LONGSLIP  Longitudinal Slip                    %          100.
FYH       Lateral Force                        N
FX        Longitudinal Force                   N
MZH       Aligning Moment                      Nm
**MEASURDATA
1         0.        4000        0.00      0.00      0.        0.       0.
2         1.        4000        0.02     -0.01     -200      -100.     20.
3         2.        4100        0.04     -0,02     -400.      0.       40.
**MODELDEFINITION
MODELREF  Cornering Stiffnesses                 DZ/GZ     MICHELIN
**MODELCHANNELS
CSFYH     Cornering Stiff.Lat.Force            N/deg      out
CSMZH     Cornering Stiff.Align.Moment         Nm/deg     out
FZH       Vertical Force                       N          in       500      10000
INFLPRES  Inflation Pressure                   bar        in       1.8      2.8
**MODELOUTPUT
221       5.869                  500         1.8
1255      37.42                  3000        1.8
3187      104.55                 10000       1.8
...       ...                    ...         ...
**MODELEND
**END
```

Fig. 2. The structure of the TYDEX file.

**CONSTANTS section instead of the **MEASURCHANNELS or **MODELOUTPUT sections save disk space, but they can be used as if they were measured or modelled. Comparable to section **HEADER the names of the constants are defined in the TYDEX-Format Manual and the abbreviations are fixed to be used for an automated postprocessing system. The abbreviation is in columns 1 to 8, from column 11 on the name is mentioned, from column 41 on the unit and starting with column 51 the value is shown.

The section **MEASURCHANNELS contains only measured (time dependent) data. The section gives the abbreviations (columns 1 to 8), the names of the variables (starting at column 11), the units (starting at column 41) and (if necessary) factors for the conversion into physical units (from column 51 on). Here also the abbreviations of the variables are fixed and have a definite meaning to be used for an automated postprocessing system.

In the section **MEASURDATA all measured data is given sample by sample. The measured data is listed in the same order as in the **MEASURCHANNELS section. Every new measurement sample starts in a new line. The number of values per line corresponds to the number of channels in the **MEASURCHANNELS section.

For the filing of tyre model data 6 keywords are available which, depending on the used tyre model, do not all have to appear. If sections concerning the tyre model are used they have to be pooled in one block. In the beginning of this block the section **MODELDEFINITION appears, in which the used tyre model is defined and a model reference code is shown to avoid mistakes. In the end of the block **MODELEND appears so that it is clear that all sections between **MODELDEFINITION and **MODELEND are related to one tyre model.

In section **MODELPARAMETERS model parameters appear in a formatted form. These parameters have a physical meaning. Section **MODELCOEFFICIENTS has free format and is intended for the filing of coefficients without physical meaning. Sections **MODELCHANNELS and **MODELOUTPUT are arranged corresponding to the sections **MEASURCHANNELS and **MEASURDATA. Section **MODELCHANNELS determines the sequence of the generated model data appearing in columns in the section **MODELOUTPUT. More details about the contents of model specific sections may be found in the TYDEX-Format Manual.

At the end of each TYDEX data file there appears the keyword **END. This keyword must absolutely occur in every TYDEX data file to recognize loss of data. Any information after this line is ignored.

By above described structure of data files (more details in the TYDEX-Format Manual) it is possible to put in tyre measurement data without problems, without additional co-ordination between the data supplier and data processor. As mentioned above the meaning and the designation of the parameters must be known exactly to ensure a correct data processing. This has been achieved by standardization of the parameters designation.

The Standardization of Parameter Designations
To enable data suppliers and data users to have same terms for measured quantities as well as for constants a great number of terms have been fixed in the TYDEX Format Manual. Whenever possible an already existing norm (e.g. ISO 8855, DIN 70000, ISO 3911, ETRTO-Standard or SAE J2047) was adopted. In Figure 3 a part of the TYDEX parameter list is shown. In the first column you find the above mentioned abbreviations for the parameters, which are necessary for the automatic data processing. In the second column there are the parameter terms in English, German and French. Finally there follow the units and examples. The complete TYDEX list has 13 pages so that all usual parameters are contained.

The Axis Systems
To avoid any mistakes not only terms have to be fixed but also the axis systems have to be defined exactly. According to the circumstances it is wise to use different definitions. Therefore you find in the TYDEX-Format Manual three axis system definitions: the C-axis system, the H-axis system and the W-axis system. The definition's basis is the axis system according to ISO 8855 which, in principle, corresponds to the W-axis system (see Figure 4). Contrary to ISO 8855 in the W-axis system also a sloped ground is permissible. The C- and H-axis systems are comparable in the directions of co-ordinate axes but have their origin in the centre of the WHEEL.

In Figure 4 the slip angle is exactly defined. Correspondingly the longitudinal slip is exactly determined by a formula in the TYDEX-Format Manual.

Experiences with the TYDEX-Format
Meanwhile the TYDEX-Format is used by several European companies and facilitated the data processing essentially. For example the TYDEX-Format is used at the DELFT-TYRE and at the IPG-TIRE. This makes a relatively simple application of these models possible. Above all this data format proved to be good in performing the TIME project (see Chapter 4) where data in large quantities has to be transferred between the participating companies and institutions. Data transfer without problems is requirement for this project and can only be realized by a uniform data format as the TYDEX-Format.

Abbrev. (8 char.)	Parameter (max. 29 characters)	Unit	Example	CR

D. Ambient Conditions

TRCKTEMP	temperature of track Fahrbahntemperatur température de la chaussée	deg C	22	
CURVTRSF	curvature of track surface *(the curvature of the track surface is negative for internal drum test rigs)* Krümmung der Fahrbahnoberfl. courbure de surface chauss.	1/m	-0.526	
TRCKSURF	surface of track Fahrbahnoberfläche surface de la chaussée	-	safetywalk	CR
TRCKCOND	condition of track surface Zustand der Fahrbahnoberfl. état de surface chauss.	-	wet	CR
WATDEPTH	waterfilm depth Wasserfilmhöhe hauteur de la couche d'eau	mm	3.0	
AMBITEMP	ambient temperature *(outside temperature)* Umgebungstemperatur température exterieure	deg C	20	
HUMIDITY	humidity Luftfeuchtigkeit humidité de l'air	%	60	

△ E. Wheel Forces and Moments
(forces and moments from tyre to rim)

| △ ♦ | FX | longitudinal force
(longitudinal force at WHEEL, $F_{XC} = F_{XH} = F_{XW}$)
Umfangskraft
force longitudinale | N | 2000 | |
| | FYC | lateral force
(lateral force at WHEEL in C-axis system)
Seitenkraft
force transversale | N | 1500 | |

Fig. 3.   Example from the TYDEX parameter list.

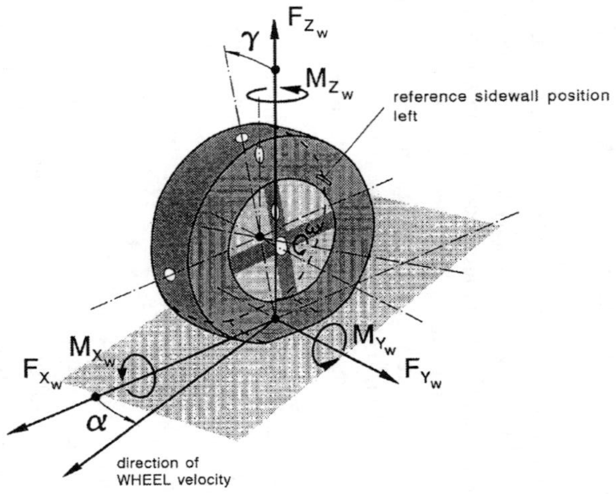

Fig. 4.   The TYDEX W-Axis system, the forces and moments act from the tyre onto the rim.

3.2 Standardized Interface between Tyre Models and Vehicle Models

The advantages of a standardized interface between tyre models and vehicle models are obvious. The industrial user can easily switch between different tyre models, as in the ideal case no modifications are necessary on the side of the vehicle model. This allows the exchange of tools in project groups, and it promotes competition between the vendors of tyre models. The advantage for all those who are offering tyre models is that their models can be integrated into different simulation environments without expensive adaptation, increasing their sales prospects.

If talking of 'interface', we should be aware that interface does not only mean the formal program interface between tyre model and vehicle model, i.e. the parameter list, but also the definition of physical units, coordinate systems and of all assumptions concerning the behaviour of the tyre model.

The physical units, which have been chosen, are SI-units. The question, which coordinate systems are most appropriate, was an intensively disputed item, which has been modified several times. At the end it has been fixed to work with an earth fixed coordinate system and with a wheel carrier coordinate system, where the transformation matrix between these two systems has to be provided by the vehicle model.

The interface has been defined as universally valid as possible. Thus only physically unambiguous parameters are exchanged, not those which can be defined in different ways. Furthermore, it has been taken into account that the tyre model can obtain dynamical degrees of freedom which have to be integrated by the solver together with the state variables of the vehicle model. In addition, no restrictions have been made with respect to road properties, and it has been considered, that a vehicle can be equipped with an arbitrary number of different tyres.

In this publication the interface shall not be described in detail. This is the aim of the documentation which can be obtained at the contact addresses given at the end of the article. The goal is rather to give a basic understanding of the definitions. Formally, the interface is the parameter list of a FORTRAN77 subroutine, as FORTRAN is at the moment and also during the next years the standard programming language for external and user written subroutines in multi body simulation environments. It has been decided to have a single precision version as well as a double precision version, which differ with respect to their names and to the type of the parameters, but are, in principle, identical.

Part of the input to the tyre model are some control parameters, which state, if the actual call to the model is during initialization, during normal simulation or if it is the final call to the model. Another parameter which is important for dynamic tyre models is the simulation time.

The next group of input parameters contains information about the kinematic state of the rim. Transferred are values defining displacement, orientation and velocity of the rim in an unambiguous way. Using these parameters, the tyre model can calculate internally such parameters as the longitudinal slip, which can be defined in slightly different ways according to the tyre model.

Next group of input parameters are the state variables of the tyre model which have been integrated by the solver.

Furthermore, the tyre parameters of the actually called tyre are given to the tyre model. Here the interface is built in such a way, that the parameters can be transferred using the interface or can be read in from a file. Finally follows a group of parameters which have to be transferred to the road model.

Primary output of the tyre model are the forces onto the rim at the center of the wheel and the torques applied from the tyre onto the rim. With this definition all problems have been avoided which would occur if referencing e.g. a 'contact point' or a 'contact area', which are very specific for different tyre models.

Next group of output parameters are the state variables of the tyre model, which have to be transferred to the solver.

In an info-array, any information about internal variables of the tyre model can be put out to be stored in the result file. A first part of this info-array is reserved for arbitrary values like slip angle, longitudinal slip or camber angle. The second part of the array is up to users needs.

Finally, the parameter list contains some work arrays for saving internal values which are needed when the same tyre is called the next time, and an error parameter.

Beside the interface between tyre model and vehicle model, also a proposal for an interface between tyre model and road model has been worked out. Basic assumption is that the road model or a special module of the road model is called by the tyre model. For a point (x,y), given by the tyre model, the road model has to deliver different information about road properties like the altitude $z(x,y)$ and their derivatives at this point.

Both interfaces have been used in daily work by different members of the TYDEX group just for a longer period, and all experiences have been incorporated into the definitions. Thus the interface between tyre model and vehicle model has reached a very mature and stable state. The interface between tyre model and road model should be seen as a well thought-out proposal, which, at the moment, might not cover all possible needs, but which should be used to gain further experiences.

The present publication shall encourage all colleagues all over the world to use these interfaces in their own simulation environments. Detailed specifications of the interfaces can be obtained at the contact points given in the Appendix.

4. TIME: STANDARD TYRE TEST PROCEDURE

Background
In order to develop vehicles which have maximum active safety, car manufacturers have an increasing need for information about the so-called force and moment properties of tyres. Vehicle manufacturers, tyre suppliers and automotive research organisations have advanced test equipment to measure the forces between a tyre and a road surface under a variety of loading conditions. However, because of the large differences in test equipment and the measurement procedures used, the consistency of the tyre force and moment properties determined with the different test devices is a major problem. As mentioned in Section 2, for an arbitrary tyre property like cornering stiffness one can find large differences when comparing this parameter derived from different tyre test devices. Differences can be caused by:
- differences between the tyre test devices, like surface friction, surface curvature, ambient conditions, etc;
- inconsistencies between the measurement signal acquisition and data postprocessing;
- the measurement procedure itself including warming up.

The differences between the tyre test devices must be seen as boundary conditions, which are difficult to change. One could look for a correlation between the devices by investigating the effect of the differences. But knowledge about these effects and standardization on signal processing and the measurement procedures would improve the comparability of tyre test data of various test devices considerably. Not forgetting that the starting point for tyre testing guidelines should always be the real life conditions of a tyre during use under a vehicle.

In the TYDEX Workshop a discussion started on the development of guidelines for tyre testing, resulting in the initiative to start a common research project called TIME (TIre MEasurements, forces and moments) within the scope of the Standards, Measurements & Testing programme of the EC (DG XII) [5]. The consortium consists of the six major European tyre manufactures, three vehicle manufactures, four research organisations and one commercial automotive company.

Objectives

The TIME project aims at the development of a common tyre measurement procedure that will be reliable and consistent with realistic driving conditions.
The research will concern passenger car tyres under steady state cornering conditions (from low to high lateral acceleration levels up to the safety limits of the vehicle).
The general objective will include following items:
- To set up a correlation between each of the main different tyre test devices and the reference test conditions;
- To explain the differences in measurement results from different tyre testing devices;
- To identify the conditions a tyre experiences in real life use;
- To develop running-in and warming-up procedures for each test device before testing a tyre;
- To establish the requirements for a tyre testing laboratory (as a data supplier) in terms of measurements quality insurance, testing procedure, data acquisition and processing;
- To specify the validity range of tyre data coming from a given procedure with respect to actual driving conditions on a vehicle.

Approach

The activities of the project have been defined in five major workpackages.

In workpackage I, the differences between results obtained with the main European tyre test devices will be investigated. Six different tyre types will be tested at each test device using exactly (if possible) the same measurement procedure. Although the test programme should be limited in order to compare the most important tyre properties, care must be paid to:
- the running-in and warming up procedures;
- the sequence of the tests;
- the selection of tyre types and sizes;
- the number of tyres required to execute the test programme;
- the definition of variables to be measured;
- the description for the processing of the data.

The eleven test devices involved concern the major European internal (1) and external drums (6), flat-track machines (2) and road test devices (2).

Workpackage I will be followed by a parameter sensitivity study in Workpackage II in order to explain the differences between the results obtained with the different devices. The influence of parameters like surface friction, test speed, tyre inflation pressure, ambient temperature, tyre type, etc. are investigated in order to derive a first correlation between the different measurement devices.

One of the most important items is the establishment of reference measurements for force and moment properties under realistic vehicle driving conditions in Workpackage III. Three different types of passenger cars will be used to record time histories of tyre forces and moments under different conditions like track surface, ambient temperature, etc. Using the time histories four tyre test devices are involved to develop the reference measurements. For a most accurate correlation, tests with the vehicles and tyre test devices will be carried out under most similar conditions.

Taking the reference measurements as a basis a (draft) common tyre test procedure will be designed in Workpackage IV. Knowing the specifications of the different tyres an attempt is made to define one procedure for all test devices.

In the first part of Workpackage V the draft procedure is tested at all tyre test devices, evaluated and modified if needed. The second part of the research concerns the validation of the tyre measurement procedure in a vehicle dynamics simulation environment because such applications are the main reason to look for realistic and consistent tyre measurement data.

Present status
The TIME project started in February 1996 and will conclude with a standard tyre measurement procedure in January 1999. At present the measurements of Workpackage I have been performed and are evaluated resulting in a programme for Workpackage II.

5. SUMMARY AND CONCLUSIONS

The TYDEX Workshop started in 1991 on the standardisation of data exchange in tyre testing and tyre modelling. Taking into account that a wide range of companies and institutes participated and cooperated intensively during five years, one can conclude that there is a need for improvement on the communication between vehicle manufacturers, suppliers and research organisations with respect to tyre modelling and tyre testing.

Now the TYDEX Workshop presented their results, two standard interfaces:
- an interface between vehicle models and tyre models: the *Standard Tyre Interface*;
- an interface between tyre models and tyre measurements: the *TYDEX-Format*.

The STI is already available in the major commercial multi-body simulation software packages; the TYDEX-Format has proven its usefulness in the TIME project and tyre model related software.

Another result of the standardisation efforts in the TYDEX Workshop is the development of a standard tyre measurement procedure within the European *TIME* project.

The participants of the TYDEX Workshop aim at a world wide distribution and use of their standards. A first review and evaluation of the TYDEX standards and the TIME standard tyre measurement procedure is planned in 1999.

Until that time it is planned to sample as many experiences as possible. Then all experiences shall lead to the next releases of the interfaces. All users of the interfaces are cordially invited to influence the further development by sending their remarks to the contact addresses and by taking part in the work of the TYDEX group. They will be informed about the further process and will be invited to the next meeting in the first half of 1999.

REFERENCES

1. J. Zamow,
    Overview of typical driving manoeuvres and the requirements on the tyre model, presented at 2nd TYDEX Workshop, September 1992.

2. J. Zamow,
    Messung des Reifenverhaltens auf unterschiedlichen Prüfständen, VDI Berichte nr. 1224, 1995.

3. H.-J. Unrau, J. Zamow,
    TYDEX-Format, Description and Reference Manual, Release 1.3,
    Initiated by the TYDEX Workshop, 1996.

4. A. Riedel,
    Standardized Interface Tyre Model - Vehicle Model, Release 1.3, TYDEX Workshop, 1996.

5. J.J.M. van Oosten, e.a.,
    TIME, Tire Measurements, Forces and Moments, proposal, Scientific and Technical Content, part A, TNO, Delft, 1995.

Appendix

CONTACT ADDRESSES TYDEX

General:
TNO Road-Vehicles Research Institute
J.J.M. van Oosten
P.O. Box 6033
2600 JA Delft   phone: +31 15 2696420
The Netherlands   fax: +31 15 2697314

TYDEX-Format:
Universität Karlsruhe
Institut für Mkl and Kfz. Bau
H.-J. Unrau
Kaiserstrasse 12
D-76128 Karlsruhe phone: +49 721 6083795
Germany   fax: +49 721 6086051

Standard Tyre Interface:
IPG
A. Riedel
P.O. Box 210522
D-76155 Karlsruhe phone: +49 721 9852013
Germany   fax: +49 721 9852099

On Modeling Contact and Friction
Calculation of Tyre Response on Uneven Roads

CH. OERTEL

ABSTRACT

In this paper, several alternatives in modeling contact and friction within tyre models are discussed. Focusing on existing models for the calculation of ride comfort, the different approaches concerning contact and friction have been used in serveral combinations. These models do not differ very much in results, but in computational effort, which can be influenced by choice of modeling methods. As an example, a tyre model for computation of ride comfort is introduced, in which the model's dynamic characteristics (modes to be modeled) and geometric characteristics (wavelength of road disturbance) are decoupled.

INTRODUCTION

A great number of tyre models have been developed in the past to meet different objectives. In the field of vehicle dynamics, the range of requirements reaches from steady state driving conditions to acustics. Although contact and friction are one of the fundamental tasks in modeling tyre behavior, some models do not need to deal with, for instance the wellknown *magic formula* [1], because an empiric approach is used. All physical models need to deal with contact and friction, at least because of the nonlinear characteristics of tyre forces with respect to the kinematic inputs.

In first part of this paper, some alternatives in modeling contact and friction are investigated by example of existing tyre models. In the second part, a tyre model is presented, which allows the calculation of tyre response on uneven roads, especially discrete obstacles.

Modeling contact and friction has to be seen in relation to

- the desired range of wavelength taken into account for disturbances (steering, breaking) and road unevenness,
- the accuracy of the model concerning tyre dynamics and -of course-
- the amount of computational effort one can spend for calculating tyre forces.

To give a classification, models may be devided by the type of discretisation used.

a) Assuming a wavelength more than ten times the length of the contact area, no discretisation of the structure or the contact area is made. These models use the contactlength, the normal load distribution and the friction coefficients as parameters. Integration over the contact area leads to ordinary differential equations for the calculation of dynamic tyre forces. When sliding in the contact area is neglected, the equations are linear [2], otherwise nonlinear [3].

b) Shorter wavelengths - close to the contact length - demand a discretisation of the contact area. The structure is represented by *rigid body modes* of the belt [4] and the shape of the contact area as well as the normal load distribution are taken from parametric functions [5]. Contact is *simulated* in detail, but not calculated. The equations describing the displacements in the contact area are partial differential equations and friction is used to limit the forces calculated from the displacements.

c) When obstacles on the road are taken into account, a discretisation of the structure is needed. There is no longer a single contact area but two or three smaller areas. The shape of these areas and corresponding normal load distributions have to be results from computation. In some models, the vibrations of tread are integrated to calculate the contact forces, other use the force limitation method.

From class a) to class c), the representation of contact and friction changes from approximation to calculation. Figure 1 gives an illustration of the different classes from two degrees of freedom for the displacement of the contact area to about 400 degrees of freedom for the structure dynamics.

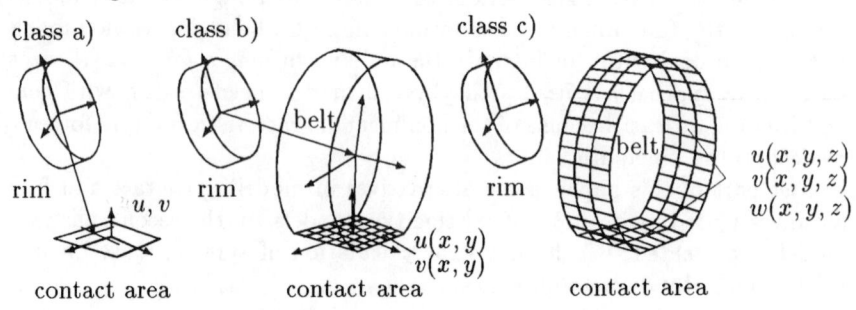

Fig. 1: Classification of tyre models

1. Friction forces in tyre models

On one hand, friction may be understood as a limitation of contact force, expressed by a friction function as ratio between normal- and tangential force with the sliding velocity as independent variable. On the other hand, friction causes selfexcited vibrations, known as stick-slip.

1.1 Simplified Description

In some models, the displacements in the contact area are calculated from partial differential equations that describe the deformation under the assumption of total sticking. For instance - in a model of class b) -

$$V_T \frac{\partial u(x,y,t)}{\partial x} + \frac{\partial u(x,y,t)}{\partial t} = V_{S_x} \qquad (1.1a)$$

$$V_T \frac{\partial v(x,y,t)}{\partial x} + \frac{\partial v(x,y,t)}{\partial t} = V_{S_y} \qquad (1.1b)$$

holds for the deformations with the displacements u, v, the transport velocity V_T and the slip velocities V_{S_x}, V_{S_y} containing the excitations. Within these equations, no limitation of tangential contact forces is made, displacements become large at large slip velocities. To reduce the tangential contact forces, in each point x_i, y_j of the mesh, limitation is done by testing

$$\sqrt{(c_x u(x_i, y_j, t))^2 + (c_y v(x_i, y_j, t))^2} \leq \mu F_N(x_i, y_j, t), \qquad (1.2)$$

wherein c_x and c_y denotes the tread stiffness and μ is the friction coefficient. Because (1.2) can not be solved for the displacements u and v, one has to add an equation for the computation of the tangential forces in case of violation of (1.2) at a gridpoint. While the magnitude is given by $\mu F_N(x_i, y_j, t)$, the direction $\vec{e}_F$ is not and may be taken from the direction of the slip velocity

$$e_x(x_i, y_j, t) = \frac{V_{S_x}}{\sqrt{V_{S_x}^2 + V_{S_y}^2}} \qquad (1.3a)$$

$$e_y(x_i, y_j, t) = \frac{V_{S_y}}{\sqrt{V_{S_x}^2 + V_{S_y}^2}} \qquad (1.3b)$$

or from the direction of the displacements

$$e_x(x_i, y_j, t) = \frac{u(x_i, y_j, t)}{\sqrt{u^2(x_i, y_j, t) + v^2(x_i, y_j, t)}} \qquad (1.4a)$$

$$e_y(x_i, y_j, t) = \frac{v(x_i, y_j, t)}{\sqrt{u^2(x_i, y_j, t) + v^2(x_i, y_j, t)}}. \qquad (1.4b)$$

In the sliding area of the contact patch, there is no equation of motion left when the simplified friction description is used. Not only the forces in the sliding have to be limited, but also the displacements because of the character of transport equations (1.1a,1.1b): a solution at t is shifted from the leading to the trailing edge of the contact area with increasing time. If the displacements remain unlimited, these large displacements will be shifted. The resulting error may be large, especially while steering a resting wheel. The simplified description may be called *static friction*, because stick-slip vibrations are not calculated.

1.2 Stick-slip effects

When - instead of (1.1a,1.1b) - the equations of motion for the tread elements are derived from the force balance, they hold for the sliding area also. In consequence, the Lagrange formulation has to be used, wherein the equations describe the motion of material points. Figure 2 shows the one dimensional case

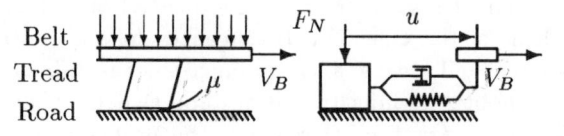

Fig. 2: Tread element

as to be used in models of class b), wherein V_B is the belt velocity, F_N the normal load, u the displacement, x the position of the element with respect to the contact area's origin and m, c and k are parameters of the tread. The equation of motion then is

$$m\ddot{u}(t) + c\dot{u}(t) + k\,u(t) = F_N(x+u,t)\,\mu(V_B(x+u,t) - \dot{u}(t)). \tag{1.5}$$

$\mu(V)$ is the friction function, that gives the friction coefficient purely as a function of velocity. Dependency on normal load or other dependencies are neglected here. Equation (1.5) holds for sliding as well as for sticking. When the relative velocity

$$V_R(t) = V_B(t) - \dot{u}(t) \tag{1.6}$$

vanishes, numerical integration may switch to

$$u(t) = u_o + \int_t^{t+\Delta t} V_B(x+u,t)\,dt \tag{1.7}$$

controlled by the sticking condition

$$F_K(x+u,t) = \mu_0\,F_N(x+u,t) > k\,u(t) + c\,\dot{u}(t). \tag{1.8}$$

In the two dimensional case, the equations of motion look similar, but the criteria for stability of steady state motions are quite different [6]. Switching between stick and slip - between (1.5) and (1.7) - can occur at any time, so within the integration of motion, the algorithm has to find the transition point and switch when the transition occurs. Other formulations of the problem, for instance using controlled constraints [7], are of the same effort.

In this modeling the assumption is made, that a transition from stick to slip and vice versa occurs, the model contains *dynamic friction* with a discontinuous friction function. This method can be used in models of class b) and c), but the integration algorithm must be capable of switching at arbitrary points in time, which may result in large computational or algorithmic effort. Another approach uses a regularized friction function [8].

1.3 Creep functions

In a surrounding V_ϵ around $V_R = 0$, the regularized function is a continuous function as shown in figure 3. This causes a creep motion instead of a stick-slip motion. When the gradient of the creep function reaches infinity, both motions are identical. On one hand, using a creep function, the numerical integration can be done without searching transitions and switching, any integration algorithm can be used. On the other hand, the problem beco-

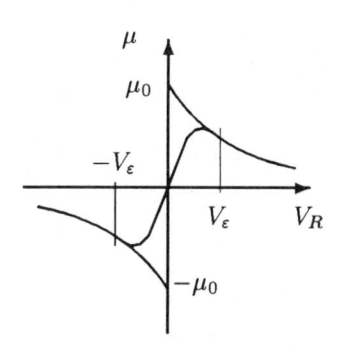

Fig. 3: Friction functions

mes numerically stiff whenever V_ε is chosen small: instead of sticking, small oscillations occur around $V_R = 0$ with high frequency, which - at the average - result in a motion similar to sticking, but must be integrated with a very small stepsize. Chosing V_ε larger, the creep motion can not longer be seen as an approximation of the stick-slip motion and the resulting tangential contact forces due to a given kinematic excitation appear to be too small. Another difference between both methods lies in the stability of the equilibrium. When $V_R < V_\varepsilon$, the creep function leads to a stable equilibrium point while the discontinuous function can - corresponding to the damping k - result in an unstable equilibrium.

2. CONTACT IN TYRE MODELS

The contact problem of a rolling tyre can be reduced to the contact of an elastic body on a rough rigid surface, because the stiffness of road and tyre are quite different. While on an even road, approximation of normal load distribution is sufficiently accurate, but it is not when obstacles or other road irregularities are considered. In consequence, the simulation of a tyre rolling over road disturbances of short wavelength (smaller than the length of the contact area) has to be done with a model from class c). The discretisation of the structure can be done with FEM, but solving the dynamic problem for about ten seconds realtime is of such an extreme large computational effort that these models are not used in vehicle dynamics. Therefore, models from class c) often discreties the belt as a three dimensional mass point system with connecting spring-dampers. Two approaches are used to solve the tyre contact problem, one using an elastic layer to support the belt and the other using constraints for the belt discretisation points.

2.1 Contact layer stiffness

As shown in figure 4, the tread is modeled by springs, that can come into contact with the road - unilateral stiffness K_c. The parameters can be obtained from geometry and rubber material. The contact condition is tested for each element in a surrounding of the road surface and in case of contact, the spring force is computed and applied to the structure. Summation over the element forces leads to the normal load, the shape of the contact area is given by the location of the elements with contact. Because of no degrees of freedom in the contact layer, the direction of element forces is given by the normal vector of a beltsegment between

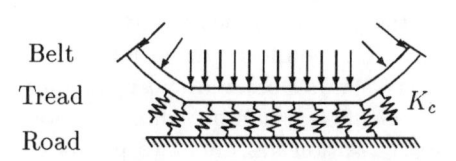

Fig. 4: Contact stiffness

two discretisation points. The tangential forces can be calculated by one of the methods mentioned above, by *static* [9] or *dynamic* friction. When rolling, the model produces a force variation in dependence of the rotational velocity and the number of springs on the circumference. To get the variation sufficiently small, a large number of tread elements have to be used, independent from the discretisation of the structure. Due to the high stiffness of the tread in comparison with the global radial stiffness of the tyre, the problem becomes numerically stiff.

2.2 Constraint displacements

To avoid the numerical stiffness, a formulation with constraints can be used for the contact. Then, instead of constraining the tips of the contact springs, the motion of the masspoints is limited by the ground. For each masspoint, the position $\vec{r}$ is constrained by

$$\vec{r}(t) \cdot \vec{e}_z \geq z_B(x,y) \tag{2.9}$$

wherein $z_B(x,y)$ denotes the road surface, see figure 5. Testing (2.9) together with the contact force obtained from the force balance, the constraint

$$\vec{r}(t) \cdot \vec{e}_z = r_z(t) = z_B(x,y) \tag{2.10}$$

can be set (component of the force normal to the surface negativ) or released. The constraint changes the structure's number of degrees of freedom. Instead from the differential equation for $r_z(t)$, the force component normal to the road surface is computed in case of contact from the force balance, the degree of freedom is frozen for the time of contact. The position $r_x(t)$ and $r_y(t)$ on the road surface is to be searched by iteration or - as a simplification - found from a projection, which in case of violating (2.9) ensures (2.10). For the computation of tangential force, one can also use both these methods concerning friction force. The constraint method ensures in any case, that penetration of the road surface is impossible, while the contact layer method does not.

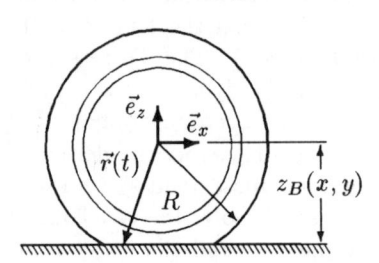

Fig. 5: Constraint motion

2.3 Road disturbance wavelength and discretisation

From the viewpoint of structural dynamics, the number of masspoints depends on the number of modes taken into account. Ride comfort calculations in vehicle dynamics have different criteria. Figure 6 shows a schematic drawing of a discretisation of the belt on an obstacle for two different numbers of masspoints and Lagrangian coordinates. When the discretisation has only two points on the cleat, the normal load distribution is not accurate enough and - following from that - the tangential forces are not too. To get reasonable results, the number of points on the cleat should be eight to ten in this example. Therefore, a connection between cleat length l_c or disturbance wavelength λ and the discretisation - expressed by the number of masspoints N_m - is given by $N_m \geq 20\pi R/l_c$ or $N_m \geq 20\pi R/\lambda$ with R from figure 5. This may be called the *geometric spacing* in opposite to the *modal spacing*, which describes the desired highest mode the model should contain and gives a bound to the upper frequency. In a three dimensional model, the number of equations of motion is $3N_m$ and may go up to 600 for a wavelength around $10mm$. Decreasing wavelength demands an increasing number of points. As a result, the computational effort is increasing and implicit integration methods are no longer useable. When numerical integration is done with an explicit formula, because of stability considerations the stepwidth has to be chosen in relation to the highest frequency in the system: an increasing number of points as a result of short wavelength leads to high frequencies and a very small stepsize, there is a coupling between road geometry (disturbance wavelength) and computational effort (stepsize).

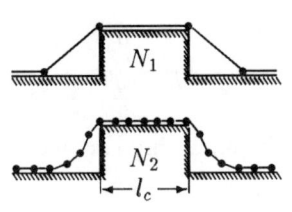

Fig. 6: Belt on obstacle

2.4 Decoupling of wavelength and stepsize

To overcome this coupling, geometric spacing and modal spacing must be seperated. This can be done, when special points - called *sensorpoints* - are introduced as massless points. Sensorpoints are interconnected to other sensorpoints or masspoints with springs (translational and rotational) to represent the tyre structure. The geometric spacing is given by the number $N_m + N_s$ of sensor- and masspoints, the modal spacing by the number N_m of masspoints. As a result, the system of equations is a system of differential algebraic equations (DAE), containing the nonlinear differential equations for the masspoints and the nonlinear algebraic equations for the sensorpoints. The set of equations can be derived from the principle of virtual work. The location of a point is given be $\vec{r}$, the point is loaded by the forces $\vec{F}_{ij}$ from springs and dampers (structure) and $\vec{F}_i$ from contact and tyre pressure. By

$$\delta A = \sum_i \left(\sum_j \vec{F}_{ij} + \vec{F}_i - m_i \ddot{\vec{r}}_i \right) \delta \vec{r}_i + \sum_k \left(\sum_l \vec{F}_{kl} + \vec{F}_l \right) \delta \vec{r}_k = 0, \quad (2.11)$$

a coupled system of $3N_m$ ordinary differential equations and $3N_s$ algebraic equations is given. The structure dynamics' upperbound frequency as well as the contact wavelength can now be chosen independently from each other. Seperating degrees of freedom of mass- and sensorpoints into the state vectors

$$\underline{q}^m = \begin{bmatrix} q_1^m \\ \vdots \\ q_{N_m}^m \end{bmatrix}, \quad \underline{q}^s = \begin{bmatrix} q_1^s \\ \vdots \\ q_{N_s}^s \end{bmatrix} \quad (2.12)$$

equation (2.11) can be rewritten as

$$\dot{\underline{z}} = \underline{h}(\dot{\underline{q}}^m, \underline{q}^m, \underline{q}^s) + \underline{R} \qquad (2.13a)$$
$$\underline{0} = \underline{g}(\underline{q}^m, \underline{q}^s) \qquad (2.13b)$$

wherein it is

$$\underline{z} = \begin{bmatrix} \underline{q}^m \\ \dot{\underline{q}}^m \end{bmatrix}, \quad \underline{h} = \begin{bmatrix} \dot{\underline{q}}^m \\ -\underline{M}^{-1}\underline{F}^{(i)}(\dot{\underline{q}}^m, \underline{q}^m, \underline{q}^s) \end{bmatrix}, \quad \underline{R} = \begin{bmatrix} 0 \\ \underline{M}^{-1}\underline{F}^{(o)}(\dot{\underline{q}}^m, \underline{q}^m, \underline{q}^s) \end{bmatrix},$$

$\underline{F}^{(i)}$ containing spring- and damper-forces and $\underline{F}^{(o)}$ containing contact- and tyre pressure forces. With given initial conditions

$$\underline{q}^m = \underline{q}_0^m, \quad \dot{\underline{q}}^m = \dot{\underline{q}}_0^m, \quad \underline{q}^s = \underline{q}_0^s. \qquad (2.14)$$

(2.13b) can be solved using for instance the Newton method

$$\underline{q}_{[\nu+1]}^s = \underline{q}_{[\nu]}^s - \underbrace{\left(\frac{\partial \underline{g}(\underline{q}_0^m, \underline{q}_{[\nu]}^s)}{\partial \underline{q}^s}\right)^{-1}}_{\underline{J}_{gs}} \underline{g}(\underline{q}_0^m, \underline{q}_{[\nu]}^s). \qquad (2.15)$$

starting with $\nu = 0$ for the equilibrium position of the sensorpoints. Then, an integration step with (2.13a) is made and (2.15) is solved again. This leads to an integrationscheme like

$$\underline{z}^P(t + \Delta t) = \underline{z}(t) + \Delta t \left(\underline{h}(\dot{\underline{q}}_m(t), \underline{q}_m(t), \underline{q}_s(t)) + \underline{R}(t)\right) \qquad (2.16)$$

to predict the solution at $t + \Delta t$ and - after solving (2.15) - correct with

$$\underline{z}^C(t + \Delta t) = \underline{z}(t) + \Delta t \left(\underline{h}(\dot{\underline{q}}_m(t), \underline{q}_m(t), \underline{q}_s(t + \Delta t)) + \underline{R}(t)\right), \qquad (2.17)$$

wherein evaluation of (2.13a) can be done also alternatively. This scheme has the character of a semi implicit integration method and is stabel even at a large stepwith.

Contact of mass- and sensorpoints and friction can be modeled with one of the methods mentioned above. When the constraint method is used, the size of $\underline{J}_{gs}$ changes due to contact and the normal force is to be taken from the corresponding row of (2.13b). Hereby, when points have contact, the number of equations to be solved decreases. Using the contact layer method, the number of equations remains unchanged in case of contact.

The friction force at the sensorpoints can not be calculated directly, because no information about the velocity of such points is available. Velocities can be interpolated, when velocities of *near* masspoints are used.

As a physically interpretation of the decoupling, the sensorpoint's degrees of freedom can be understood as *inner variables* of the structure. Instead of a simple rod with stiffness k connecting two masspoints, a more complex approximation of the structure is given. In figure 7, an example is shown. With a given masspoint density, the deformation of the belt is approximated better by using sensorpoints. Adding rotational springs, the element has k_t and k_r as parameters and deforms more like a simplified beam. The forces between mass- and sensorpoints are inner forces of the structure. Expanding the idea from discrete elements to beams or other elements leads directly to FEM, but increases the computational effort dramatically [10], when investigating obstacles with short wavelengths.

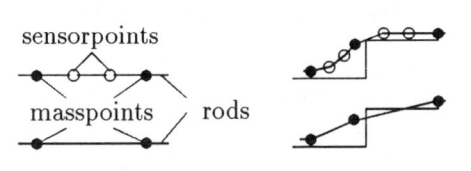

Fig. 7: Sensorpoints

3. TYRE MODEL FOR RIDE COMFORT CALCULATIONS

In the following section a tyre model is presented, which contains sensorpoints, constraints for contact and a creep function for friction. Assuming a tyre rolling over a cleat without large steering- and camber angles, a two dimensional model is build to save computation time compared to a three dimensional. The model's degrees of freedom are the displacements in the rim plane, for the mass- and sensorpoints. Between two masspoints, two sensorpoints are located at the circumference of the belt. The points are interconnected with translational and rotational springs, in radial direction with corresponding points at the rim, in circumferential direction with each neighbour. Damping is modeled between masspoints only. The parameters of the tyre structure can be obtained from measurement following [11]. The motion of the rim is assumed to be given by an MBS-model of the vehicle. Numerical and mechanical coupling between the tyre model and the MBS then is done according to [12], where $<\vec{e}_1, \vec{e}_2, \vec{e}_3>$ is called the tyre base, which is connected to the rim, but does not rotate with Ω. Transformation $\underline{T}_{IR}$ between $<\vec{e}_1, \vec{e}_2, \vec{e}_3>$ and $<\vec{e}_x, \vec{e}_y, \vec{e}_z>$ here contains steering and camber only.

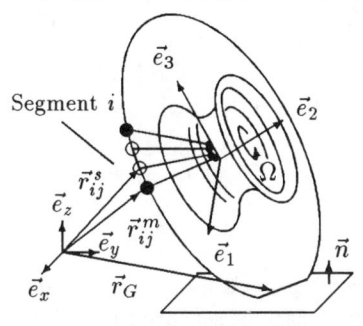

Fig. 8: Rim and plane

3.1 Model equations

Because of the topological scheme from figure 8, the Jacobian in (2.15) can be splitted into [4x4]-matrixes. If $\underline{q}_i^m$ denotes the degrees of freedom of the two masspoints that surround the two sensorpoints, $\underline{J}_{gs_i}$ - the Jacobian of the segment i - depends on these states of the masspoints state vector only. If the states $\underline{q}_i^m$ are given (according to the integration scheme), (2.15) is a separated system of four equations. Then, a symbolic calculation of $\underline{J}_{gs_i}^{-1}$ is done and evaluation of (2.15) is of small effort. With the degrees of freedom $\underline{q}_i^s$ of the sensorpoints, a number of $N_s/2$ nonlinear systems

$$\underline{g}_i(\underline{q}_i^s, \underline{q}_i^m) = \underline{0} \tag{3.18}$$

is to be solved for the displacements of sensorpoints. With a point on the road surface $\vec{r}_G$ and the rim position $\vec{r}_R$

$$\vec{r}_G = x\,\vec{e}_x + y\,\vec{e}_y + z_G(x,y)\,\vec{e}_z \tag{3.19}$$
$$\vec{r}_R = x_R\,\vec{e}_x + y_R\,\vec{e}_y + z_R\,\vec{e}_z \tag{3.20}$$

the location of mass- or sensorpoints is given by

$$\vec{r}_{ij}^s = \vec{r}_R + q_{ij_1}^s \vec{e}_1 + q_{ij_3}^s \vec{e}_3 \tag{3.21}$$
$$\vec{r}_{ij}^m = \vec{r}_R + q_{ij_1}^m \vec{e}_1 + q_{ij_3}^m \vec{e}_3. \tag{3.22}$$

When contact occurs

$$\vec{r}_{ij}^s \cdot \vec{n} = z_{ij}^s \leq z_G(x,y) \tag{3.23}$$
$$\vec{r}_{ij}^m \cdot \vec{n} = z_{ij}^m \leq z_G(x,y). \tag{3.24}$$

the displacements are constrained and (3.18) change from [4x4] to [2x2] if both sensorpoints are in contact. In case of violation of (3.23) or (3.24), the point is set to the road surface with

$$z_{ij}^s = z_G(x,y) \tag{3.25}$$
$$z_{ij}^m = z_G(x,y) \tag{3.26}$$

The remaning equations of (3.18) are solved for the point's location normal to the road surface in the line of contact iterativly in case of sensorpoints. After the solution is achieved, the positions of the masspoints are computed from the (remaining) equations of motion. In these equations as well as in (3.18), friction forces are taken from

$$\vec{F}_f = F_n\,\mu(|\vec{v}_c|)\frac{\vec{v}_c}{|\vec{v}_c|}, \tag{3.27}$$

with the normal load F_n (constraint force) and a continuous friction function

$$\mu(|\vec{v}_c|) = a\arctan(b\,|\vec{v}_c|) \tag{3.28}$$

with parameters a, b for the shape and $\vec{v}_c$ as the sliding velocity, in case of sensorpoints obtained from interpolation.

3.2 Variation of discretisation and stepsize
As an example, the tyre reaction crossing an obstacle

$$z(x) = z_0\,\sin\left(\frac{x - x_0}{l}\right) \tag{3.29}$$

MODELING CONTACT AND FRICTION

($z_0 = 5$ mm, $l = 150$ mm, $V = 20$ km/h) is computed with a different number of points. According to the obstacle's length, about 100 points are needed at the circumference to get about ten points on the obstacle. Figure 9 and 10 show the rim force results, when - starting with 100 masspoints - the number is reduced in three steps to 40: no significant differences are to be found between the four results. Because of the use of inner variables, even with a relativ small number of masspoints, the deformation of the structure is described with sufficient accuracy and the stepwith can be chosen nearly independent from the wavelength of road disturbance. The choice of the stepsize becomes important,

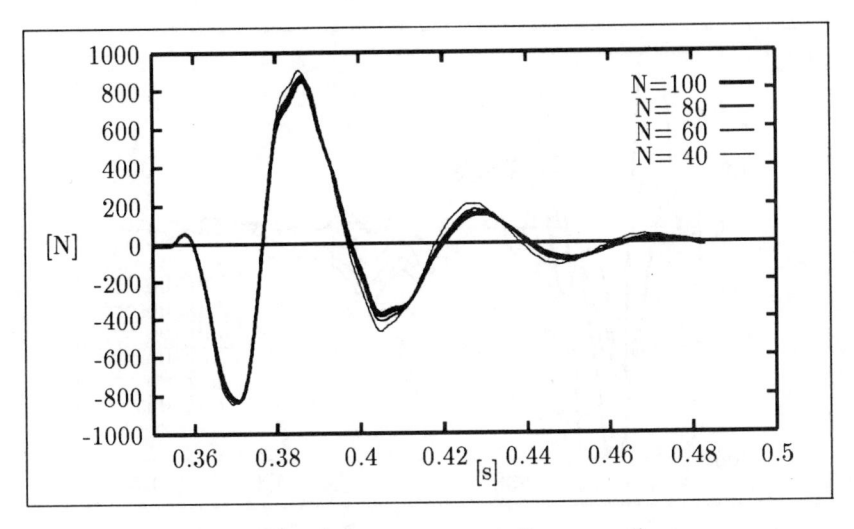

Fig. 9: Rim force component F_x versus time

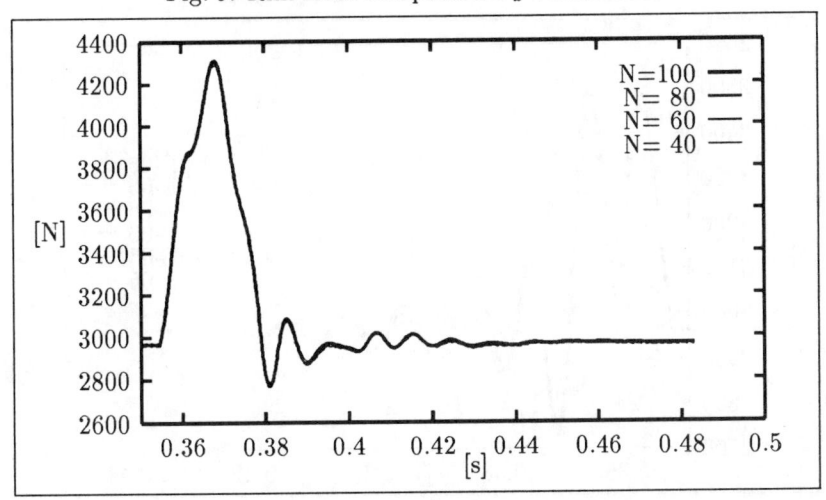

Fig. 10: Rim force component F_z versus time

when obstacles with short wavelength are investigated. In figure 11 and 12, the results from calculations with an obstacle of 15 [mm] length with varied stepsize are shown. Because of the use of a regularized friction function, the tangential force (F_x) changes with the stepsize while the normal force (F_z) does not. Therefore, the comparison between measurement and calculation has to focus on the tangential contact force.

In the following figures, the comparison between measurement on a drum and corresponding calculations are shown for the tangential contact force. The cleat length was 15 [mm]. Two different cleat hights were tested. In both results

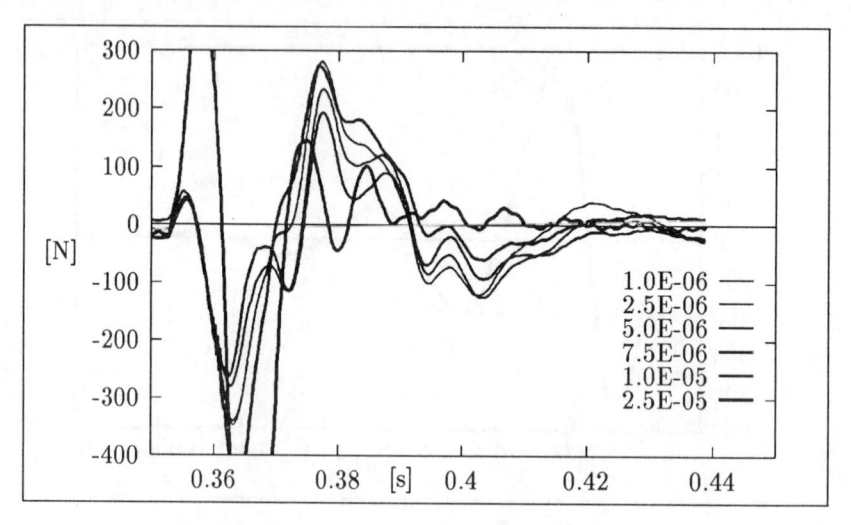

Fig. 11: Rim force component F_x versus time

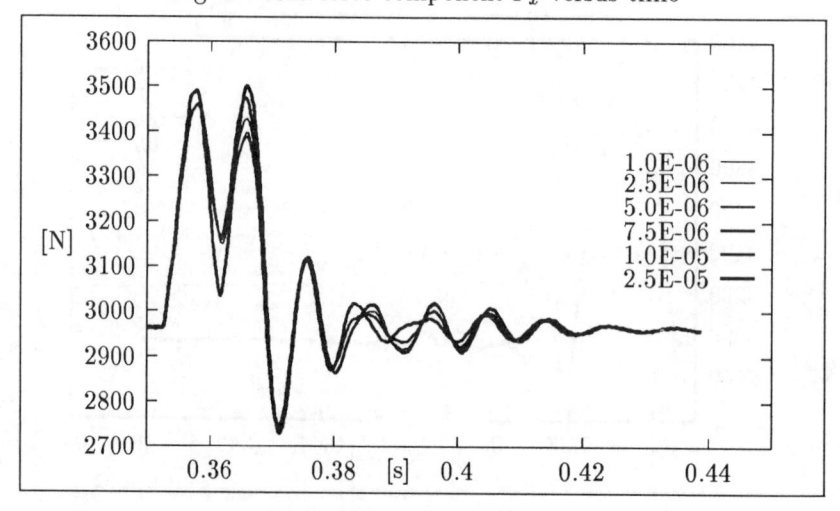

Fig. 12: Rim force component F_z versus time

(figure 13 and 14), the measured impact reaction is reproduced well by the calculation. Calculations have been done with 150 masspoints and 10^{-5} [s] as stepsize. As the results show, because of the use of viscous damping only, the results differ after crossing the cleat, inner friction has to be introduced to improve the model's accuracy.

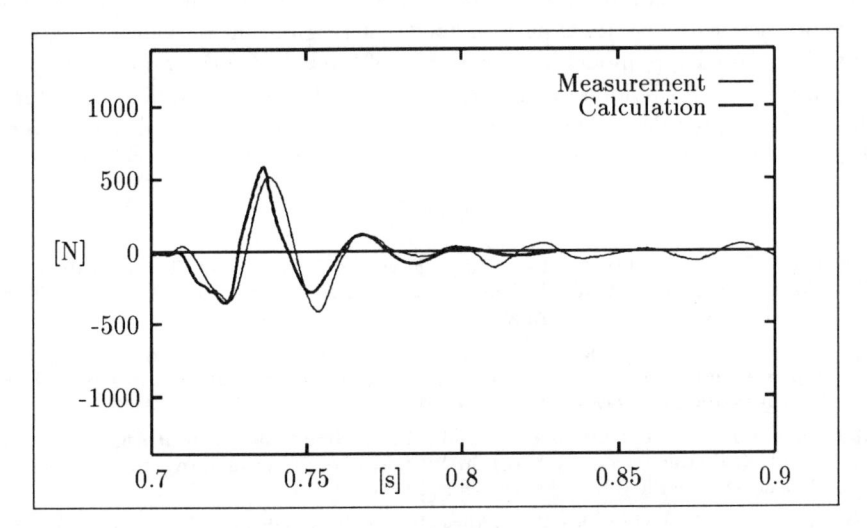

Fig. 13: Rim force component F_x versus time

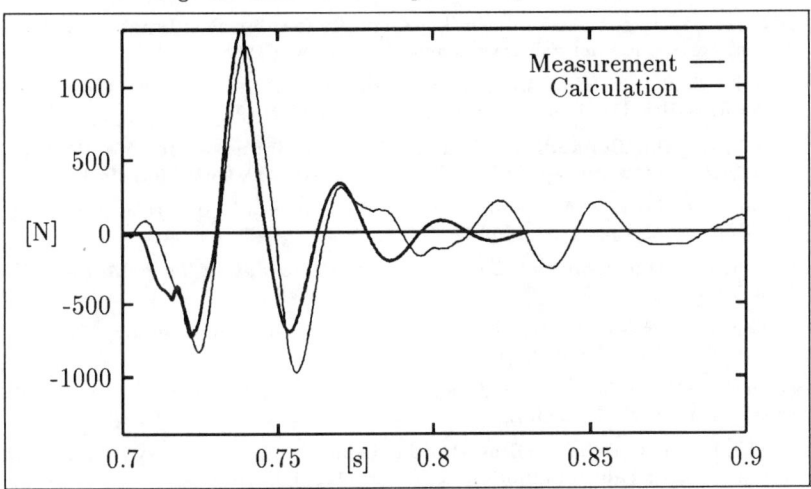

Fig. 14: Rim force component F_z versus time

SUMMARY

Some alternatives in modeling contact and friction for the prediction of tyre response on uneven roads are discussed. They appear to converge to a common behavior, if taken to the limits: the creep motion reaches a stick-slip motion and the contact layer becomes a constraint. Therefore - for the choice of a method - the computational performance is the significant measure.

In the presented tyre model inner variables of the belt deformation are used to decouple frequency bounds (vibrations of the structure) and wavelength bounds (form of the obstacles). The results show good agreement with measured tyre response, but the damping mechanism has to be expanded to inner friction.

REFERENCES

[1] Pacejka, H. B., Bakker, E.: **The Magic Formula Tyre Model. Tyre Models for Vehicle Dynamics Analysis**. Proceedings of the 1st Internation Colloquium on Tyre Models for Vehicle Dynamic Analysis held in Delft, Oct. 21-22, 1991

[2] Kollatz, M.: **Kinematik und Kinetik von linearen Fahrzeugmodellen mit wenigen Freiheitsgraden unter Berücksichtigung der Eigenschaften von Reifen und Achsen**. Fortschritt-Berichte, VDI-Reihe 12 Nr. 118, Düsseldorf 1989

[3] Schulze, D. H.: **Instationäre Modelle des Luftreifens als Bindungselemente in Mehrkörpersystemen für fahrdynamische Untersuchungen**. Fortschritts-Berichte, VDI Reihe 12 Nr. 88, Düsseldorf 1987

[4] Böhm, F.: **Theorie schnell veränderlicher Rollzustände**. Ingenieur-Archiv 55 (1985), 30-44.

[5] Böhm, F., Eichler, E. and Kmoch, K.: **Grundlagen der Rolldynamik von Luftreifen**. Fortschritte der Fahrzeugtechnik, Nr.8, Vieweg 1988

[6] Oertel, Ch.: **Untersuchung von Stick-Slip-Effekten am Gürtelreifen**. Fortschritts-Berichte, VDI Reihe 12 Nr. 147, Düsseldorf 1990

[7] Kölsch, D.: **Die Behandlung Coulomb'scher Reibung in der Kraftfahrzeug-Simulation**. Fortschritts-Berichte, VDI Reihe 12 Nr. 230, Düsseldorf 1994

[8] Oden, J.T., Martin, J.A.C. and Simoes, F.M.F.: **A Study of Static and Kinetic Friction**. Int. J. Engng. Sci., Vol 28, No. 1, 1990

[9] Eichler, E.: **Ride Comfort Calculations with Adaptive Tyre Models**. AVEC (1996) Aachen.

[10] Gipser ,M.: **DNS-Tire - ein dynamisches räumliches, nichtlineares Reifenmodell**. VDI-Berichte, Nr. 650, Düsseldorf 1987

[11] Bannwitz, P. and Oertel, Ch.: **Adaptive Reifenmodelle: Aufbau, Anwendung und Parameterbestimmung**. VDI Berichte Nr. 1224, Düsseldorf 1995

[12] Eichler, M. and Oertel, Ch.: **Eine standardisierte Schnittstelle zwischen Reifenmodellen und Fahrzeugmodellen** ATZ 96 (1994), S. 184-188

On the Roots of Tire Mechanics

by F. BÖHM

ABSTRACT

Dynamics of cables has something to do with tires because of the fact that a tire consists of 3 - 4000 cables. Rubber is a glue or adhesive to adhere the cables. After vulcanisation one gets a high elastic but strongly anisotropic structure which is very good to fill the gap between steel wheel and road. Increasing comfort is achieved when filling the structure with air. So we get a rolling spring with one end continuous changing in time. The understanding of this event lies not in the scope of classical mechanics.

INTRODUCTION

The tire is a textile product with high anisotropy. Mathematic foundation is the discovery of a mathematician from Berlin, Roth 1906, who found that a net consisting of inextensible filaments can cloth any surface! By coating of the net with rubber and by the finite thickness of the filaments an elastic locking effect occurs, so the change of tire surface and layers bending is limited. Also the relative movements of the separate sheets of filaments forming the tire membrane shell have to obey the conditions of anisotropy of all filaments. This forms the internal tire kinematic field, mainly shear restrain, heating up the rubber.

For the external tire kinematics the necessity arises that bending plus shear occurs always together and produces bridges in contact creating a filtering of short waves of the road surface.

The usual shell theory is not good enough to describe mathematically the behaviour of the tire shell. It is necessary to use a sheet theory due to Timoshenko instead of the Bernoulli-bending theory.

In engineering practice linear differential equations are used to conclude straight forward the statical and dynamical behaviour of the underlying system. In case of the tire this is generally not true. First, inflation of the tire has a non-linear start when the cords are without prestress! Second during rolling the tire surface undergoes a non-linear deflection in contact and non-linear friction forces.

Only when the tire is inflated small additional deformations can be computed by aid of linear (perturbation) theory, if the forces on its surface are given. In this case a linear differential equation system by perturbation theory can be elucidated. This theory gives insight in the mechanism of the spreading of local deformations in space and in time.

In the subsequent argumentation static and dynamic modelling ideas are presented for better understanding of tire mechanics. As engineers are familiar with the interpretation of eigenvalue-problems we will use these from analysing linearized prestressed systems.

The start of the analysis is difficult because we have to start with a non-linear problem. So we try to make a decomposition into force flow along the cord, and the bearing of inner pressure by the first stressed cords. For the first problem we use well known strip theory, for the second Euler membrane theory, despite the fact that in reality both are working non-linear together.

1. MODELLING EQUILIBRIUM SHAPE

From [1] we know that at the belt-edge an equation for maximum shear deformation exists:
$$-E_G \frac{(3d)^2}{1-v_G^2} \cot \beta \frac{d}{2} \frac{d^2\gamma}{dy^2} + dG\gamma = 0$$

with the eigenvalue $\lambda_1 = \frac{1}{3d}\sqrt{(1-v_G)\tan \beta}$

With $v_G = 0.5$, $d = 0.17 cm$ $\beta = 18.5°$ we get $\lambda_1 = 2.36 cm^{-1}$ and $\frac{1}{\lambda_1} = 0.43 cm$ as a relaxation length. So our mesh for computation should be smaller than 0.73 cm. But it is a question if this is really necessary, in practice the thickness d is increased and decouples both layers.

From [2] we know that the first belt layer resting on the elastic carcass needs only a short length of 0,5 cm to get the whole stress from carcass-cords due to elasticity theory. To improve the stress gradient at this point by a filler the thickness of rubber sheet in-between is increased for decoupling.

The equilibrium configuration of a belted tire is computed using Euler-membran theory [3]. Questionable is the choice of the belting function. For instance if one uses a given structure see Feng [4] fig 1, one gets a definite stress distribution in belt cords and carcass cords under inner pressure. Because of free ends of belt cords the stress there is zero. An analytic solution was found under the assumptions of constant radial deformation w of the belt in meridional and (naturally) in circumferential direction. One gets [2] the differential equation

$$G_{PL}\gamma - E_K h_K \frac{d^2\gamma}{dy^2} = -E_K \frac{1}{a\tan^2 \beta} \frac{dw}{dy} \doteq 0$$

from which we conclude the eigenvalue $\lambda_2 = \sqrt{G_{PL}/(h_K E_K)}$. The stress transfer zone from belt to carcass is longer than $1/\lambda_2$ that is more than 3 - 4 cm! So the solution of $\gamma(y)$ is a cosh-function which is nearly a parabola. Measured radial deformation is shown in fig. 2, it is nearly constant. Starting from side wall stress σ_{KO} there is a constant decrease of $\Delta \sigma_K = -E_K \dfrac{w}{a \tan^2 \beta}$ and the transient part is $\Delta \sigma_K$ cosh λy. Vice versa in the belt there exists a stress d cosh λy.

Lastly the curvature of the tire section is relevant for the distribution of stress between belt and carcass. For the static equilibrium we can define a belting function which regulates the bearings [4] of inner pressure p and also the shear of rubber layer:

$$\tau_{yz} = \frac{d\sigma_y}{dy} = G\gamma_{yz} = -\frac{2\kappa_o P/b^2}{\dfrac{1}{a\tan^2 \beta} + \dfrac{1}{\rho}} \cdot y$$

by use of a parabolic belting function. This function is often used but creates a tread curvature which is too round. By this roundness longitudinal slips-differences are produced and strongly shoulder wear. Increasing the exponent of parabolic improves this wear but gives rise to higher shear τ_{yz} than linear in y! Further improvement came from using nylon shrink by a third layer of nyloncords in circumferential direction. By this production method a belting function can be designed which is finite at belt corners.
With this function a flat tire tread contour is possible. Such equilibrium shapes are good for high velocity-tires.

In the case of truck-tires there is another strategy. The belt only bears in the middle part of the tread. At the shoulders the sidewall-cords bear the tread. This is due to the well known equilibrium shape after Bidermann [5]. The new super-single truck tires have low tread curvature, but therefore high shear stress in rubber between the layers. Decoupling helps but the thickness of rubber in section of shoulder is critical.

What we can learn from this belt-edge effects for our dynamic modelling of the tire? Firstly, it is not necessary to take into account the „singularity" at the belt edge, we can reduce the computed belt with. Second, we do not need to divide the beltsection into many parts because the belt stress does not change very much. Three or four parts are enough, that gives rise to four or five traces in circumferential direction. Third, because of the fact that $\tau_{yz}^1 = \dfrac{1}{2}\tau_{yz}^2$ and $\tau_{xz}^2 = 0$, the shear distribution between belt layers and belt to carcass is simple so we can use Belkin's theory [6] of tire equilibrium. Fourth, sidewall consists of membrane parts and of stiffened parts near the bead region. We can describe this by one or two traces for one sidewall. So for dynamics of rolling we get a good tire model with only eight or nine traces! We should take into account bending in meridional and circumferential direction from rubber and cord tension.

The first model to discuss is the elastic founded circular ring model. It has only one trace in the middle of the belt and two massless sidewalls. Therefore it only produces a pressure distribution on 3 lines consisting of tread pressure on ground plus sidewall pressure. So the computed creep force distribution is only correct in the non-slipping part of contact. For instance in application for truck tires contact area is splitted into three traces, so the shoulder bearing is separated from belt bearing. Here is also introduced dynamics of sidewall-waves modelling it by several point masses in contact region. This produces a good load-deflection behaviour of the model. Even for passenger car tires the deflection of tire section would improve this load-deflection characteristic, but because in this case the belt circumferential tension is also active in shoulder traces this modelling tends to a three-trace model.

2. MODELLING DEFLECTION BY VERTICAL LOAD

In circumferential direction we look now for the decay of radial deformation produced by a single force. Because of smallness of bending stiffness EI_y and small strain of the belt $\varepsilon_\rho \doteq 0$ we simply use the equation

$$\frac{T_o}{a^2} v'' - (k_a - \frac{T_o}{a^2}) v = q(\varphi)$$

The eigenvalue is $\lambda = \sqrt{\frac{a^2 k_a}{T_o}}$ or for all types of radial tires $1/\lambda = 0.2 \text{ rad} \doteq 12°$. This shows us that we need a minimum spacing in circumferential direction of 30. A spacing of 60 - 70 is really good enough.

In laboratory we measure this radial deformation also on several sections of the tire. Good tires have constant deformations over the whole width of belt section. Increasing pressure from 1 to 2 bar one gets 1.5 to 2.5 mm radial extension.

Radial deformation due to inner pressure produces a radial equilibrium of the model of

$$\frac{EF}{a} \varepsilon_{\varphi o} = pb \quad \text{with} \quad \varepsilon_{\varphi o} = \frac{v_o}{a}.$$

So by measuring v_o, f. i. $v_o = 2$ mm we get $\dfrac{EF}{30^2} \cdot 0.2 = 2 \cdot 12$ or $EF = 108000$ daN.
Parallel to this stiffness a big hysteresis is present.

Vertical deflection under load produces also circumferential deformation. Neglecting bending stiffness again we get the equation

$$\frac{1}{a}\Delta T' = k_\tau u \quad \text{with} \quad \Delta T = EF\varepsilon_y = EF\frac{u'+v}{a}$$

from which we find

$$\Delta T = \int a k_\tau u\, d\varphi + C.$$

The assumption that at the highest point of the tire $T \doteq T_o$ holds, see Feng [4], reduction of belt tension in contact area can be computed, when $u(\varphi)$ is measured. One gets a reduction of 50 % for 2 cm deflection! Because $\varepsilon_\varphi \doteq 0$ and therefore $u' = -v$ holds for radial and for tangential load.

By aid of free body diagrams I deduced [1] the eigenvalue equations of the model:

$$\begin{vmatrix} \dfrac{EF}{a^2}n^2 + k_\tau - \rho\omega^2 & \dfrac{EF}{a^2}n + \dfrac{EI_y}{a^4}(n^2-1)n \\ \dfrac{EF}{a^2}n & \dfrac{T_0}{a^2}(n^2-1) + \dfrac{EF}{a^2} + \dfrac{EI_y}{a^4}(n^2-1)n^2 + k_a - \rho\omega^2 \end{vmatrix} = 0$$

the lowest radial eigenmode $n = 1$ has an eigenfrequency which does not depend on belt tension T_0 because of term $T_0(n^2-1)$. The mode is a sin-wave with length of $2\pi a$. But we found by parameter optimisation [7] that the real figure is 1 Hz smaller than the computed one despite the fact that all other n > 1 measured modes coincide with computation. The reason is that the eigenmode n = 1 should be computed by the three-trace model which changes the belt section and is not exactly an cylinder.

Using energy-principles and variational computation with help of „Mathematica" we got another solution for the radial eigenmodes:

$$\begin{vmatrix} \dfrac{T_0+EF}{a^2}n^2+k_r-\rho\omega^2 & \dfrac{T_0+EF}{a^2}n \\ \dfrac{T_0+EF}{a^2}n & \dfrac{EF}{a^2}+\dfrac{T_0}{a^2}n^2+\dfrac{EI_y}{a^4}(n^2-1)n^2+k_a-\rho\omega^2 \end{vmatrix}=0$$

There we found another connection of T_0 with mode number! So we tried again changing from Bernoulli-bending to Timoshenko-bending:

$$\begin{vmatrix} \dfrac{T_0+EF}{a^2}+k_r-\rho\omega^2 & \dfrac{T_0+EF}{a^2}n & 0 \\ \dfrac{T_0+EF}{a^2}n & \dfrac{EF}{a^2}+\dfrac{d^2EF}{4a^4}(n^2-1)^2+ \\ & +\dfrac{T_0}{a^2}n^2+k_a v-\rho\omega^2 & -\dfrac{d^2EF}{4a^3}(n^2-1)n \\ 0 & -\dfrac{d^2EF}{4a^3}(n^2-1)n & \dfrac{d^2EF}{4a^2}n^2+GF \end{vmatrix}=0$$

There again we found the old dependency. In the two last cases the eigenvalue-problem is symmetric. So, at least, we have the following situation:

The first case was wrong because of Bernoulli-theory and nonsymmetric matrix-term (we did not use it).
The second case was wrong because of connection of T_0 with n.
The third case was good, but the first mode does not coincide with measurement.
The general conclusion is that a 2D-ring-model can not describe behaviour of a 3D-tire in respect to sidewalls and meridional deformations. Nevertheless it is a good model for numerical computations, has only a few DOF and is easy to couple with car dynamics.

Non-linear geometric behaviour during rolling needs for discretisation of the structure. The method of particle dynamics together with explicit integration in time needs to state a time step for highest frequency in the system of ordinary differential equations. The highest frequency is longitudinal belt vibration. We take n points equidistant at the circumference, so one point has mass $m = m_g/n$ where m_g is belt plus tread mass. Belt stiffness is EF and for connecting two points we get an elastic spring stiffness of $c = \dfrac{EF}{2\pi R/n}$. The highest possible mode has knots at the half distance between two masses, therefore $f_{max} = \dfrac{1}{2\pi}\sqrt{\dfrac{4c}{m}}$. The time step is then $\Delta t \leq \dfrac{1}{10 f_{max}}$ for fulfilling Shannon criterion. Measurements then show the damping in the range of 0.03 - 0.06 for the dimensionless damping factor. In most cases we use $\Delta t = 1/30000 \sec$.

3. GENERAL MODELLING AND CONCLUSION

In case of 3D-model constants for fiber-stiffness of steel cords and of rayon cords are used together with rubber elasticity modulus (table 1). Starting with a non-linear dynamic computed equilibrium shape with 19 points, fig. 3, this shape is simplified defining six traces. Four traces are used for the belt and one trace for every sidewall. These few traces are only good for modelling belt and tread rolling dynamics, while the sidewall dynamics are too simplified. For instance waves in the sidewalls disturb laterally by bending the profile elements in the tread zone along the sidewalls. For contact with the ground the profile elements are substituted by contacting members which are sticking and sliding appropriate to the vertical contact force of the contacting particle.

The highest frequency in the system is due to the steel elements of the belt. The same formula is used as before but stiffness is increased by rubber shear stiffness between the layers. The usual time step is $\Delta t = 10^{-5}$ sec. In fig. 3 tire section and in fig. 4a frequency spectra of the model are shown. The model has 324 DOF and is adapted to a car model with 114 DOF. As for the car model the frequency range is up to ~ 150 Hz. For higher frequency side-wall waves in contact region can be added. Damping factors are in the range of 0.05 to 0.10 for the belt.

The most complicate model uses profile elements resting on the membrane 3D shell, fig. 4b . It is a tire model for off-road tires and the profile elements are discretised so to produce a shear vibration of 1800 Hz and a thickness vibration of 2000 Hz. In fig. 5 rolling together with side slip on a glass-plate was computed. Fig. 6 shows dynamic behaviour of all six traces in radial direction during rolling.

Summing up all this roots we can say that a discretised 3D tire model should have a belt which is nearly 3 cm smaller as the real one at both sides. In this remaining region the deformation field is nearly linear and the stress field is a parabola with unsymmetric part. So one needs only four traces for modelling the belt. Bending of the tire shell is small and can be simulated by the rubber bending. The sidewall waves have only in the range higher than 300 Hz influence along the contact area but not at the free circumference and can be simulated by simple oscillatory subsystems. The profile elements are shorter discretisied but have only rubber stiffness. Belt system therefore the stiffness and mass of belt defines the total stiffness of the system and the time step necessary for dynamic explicit integration. The model does not include the sidewall dynamics, which would need some additional traces in the sidewall as was introduced in Diploma Thesis of Gallrein [8]. To use this model we are waiting for faster computers.

REFERENCES

[1] Böhm, F.: Mechanik des Gürtelreifens. In *„Ingenieur-Archiv"*, 35. Band, 2. Heft, pp. 82-101, Springer-Verlag 1966.
[2] Böhm,F., Duchow,A., Hahn,P.: Beanspruchung und Verformung des Gürtelreifens unter Innendruck. In *Automobil-Industrie* 3/85, Reifenmeßtechnik, pp. 317-323
[3] Böhm, F.: Zur Statik und Dynamik des Gürtelreifens. In *ATZ* 69 (1967) 8, pp.255-261, Stuttgart 1966
[4] Feng, K.: Statische Berechnung des Gürtelreifens unter besonderer Berücksichtigung der kordverstärkten Lagen. In *Fortschr.-Ber. VDI* Reihe 12 Nr. 258, VDI Verlag, Berlin 1994
[5] Bidermann, W.L.: Autoreifen (Konstruktion, Berechnung, Prüfung, Verwendung). Moskau 1963, Staatlicher Wissenschaftlicher und technischer Verlag für chemische Literatur, Kapitel 4.
[6] Belkin, A.E.: Theoretische Grundlagen der Berechnung des Spannungs- und Wärmezustandes in Radialreifen nach Modellen mehrschitiger bewehrter Schalen. and
Narskaya, N.L.: Computer realisation of radial tire calculations an the basis of shell models. *Presentations at mechanic seminar of 1. Institut für Mechanik at the TU-Berlin*, 16.1.1995
[7] Gallrein, A.: Berechnung hochfrequenter Stollendynamik am Gürtelreifen. *Diplom-Arbeit am 1. Institut für Mechanik an der TU-Berlin*, Dezember 1992

	Passenger Car Tire	Truck Tire
tread	55 - 65 daN/cm^2	55 - 58 daN/cm^2
steel-belt coating compound	75 - 80 daN/cm^2	70 - 82 daN/cm^2
carcass coating	35 - 50 daN/cm^2	45 - 50 daN/cm^2
sidewall coating	25 - 30 daN/cm^2	30 - 35 daN/cm^2
innerliner	25 daN/cm^2	20 daN/cm^2
belt-filler	40 daN/cm^2	40 daN/cm^2
filler at bead coil	70 - 80 daN/cm^2	70 - 80 daN/cm^2
ply-edge coating	80 daN/cm^2	-
reyon-cord	$9 \cdot 10^4$ daN/cm^2	-
steel-cord	$1.7 \cdot 10^6$ daN/cm^2	$1.7 \cdot 10^6$ daN/cm^2
high elongation cord	-	$1.9 \cdot 10^6$ daN/cm^2
area of belt-cord	0.22 - 0.36 mm^2	
area of carcass-cord	0.28 mm^2	

Table 1: Elastic moduli, areas of compounds, cords, steel-cords

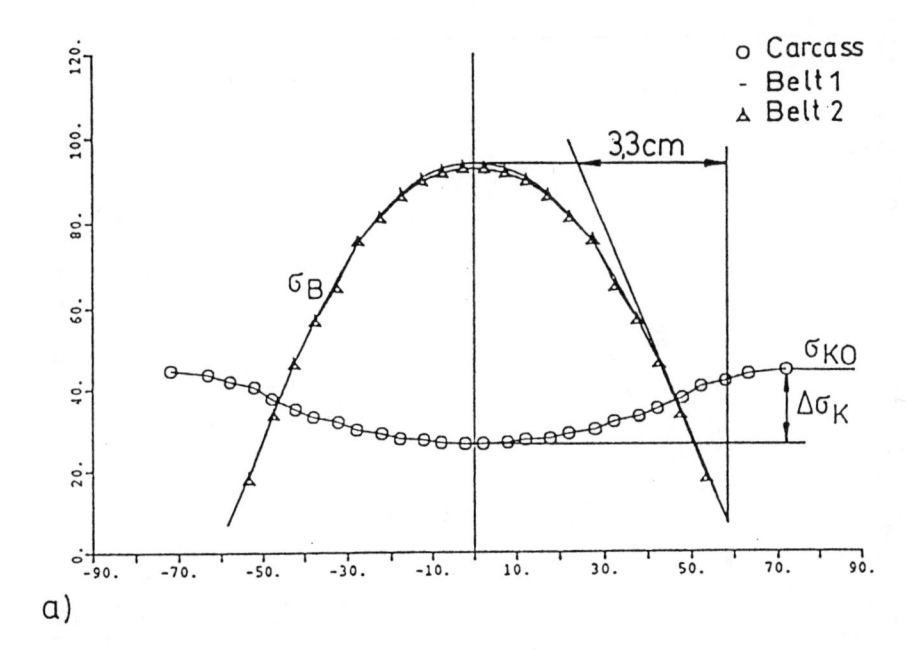

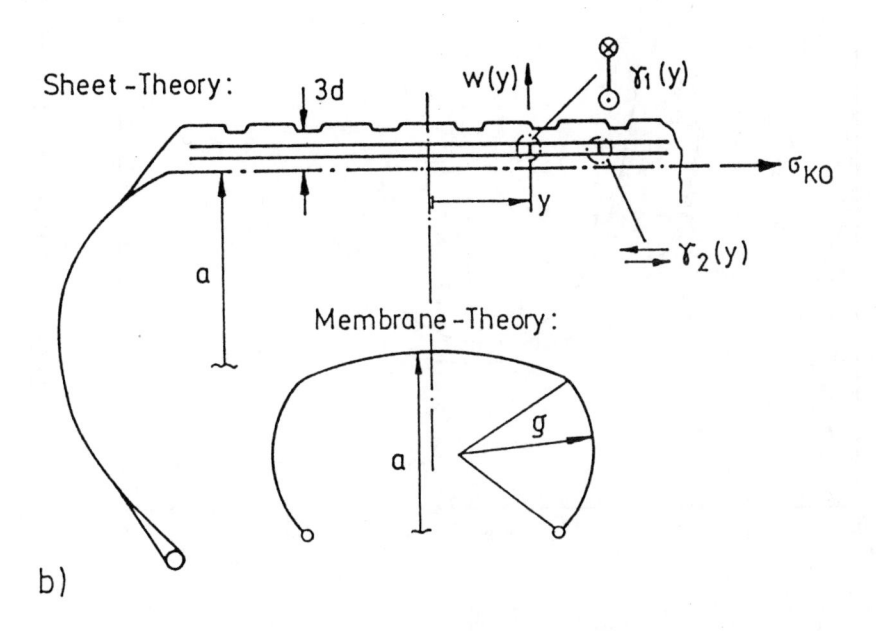

Fig. 1 a) Stress - distribution after Feng [4]
b) Sheet- and Membrane theory

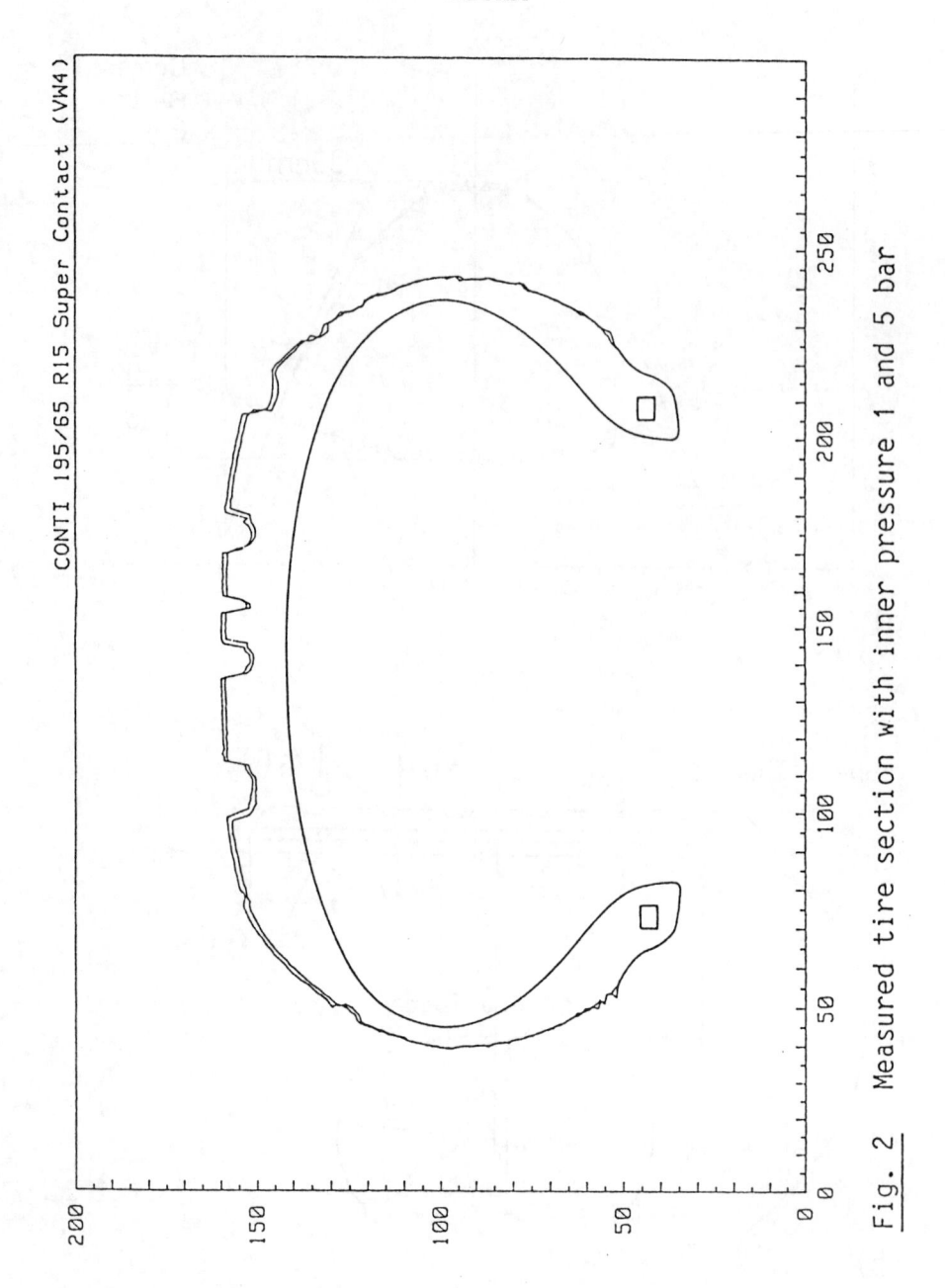

Fig. 2   Measured tire section with inner pressure 1 and 5 bar

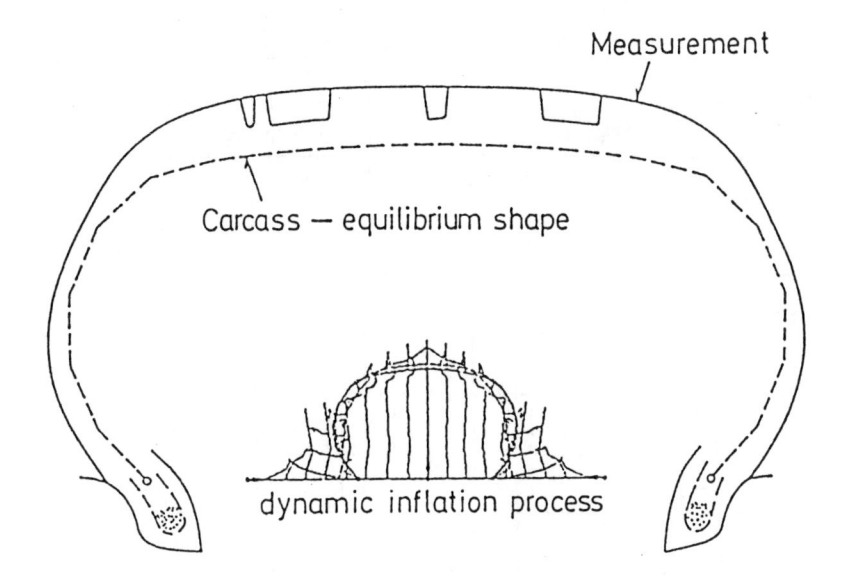

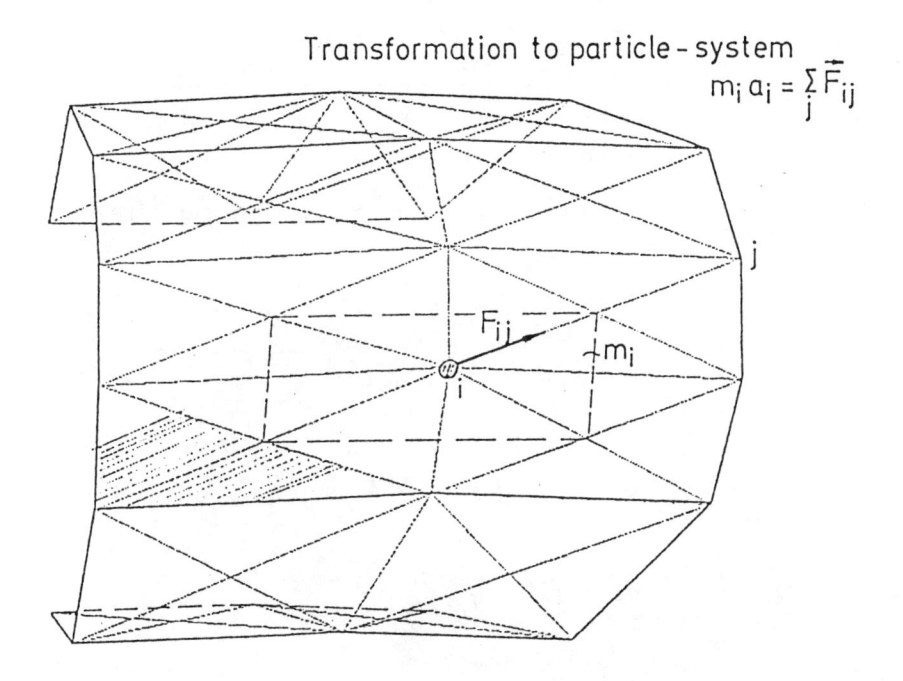

Fig. 3 Blow up process and particle system

Membrane 3D-Spectrum, computed

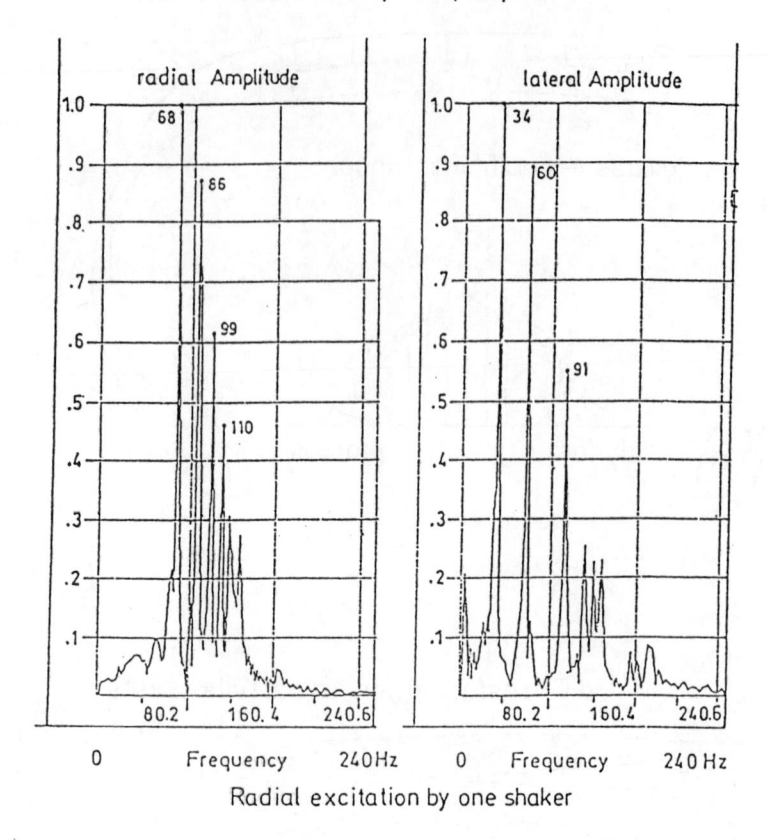

Fig. 4a   Radial and lateral Spectra of a passenger car tire

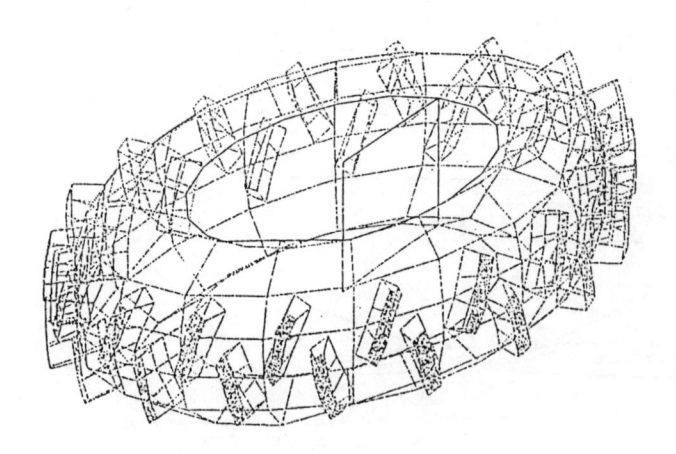

Fig. 4b   Agricultural tire under vertical load

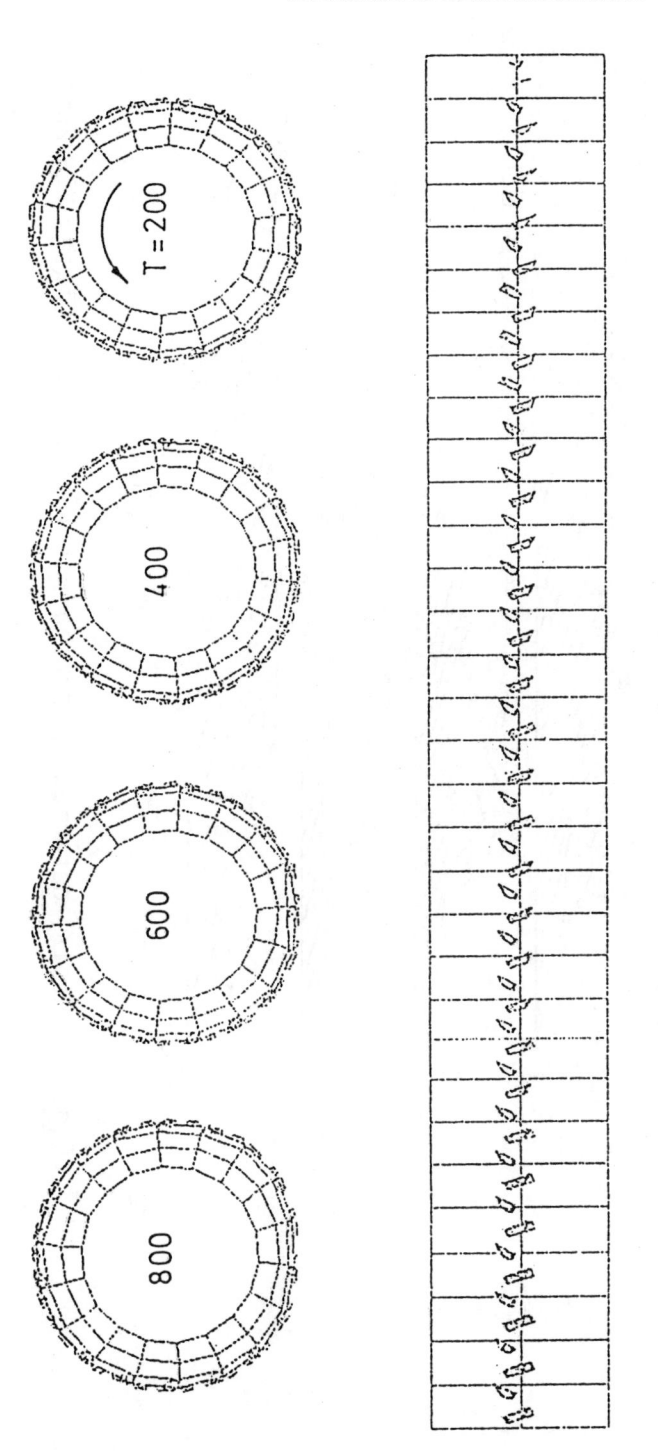

Fig. 5 Rolling of agricultural tire on rigid plate T..time step counter, $\Delta t = 10^{-4}$, $V = 20$m/s, cornering angle 5°

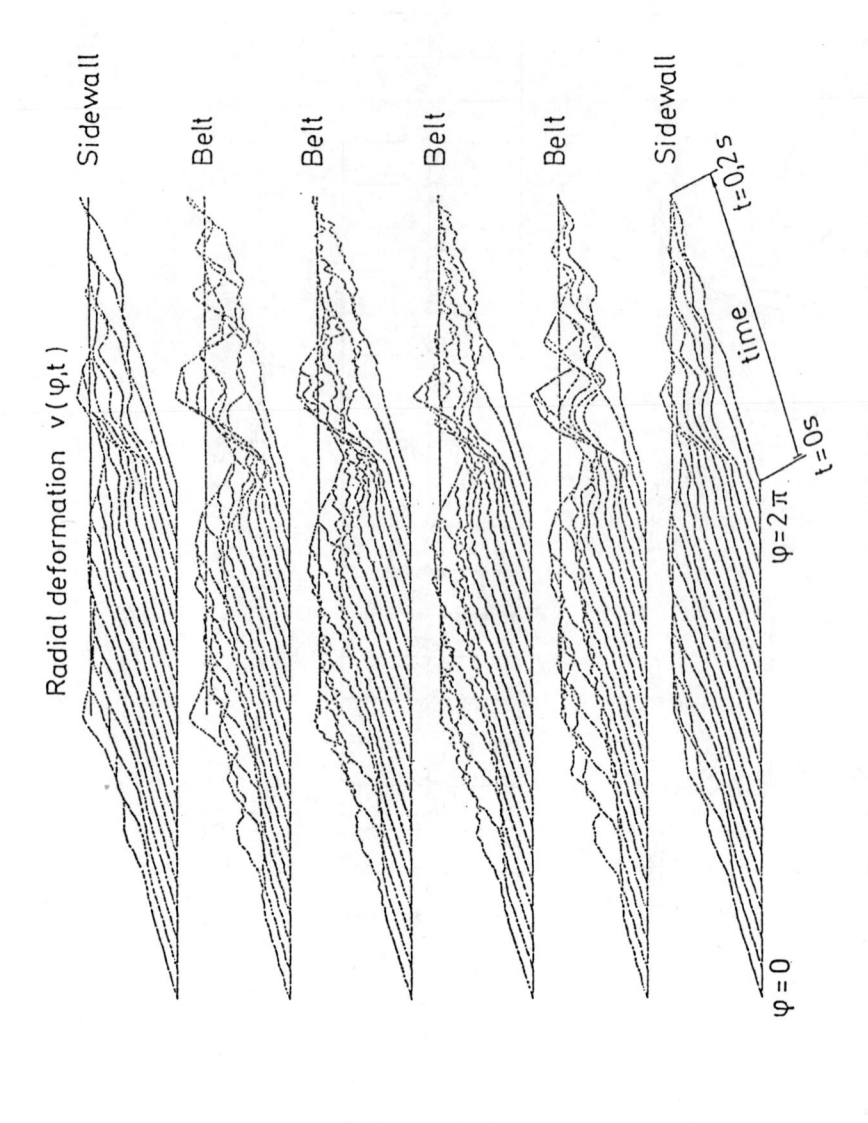

Fig. 6 Radial deformation during rolling on rigid plate, $V = 20$ m/s

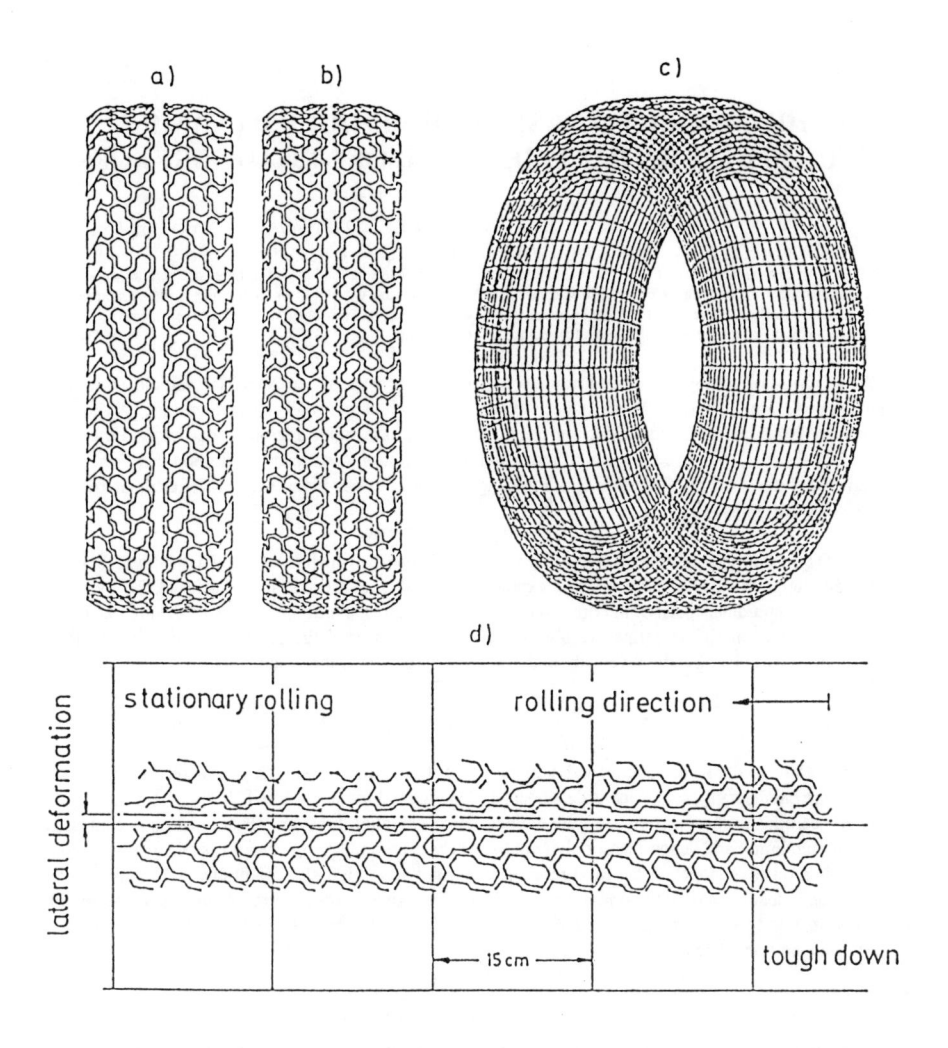

Fig. 7 a) For computation simplified tread profile
       b) Measured tread profile
       c) Net model of tire structure
       d) Tread rolled on glass-plate with
          3° cornering, small rolling velocity

A TIRE MODEL FOR USE WITH VEHICLE DYNAMICS SIMULATIONS ON PAVEMENT AND OFF-ROAD SURFACES

R.W. ALLEN, J.P. CHRSTOS AND T.J. ROSENTHAL

ABSTRACT

STIREMOD is an expanded version of an earlier vehicle simulation tire model designed for the full range of operating conditions (slip, camber, normal load) on both paved and off-road surfaces. Within a vehicle dynamics simulation tire model parameter sets can be changed dynamically to reflect the local operating condition of each tire. This paper summarizes the general tire model structure, data sources, parameter identification and model fits to tire test data.

INTRODUCTION

STIREMOD was originally developed for paved surfaces as part of the VDANL vehicle dynamics computer simulation [1] and has recently been expanded to include off-road characteristics [2]. The off-road characteristics cover the adhesion, transition regions, and saturation regions. To simulate the high side forces associated with tire plowing on loose soil, the saturation coefficient of friction can be set low for longitudinal slip, and high for side slip, resulting in an equivalent high lateral coefficient of friction (greatly exceeding unity) [3,4].

MODEL FORMULATION

STIREMOD equations are based on a composite slip formulation (σ), which is basically a quadratic function of lateral and longitudinal slip. Lateral slip is expressed as the ratio of the side slip velocity of the tire patch relative to the longitudinal speed of the tire patch, which is the equivalent of the tangent of the tire patch slip angle α. Longitudinal slip is defined as the ratio S of the differential tire patch to ground longitudinal velocity divided by the longitudinal velocity of the wheel hub relative to the ground.

$$\sigma = \frac{\pi a_p^2}{8\mu F_z} \sqrt{K_s^2 \tan^2\alpha + K_c^2 \left(\frac{S}{1-S}\right)^2} \quad (1)$$

This formulation accounts for lateral and longitudinal stiffness coefficients (K_S, K_C respectively) and changes in tire patch length, a_p, which is dependent on longitudinal force response [2].

Given a composite slip function, we now define a force saturation function. This function is expressed as a ratio of numerator and denominator polynomials with two important properties: 1) as slip increases from zero there is a positive slope; 2) the ratio reaches an asymptote at high slip conditions. The roots of the numerator and denominator polynomials can be located to give a wide variation in the shape of the saturation function to accommodate paved and off-road surfaces [2]. This function is a ratio of polynomials that define a load normalized composite force F_C:

$$f(\sigma) = \frac{F_c}{\mu F_z} = \frac{C_1\sigma^3 + C_2\sigma^2 + C_5\sigma}{C_1\sigma^3 + C_3\sigma^2 + C_4\sigma + 1} \quad (2)$$

Given the normalized composite force, we then define the normalized lateral and longitudinal forces without cambering in terms of the stiffness weighted lateral and longitudinal slip ratios. With camber (γ), the normalized side force also includes an additional camber component where $Y_\gamma = dF_y/d\gamma$:

$$\frac{F_x}{\mu F_z} = \frac{-f(\sigma)K_c'S}{\sqrt{K_s^2 \tan^2\alpha + K_c'^2 S^2}} \quad ; \quad \frac{F_y}{\mu F_z} = \frac{-f(\sigma)K_s \tan\alpha}{\sqrt{K_s^2 \tan^2\alpha + K_c'^2 S^2}} + Y_\gamma'\gamma \quad (3)$$

The primed quantities Y_γ' and K_c' account for saturation effects discussed below.

Aligning moment is considered a basic function of side force operating on a "pneumatic trail" moment arm. Under saturation (i.e. high slip) conditions aligning moment approaches zero as the pneumatic trail goes to zero. The aligning moment is expressed as the product of two functions:

$$M_z = \frac{K_m a_p^2 \tan\alpha}{\left(1 + G_1\sigma^2\right)^2} \left[\frac{K_s}{2} - G_2 K_c \frac{S}{1-S}(2+\sigma^2)\right] \quad \text{where} \quad K_m = K_1 F_z \quad (4)$$

The first function gives an initial linear slope of aligning moment as a function of lateral slip ratio which then falls off at higher slips due to the denominator quadratic in composite slip. The shaping coefficient G_1

allows fitting the peak and fall off of aligning torque at high slip. The second function and shaping coefficient G_2 allow accounting for combined cornering and braking effects on aligning moment due to tire patch lateral offset.

On paved surfaces tire forces reach a peak at relatively low slip conditions, then fall off with further increases in slip in the saturation region of force production. This peak coefficient of friction can be different for lateral and longitudinal force production, and can be interpreted in the sense of a friction ellipse. STIREMOD includes separate lateral and longitudinal slip to slide transition equations for coefficient of friction:

$$\mu_x = \mu_{px}\left(1 - K_{\mu x}\sqrt{\sin^2\alpha + S^2\cos^2\alpha}\right) \; ; \; \mu_y = \mu_{py}\left(1 - K_{\mu y}\sqrt{\sin^2\alpha + S^2\cos^2\alpha}\right) \quad (5)$$

Here μ_x and μ_y are the equivalent of "slide" coefficients of friction, and μ_{px} and μ_{py} are the peak coefficients of friction. Under paved surface conditions the limit slip coefficients of friction are referred to as slide coefficients of friction and are typically 10-30% below the peaks (μ_{px} and μ_{py}) depending on speed as defined by the $K_{\mu x}$ and $K_{\mu y}$ parameters. By setting $K_{\mu y}$ to negative values, tire forces can also be caused to increase beyond the peak transition region, which can be used to produce side forces due to surface deformation under high lateral slip conditions. The peak coefficients of friction are also a function of normal load [2].

In the slip to slide transition region two other saturation functions are defined. The longitudinal stiffness coefficient K_C merges to the lateral stiffness coefficient K_s for symmetry in the limit locked wheel condition and the camber stiffness Y_γ goes to some small value:

$$K'_c = K_c + (K_s - K_c)\sqrt{\sin^2\alpha + S^2\cos^2\alpha} \; ; \; Y'_\gamma = Y_\gamma\left[1 - K_\lambda f^2(\sigma)\right] \quad (6)$$

MODEL PARAMETER IDENTIFICATION

Tire data for paved surfaces is typically obtained on tire testing machines. Procedures have been developed using MATLAB© routines to carry out nonlinear parameter identification required by the above model equations and their subsidiary load varying coefficients [2]. Metz [5] has formulated a set of lateral force equations for off-road surfaces and provided coefficients for several soil conditions. The saturation function in equation 6 above can be used to approximate Metz's exponential response [2] given proper values for the C_i coefficients. Slide coefficients of friction for large side slip conditions have been noted by DeLeys and Brinkman [3] and measured for a variety of surfaces by Christoffersen, et al. [4].

Example pavement parameter estimation data is illustrated in Figure 1 for a General Tire "Ameritech ST" P205/65R15 tested on an MTS Flat-Trac II tire machine. Parameters were fit in a two step process. First, initial condition and load varying parameters were fit with quadratic functions [2]. The load varying parameters included longitudinal, lateral and camber stiffness, and longitudinal and lateral peak and slide coefficients of friction. The second step involved non-linear iterative estimation procedures to fit the force saturation polynomial, braking/driving force asymmetry factor, longitudinal force effects on cornering stiffness, and aligning moment shaping parameters. The pavement characteristics are modeled quite well, including the cornering stiffness at high longitudinal slip (i.e. accelerating or braking during cornering).

The influence of off-road parameters as measured by Metz [5] and Christoffersen, et al. [4] are illustrated in Figure 2. The effect of the this loose soil condition is to give a low longitudinal coefficient of friction, a high lateral coefficient of friction under high side slip conditions. The development of the high lateral forces under high side slip conditions depends on the soil shear strength, and the shear stress exerted by the tire. This process can be approximated by a shear force development time constant similar to a rolling tire's relaxation time constant [2].

REFERENCES

1. Allen, R.W., Szostak, H.T. and Rosenthal, T.J., "Steady State and Transient Analysis of Ground Vehicle Handling," SAE Paper 870495, Society of Automotive Engineers, Warrendale, PA.
2. Allen, R.W., Chrstos, J.P. and Rosenthal, T.J., "A Vehicle Dynamics Tire Model for Both Pavement and Off-Road Conditions," 1997 SAE International Congress (in press), Society of Automotive Engineers, Warrendale, PA
3. DeLeys, N.J. and Brinkman, C.P., (1987), "Rollover Potential of Vehicles on Embankments, Sideslopes, and Other Roadside Features," SAE Paper 870234, Society of Automotive Engineers, Warrendale,PA.
4. Christoffersen, S.R., et al., (1995), "Deceleration Factors on Off-Road Surfaces Applicable for Accident Reconstruction," SAE Paper 950139, Society of Automotive Engineers, Warrendale, PA.
5. Metz, L.D., "Dynamics of Four-Wheel-Steer Off-Highway Vehicles," SAE Paper 930765, Society of Automotive Engineers, Warrendale, PA.

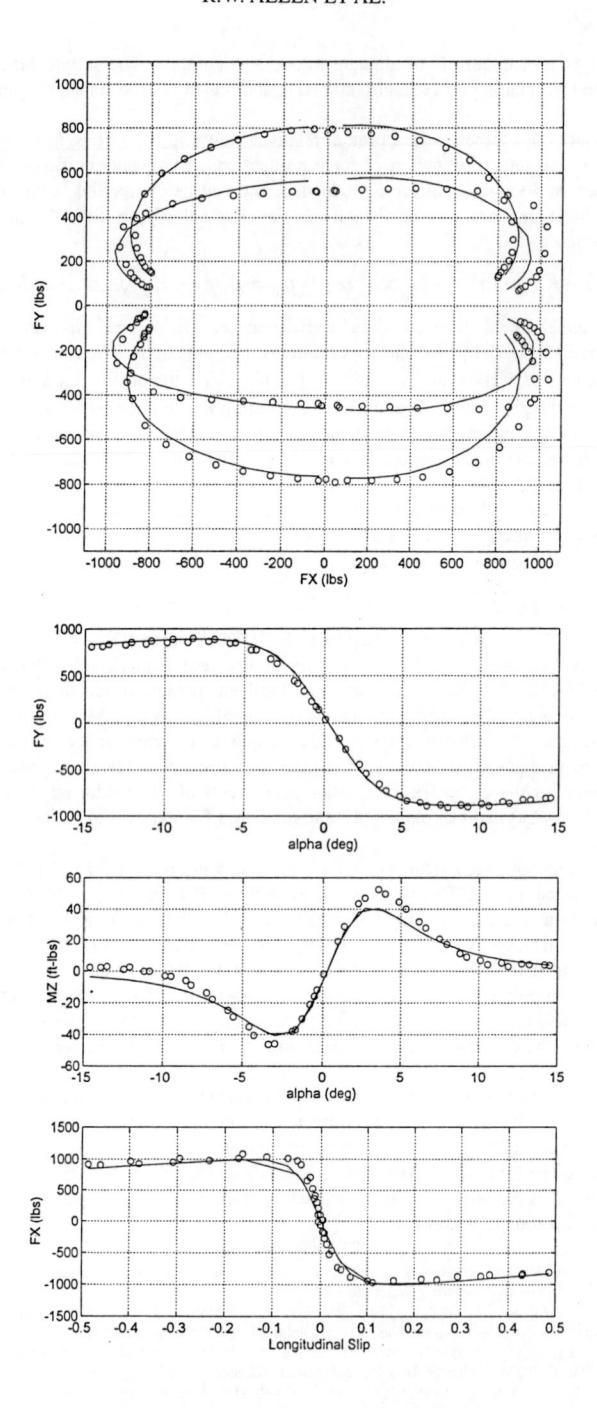

Figure 1. STIREMOD validation data for a General Tire Ameritech ST P205/65R15: Normal load 930 lbs., Slip Angles of -4,-2, 2 and 4 degrees in F_x, F_y plot.

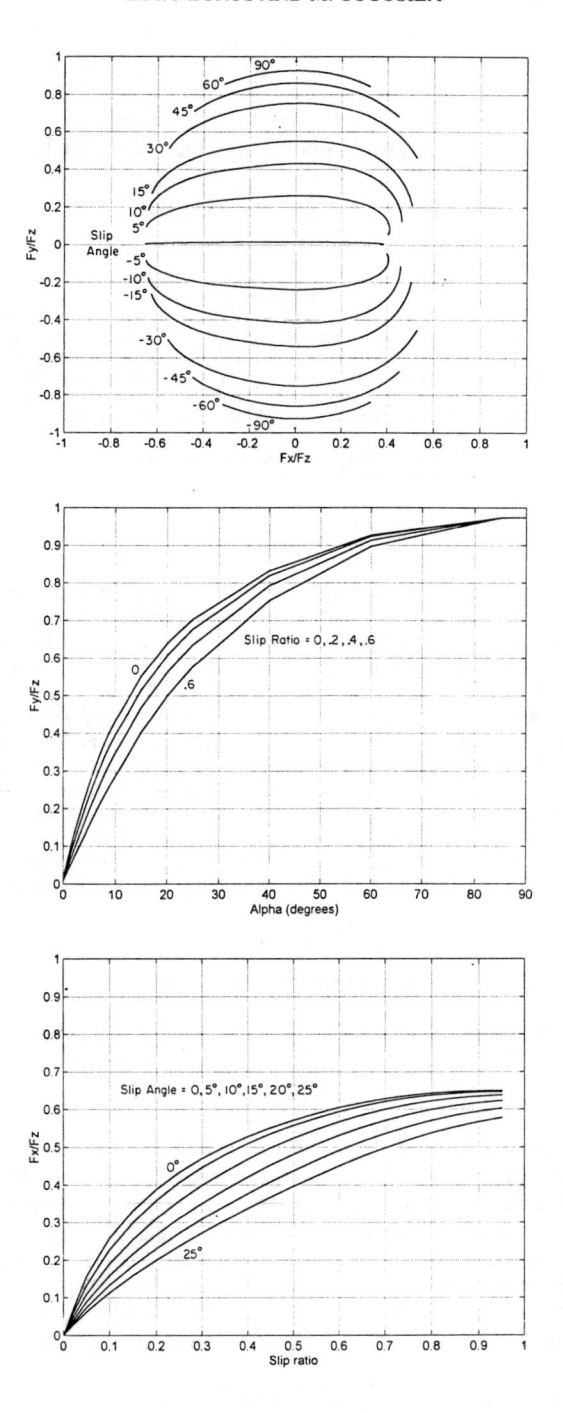

Figure 2. STIREMOD response under off-road conditions: Metz [5] `plowed field' condition for adhesion and transition; Christoffersen, et al. [4] `bedded corn rows' for high slip angle ($K_{\mu y}$=-0.5).

Flexible Model to Simulate Wheel Pass Over Singular Road Obstacles

E.M. NEGRUS and M. COCOSILA

ABSTRACT

The paper presents a new approach regarding the study of the dynamics of the tire-suspension-vehicle system during wheel pass over short road obstacles. The proposed mathematical model suitable for the above study assumes the motor-vehicle to be a system composed of rigid bodies with multiple connections and utilizes matrix calculations. This allows the model to be extremely flexible and adaptable to changes to various concrete situations. In this paper the above model is applied by means of a computer program to simulate the car wheel pass over singular road obstacles. The simulation is based upon measurement data for the tire enveloping process found in the available literature which allow a black-box model for the tire in these circumstances. The remote target of the simulation is to try to improve vehicle road safety on uneven roads. Current studies are done in order to verify the suitability of the approach presented in this work to various practical situations.

INTRODUCTION

Studying the dynamics of a road vehicle by taking into account its constitutive components is not an easy task and that usually leads to a complicated mathematical problem. Any changes, even small, to be applied to the system would cause difficult to handle changes to the system of differential equations which describe the dynamic behavior of the system. Therefore a different approach would be to consider the vehicle as a system composed of several rigid bodies. Between these bodies are connections involving various damping and stiffness characteristics. The global equation of the vehicle as a system can be then written in the matrix form and the same for the equation describing the behavior of any smaller system (as tire-suspension for example) included in the vehicle system.

This paper actually applies the matrix model philosophy to a tire-suspension-vehicle system during wheel pass over short road obstacles. The target of the computer simulation based on the matrix model theory is to find the best suspension damping characteristics which would lead to the smallest wheel center displacement during obstacle encountering. The reason of this target is to provide the best road safety conditions from the wheel displacement point of view during a road obstacle pass.

PRINCIPLE OF THE MATRIX MODEL

According to Newton's law the movement of any rigid body can be described by the following equation:

$$[M]\{\ddot{\delta}\} = \{F\} \qquad (1)$$

where $[M]$ is the mass matrix, $\{\ddot{\delta}\}$ is the accelerations vector and $\{F\}$ is the applied forces vector. The displacement of the center of gravity of the body is expressed as a vector with 6 components (3 translations and 3 rotations): $\delta = \{\delta_x, \delta_y, \delta_z, \delta_{xx}, \delta_{yy}, \delta_{zz}\}$. The same equation applies to a system composed of at least two rigid interconnected bodies with the difference that in this case the general equation contains also forces due to the connection between bodies (stiffness, damping and constraints). For example in case of a rigid body of a system the connection displacement (considered, for simplicity reasons, as a point) $\{\delta_C\}$ can be expressed in terms of the displacement of the center of gravity of the same body $\{\delta_{CG}\}$ as:

$$\{\delta_C\} = [R]\{\delta_{CG}\} \qquad (2)$$

where $[R]$ is a square matrix of fixed dimension (6 x 6) which characterizes the space position of the connection point with respect to the center of gravity of the body. One can also prove that the vector of forces and moments applied to the center of gravity of the body $\{F_{CG}\}$ can be expressed in terms of the vector of forces applied to the connection point with another body $\{F_C\}$ by the relation:

$$\{F_{CG}\} = [R]^T \{F_C\} \tag{3}$$

One should notice that the above relations (2) and (3) assume small relative displacements of bodies what is satisfactory for usual situations which consider that the time evolution of the system is investigated through numerical methods which allow small time steps.

THE DYNAMIC MODEL TIRE-SUSPENSION-CAR AS A SYSTEM OF BODIES

A motor-vehicle can be assumed to be a system of rigid interconnected bodies. After expressing the displacements of the connection points of the bodies in terms of the displacements of the centers of gravity and of the positions of the connection points with respect to the centers of gravity, the general movement equation (1) becomes:

$$[M]\{\ddot{\delta}\} + [RCR]\{\dot{\delta}\} + [RKR]\{\delta\} + [RA]\{F_{CON}\} = \{F_E\} \tag{4}$$

where the aggregated matrices $[RCR], [RKR], [RA]$ describe the position and magnitude of the damping, stiffness and constraints of the system respectively whereas $\{F_{CON}\}$ is the vector of the constraint forces in the system and $\{F_E\}$ is the vector of the external forces applied to the system.

Due to the matrix approach the mathematical model proves to be very flexible. Any modification regarding the data of the investigated body system (masses or dimensions) can be performed in an easy way by changing some elements of some matrices.

TIRE-SUSPENSION-CAR SYSTEM SIMULATION ON UNEVEN ROADS

The mathematical matrix model and the corresponding software were applied to a quarter-car model [4] which consists of two masses connected by a suspension with certain stiffness and damping characteristics and supported above ground by a tire having, similarly, certain stiffness and damping characteristics. The target of this simulation is to investigate the behavior of the above dynamic system when encountering a road obstacle with a view to improve car safety in these conditions. Road holding (the capability of the tire to follow as closely as possible the road contour) was used as a criterion for car safety.

A key problem of the simulation was how to model the tire while enveloping an obstacle. The main idea of this paper is to model the tire subjected to an enveloping process as a black-box with known output for a known input. The model built is discrete for a large number of points and it refers to a tire passing over a short road obstacle for which the variation of the vertical load is fully known from experiments performed with a constant wheel axle height, for low speeds and usual loads and tire inflation pressures [1] [3] (Figure1). These tests showed a wheel normal load variation with two maxims (corresponding to the obstacle "attack" and "release" moments, respectively). Since the tire is isolated from other suspension's effects one may assume that the vertical load variation is, in fact, a variation of the vertical stiffness coefficient of the tire. This happens in dynamic conditions at low speeds while "swallowing" the obstacle. The variable stiffness coefficient which can be calculated step-by-step is then used for a dynamic simulation of a tire-suspension-vehicle system while passing over a short road obstacle.

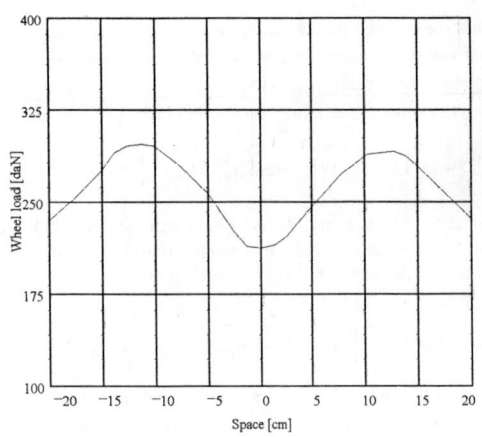

Figure 1. Variation of axle normal load as a function of obstacle position with respect to wheel axle

The model is an extension of the model presented in [2] and based upon data of [4]. The model consists of two masses (m_1 and m_2 which represent the unsprung and sprung mass, respectively) connected by a suspension with the stiffness coefficient k_s and the damping coefficient c_s, respectively. The tire is assumed to have a constant damping c_t but a variable stiffness k_t.

The data used for the simulation have been as follows: m_1=33 kg; m_2=200 kg; k_t= variable; k_s=9000 N/m; c_t=1000 Ns/m; c_s=-5000...5000 Ns/m. The system is assumed to translate with constant low speed (8 m/s) causing the tire to encounter a short rectangular obstacle (20 cm long and 2.5 cm high, wider than the contact patch). As a reasonable solution to diminish the vertical motion of the wheel, the suspension's damper was considered adjustable.

RESULTS

The simulation was performed in Matlab for Windows and is an improvement of the method and computer software presented in [2]. After several attempts it appeared that high suspension damping leads to a small wheel vertical displacement but with rather long settling time. It is likely that variation of the suspension damping coefficient in smaller steps during obstacle swallowing and some time after that moment would produce an even smaller variation of the wheel center height. Some results of the simulation are displayed in Figure 2 (the obstacle is "attacked" at time 0.06 sec).

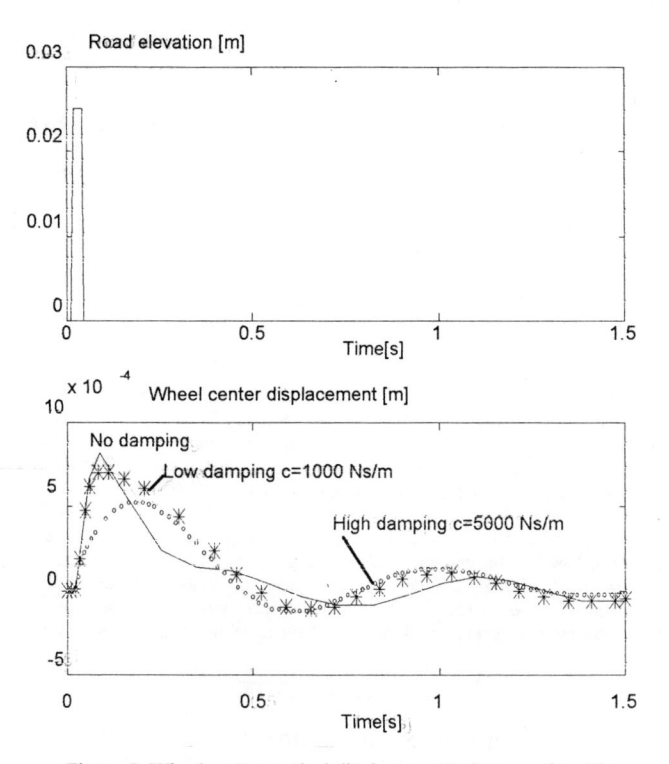

Figure 2. Wheel center vertical displacement [m] versus time [s]

CONCLUSIONS

This paper has presented a new approach used to study the motor-vehicle dynamics during wheel pass over short road obstacles. The flexible matrix model offers clear advantages in terms of model handling and updating. Using the matrix model approach for a quarter-car vehicle model which considered the tire as a black-box with output known from experimental data gave some promising results for the study of the system's behavior while encountering a road obstacle. However, extensive studies trying to apply this model as well as experimental validation of the results is mandatory before concluding upon the theoretical approach proposed in this paper. If validated by experiments this theoretical approach might be useful in studying the dynamic behavior of more refined 3D vehicle models in similar conditions.

REFERENCES

[1] Clark, S.K.,-editor, "*Mechanics of Pneumatic Tires*", University of Michigan, 1981;
[2] Cocosila, M., Stancioiu, D., "*Simulation of the Behavior of the Tire-Suspension-Vehicle System at Wheel Pass Over Singular Road Obstacles*", ESFA '95, Bucharest, 1995;
[3] Pottinger, M. G., et al, "*A Review of Tire/Pavement Interaction Induced Noise and Vibration*", The Tire Pavement Interface, ASTP STP 929, Philadelphia, 1986;
[4] Venhovens, P.J., "*Optimal Control of Vehicle Suspensions*", Delft University of Technology, Delft, 1993.

A tyre model for interactive driving simulators

Sylvain DETALLE, Julien FLAMENT & Franz GAILLIEGUE
(GIE SARA)

ABSTRACT

This paper takes place in the general context of interactive driving simulation and is based on development works made on the simulators used by INRETS, PSA and RENAULT, the partners of the SARA (Advanced driving Simulator for Automotive Research) project. There is a need for simulators that accurately reproduce the driving cues necessary to enable handling analyses with virtual prototypes of light and heavy vehicles. For this purpose, the SARA modelling team is developing ARHMM (Advanced Road Handling Modular Model). The present paper explains current characteristics of the ARHMM [tyre] module and gives an overview of incoming evolutions. The ARHMM [tyre] module, enabling the vehicle to run on three dimensional ground, is based on the coupled Michelin Magic Formula. Owing to numerical improvements, this formulation has been adapted to low velocities, important loads and high side slip angles. The functioning area has therefore been enlarged, although it is unsure by how much the validity domain has also been increased. So transient phenomena and friction still have to be introduced to render tyre dynamics.

INTRODUCTION

The SARA (**A**dvanced driving **S**imulator for **A**utomotive **R**esearch) project is aimed at building an interactive driving simulator usable during the design and development cycle of both light vehicles and trucks [3]. The vehicle model software for SARA is called ARHMM (**A**dvanced **R**oad **H**andling **M**odular **M**odel) [1] [2]. Requirements for the ARHMM project are derived from the SARA project. ARHMM should be an accurate vehicle handling model respecting real time constraints and high software quality standard.

A prototype version of the vehicle dynamic model is now finalized. It is composed of seven vehicle modules like [chassis + axles], [steering system], [drive line + front wheels], [rear wheels] or [tyre]. The tyre representation quality accounts for a major part in the global fidelity of a driving simulator model. The current tyre model is quasistatical and adapted from the formulation proposed by Michelin in [4] and hereafter named Michelin Magic Formula. The tyre vertical force F_z^{tyre} is calculated according to a linear spring radial model of the tyre. The other tyre efforts are computed as a function of F_z^{tyre} and three kinematic variables: longitudinal slip S_x, side slip angle δ and wheel/road camber angle γ. The Michelin Magic Formula provides contact effort calculations close to the measured ones in a "classical" domain of the model variables. Nevertheless, the tyre model should enable an effort computation defined in the whole functioning domain with a validity domain as large as possible. Thus, some adaptations have been made to improve the representativity of the tyre model while still insuring the stability of the complete model.

The quasistatical approximation, still used in the tyre model, is not accurate enough in some conditions (especially for low velocity). So, we decided to study different ways to introduce transient phenomena and friction in the tyre behaviour for further development.

TYRE INFLUENCE IN VEHICLE MODEL STABILITY

In case of low velocities the [tyre] module generates instabilities in the numerical integration of modules that are coupled with it (see following figure).

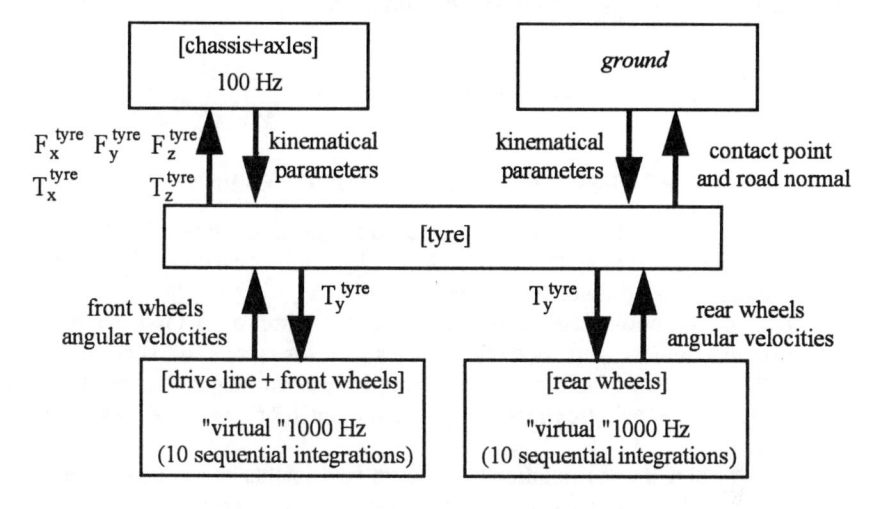

These instabilities are due to the longitudinal slip and side slip angle formulations which use velocities in the denominator. A stability analysis of the differential equations contained in the [chassis + axles], [drive line + front wheels] and [rear wheels] modules leads to a minimum condition on velocities. The solution to this difficulty is to impose a bottom value to the velocities used in the denominators. These thresholds depend upon vertical load, wheel rotation and integration step. In this way, a "safe" functioning of the [tyre] model is guaranteed. This will now be illustrated on a rear wheel.

The generic differential equation that modelizes rear wheels dynamics is:

$$I^{wheel} \cdot \dot{\omega} = T^{brake} + T_y^{tyre} \tag{1}$$

where T^{brake} is the brake torque, T_y^{tyre} is the tyre torque around wheel rotation axis, and ω is the wheel angular velocity.

The Euler method is used to resolve (1) and its linearization gives:

$$\dot{\omega} = K \cdot \omega + H \quad \text{with} \quad K = \frac{1}{I^{wheel}} \cdot \left(\frac{\partial T_y^{tyre}}{\partial \omega} \right)$$

Indeed, only T_y^{tyre} depends on ω through the longitudinal slip S_x.

If h^rear is the integration step for [rear wheels], the stability condition is:

$$\left|1+K\cdot h^{rear}\right|<1$$

It leads to $\left|V_x\right|>V_{xlimS_x}$ where V_x is the longitudinal velocity of the wheel and

$$V_{xlimS_x}=\frac{50\cdot Rr^2\cdot h^{rear}\cdot\left|BCD_{F_x}\right|}{I^{wheel}}$$ (Rr is the wheel roll radius)

Finally, while $\left|V_x\right|>V_{xlimS_x}$ stability is guaranteed and $S_x=\frac{Rr\cdot\omega-V_x}{V_x}\cdot 100$,

but when $\left|V_x\right|<V_{xlimS_x}$, the formulation becomes $S_x=\frac{Rr\cdot\omega-V_x}{sign(V_x)\cdot V_{xlimS_x}}\cdot 100$.

TYRE REPRESENTATIVITY IMPROVEMENT

The analysis of the efforts obtained from the Michelin Magic Formula shows that some of its macro-coefficients can produce physically aberrant values, when variables they are depending on are extreme.

For this reason, the tyre representation has also been improved for high vertical loads, extreme cambers and side slip angles. In the scope of such extreme conditions the Michelin Magic Formula curves have been adapted to be closer to these obtained from a real tyre measured on a trial bench or a car. This approach is only qualitative.

The parameters and the formulations of macro-coefficients are, when necessary, corrected in order to respect a series of physically consistent behaviour types of the tyre. The following figure presents, for a fixed vertical load, typical curves of lateral force and self-aligning torque as function of side slip angle. They correspond to real tyre behaviour, measured on trial bench. The corrected tyre model representation enabled us to produce such curves.

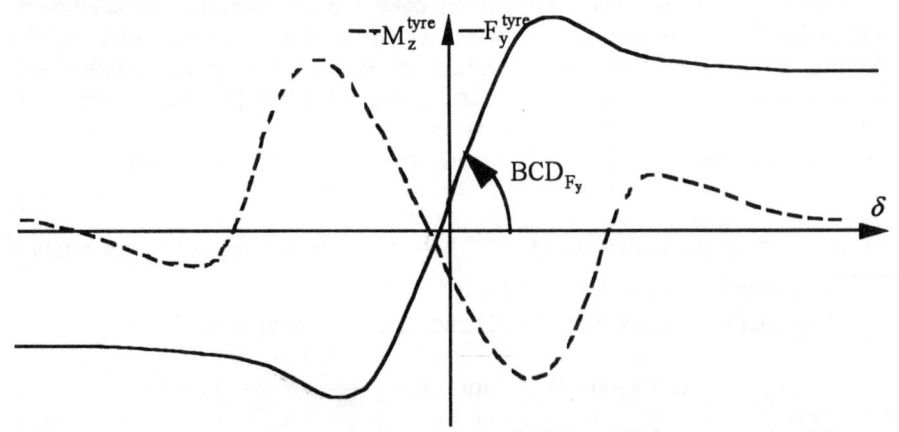

The first kind of correction consists in truncating the value of macro coefficients which are not in pre-defined interval. For example, the form factor must be less than 1 otherwise its influence on slip parameters may produce effort corresponding to a slip of opposite sign. Such corrections are applied to the all coupled coefficients.

The second correction possibility is to truncate the value of F_z^{tyre} which is used to determine the macro-coefficients. Indeed, the lateral force sliding rigidity, BCD_{Fy} should be positive and increase with F_z^{tyre}. When this is not the case, this macro-coefficient is calculated as $BCD_{Fy}=BCD_{Fy}(F_{zlim})$, where F_{zlim} is the greatest value of the vertical load for which BCD_{Fy} is still positive and increasing. Similar treatments are applied to longitudinal force, lateral force and on self-aligning torque extreme values. A complementary truncation is made on camber angle for the lateral force maximum and self-aligning torque minimum. Furthermore, the self aligning torque formulation is modified for great side slip angle values in order to reach a "null" asymptote.

When the vertical force F_z^{tyre} is zero or less, the formulation is undefined. This situation corresponds with the wheel lefting off. In that case, the "contact" efforts are forced to zero.

For the validity domain to be as large as possible, the tyre model must be able to take into account variation of the road friction conditions. At first glance, we supposed that there was a linear relation between tyre efforts and friction values. Yet, the further the friction is from the reference friction, the worse is the approximation. So, two sets of parameters are employed corresponding to two adhesion conditions: μ_0 and μ_1. In general, μ_0 corresponds to low friction value and μ_1 is close to 1. We define μ_m as the geometric mean of μ_0 and μ_1. For friction below μ_m a linear interpolation is carried out with the parameter set corresponding to μ_0 and for values above the interpolation is based on the μ_1 set.

CONCLUSION

The classical Michelin Magic Formula model has been significantly improved to allow a large and physically coherent use of interactive driving simulators. But the representativity of the current solution is not completely satisfactory for low velocities and transient domains. In order to improve vehicle design possibilities, current investigations on ARHMM consist in defining the way to efficiently represent the tyre dynamics. The most important constraints are the respect of real time computation, the compatibility with characterization standards and the need for means of validation.

REFERENCES

[1] Sylvain Detalle, Jean-Marie Rousset, Thierry Voillequin, "SARA: Real-time dynamic vehicle models and generation tools", Real Time Systems, "Driving Simulation" conference, Paris, January 11-14 1994
[2] Sylvain Detalle, Laurence Duverly, Jean-Marie Rousset, Bernd Weber, "Automatic Equation Generation for Real-Time Dynamic Vehicle Models on SARA Project", Driving Simulation Conference, Sophia Antipolis, September 12-13 1995
[3] Sylvain Detalle, Jean-Marie Rousset, "SARA : un simulateur de conduite automobile", Entretiens de la Technologie, Paris, March 27-28 1996
[4] P.Bayle, JF Forissier, S. Lafon, MICHELIN, "A new tire model for vehicle dynamics simulations", Automotive Technology International 93.

The System Tractor - Tire under the Influence of Tractor Development

Chr. von Holst[*] and H. Göhlich[*]

ABSTRACT

Tractor tires are noncircular and although the noncircularity is small, it has considerable influence on the stimulation of vibrations. In a research work at the TU - Berlin such tires are measured and a method was found to eliminate the noncircularities very easy and fast. Tires of the dimension 650 / 75 R 32 are explored and it was determined, that a noncircularity of 7 mm enlarges the amplitude of the tread bar gearing acceleration up to 25 %. With outlook to the abound use of numerical simulation, it is more then ever necessary to describe such effects for self stimulation of vibration and exact modelling of tires as decisive element of ride dynamic properties.

INTRODUCTION

In the development of modern tractors a remarkable trend to front axle suspensions is to observe. In this manner higher speed at transportation rides is possibel. By the dynamic decoupling of the axle and the vehicle the tires have to be more critical examined considering their influence to safety, comfort and soil compression. For the designer of such vehicles it is necessary to know the dynamic properties of the tires in order to design an optimal axle suspension. For the producer of tires it is necessary to consider the changing demands in ride dynamics. A research work at the TU - Berlin, Dept. of Agricultural Engineering, was concerned to determine the dynamic properties of tractor tires. In addition to spring rates and damping rates the rolling resistance, absorption capacity and the noncircularity have been explored.

During the developing of an axle suspension it is useful to apply the numerical simulation. To describe the modells as exact as necassery the knowledge about the geometric -, the mass - and the vibration properties of the system tire - vehicle is urgent. The geometric properties and masses of every part of the vehicle exist by CAD - construction. To describe the dynamic behavior of tires in numerical simulation process, it is necessary to examine the influence of every parameter to build an exact modell [1]. In this section the influence of noncircularity of tires on their dynamic properties at higher frequencies was examined. Such higher frequencies become more important, because spring mounted tractor axels have resonance frequencys between 6 Hz and 10 Hz. During the research a method was

---

[*] Department of Agricultural Engineering and Hydraulics, Technicale University Berlin, G - 14199 Berlin, Germany

developed, to measure the noncircularity and to grind the tire with high shape tolerance. The contour of the running tread was completly preserved.

DETERMINATION OF THE DYNAMIC QUALITIES OF TIRE TREAD BARS

The tire is mounted in a flat track test stand (figure 1) on a drivable and stearable tractor front axle. Two hydraulic cylinders are pressing the tire with the proof force on the steel band and are able to vary the wheel inclination. The steel band is driven by a hydraulic motor to simulate a ride velocity to the tire. The proof force is equal to a static axle load. With this test stand stearing angles between ± 5°, inclinations between ± 10°, proof forces up to 40 kN and simulated ride velocitys up to 40 km/h are applicable. This parameters are conditional to tire diameter, tire pressure and simulated ride velocity. To determine the dynamic qualities of tire tread bars, the vertical acceleration of the wheel is measured on the axle in the projection of the foot print. This experimental arrangement impedes the tire to do free oscillations. The tread bar influence to the tire dynamic properties related to the tire pressure, the ride velocity and the proof force is identifiable. The tread bar gearing frequency is detected by frequency analysis and proved by a simple formula:

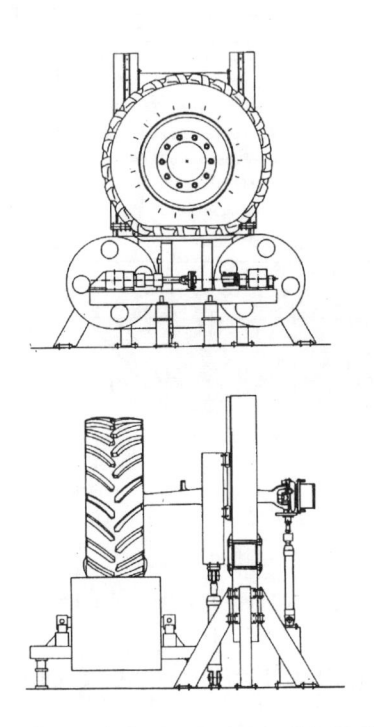

Figure 1: Flat - track tire test stand at the TU - Berlin [2]

$$f_{Otbg} = \frac{Zv_{Fahr}}{U_{Reifen}} \qquad (1)$$

With: f_{Otbg} = tread bar gearing resonance frequency (Hz), Z = number of tread bars, v_{Fahr} = ride velocity (m/s) and U_{Reifen} = tire circumference (m).

The result of such an experiment is shown exemplary in figure 2. Very pronounced vibrations are observed at 15 km/h and 30 km/h. The half tread bar resonance frequencies are dominating. The reason is the difference of tread bar hight in the middle of the foot print and the better conduction of vibration of the tire sidewalls.

The maximum acceleration peaks could be found at lower frequencies than expected. Particulary at low ride velocities such vibrations may have an effect to spring mounted front axles. The resonance frequencies of spring mounted front axles are in the range of approx. 6 Hz up to approx. 10 Hz. Additional the vibrations, stimulated by tread bar gearing, affect ride comfort and may demage vehicle components (i. e. mirrors o. e.).

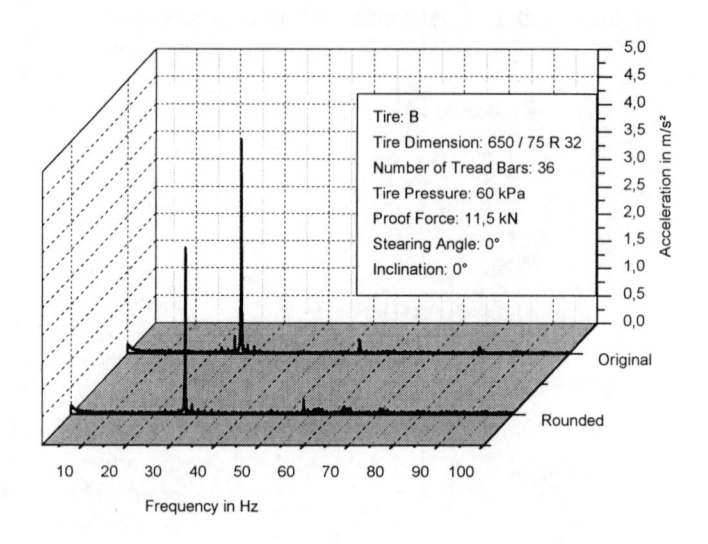

Figure 2: tire tread bar resonance frequency at 30 km/h of the original and the rounded tire

DETERMINATION AND ELIMINATION OF NONCIRCULARITY

The noncircularity of the unloaded tires and the contour of the tread bars was measured by a laser - distance sensor. The tire was mounted in a rocker. The laser - distance sensor was displaced transversaly to the running tread in different defined positions, the tire was turned and the distance between laser - sensor and tire tread surface and the rotation angle are measured. Also the bead of the rims noncircularity was measured, because the noncircularity of the rim has experiencely a remarkable influence on the wheel noncircularity [3]. The results are exemplary shown in figure 3. Then the tire was rounded by grinding. A special link was constructed, which allowes a grinding machine to follow exactly the former tire tread surface contour. After rounding the tire the acceleration by tread bar gearing was measured again.

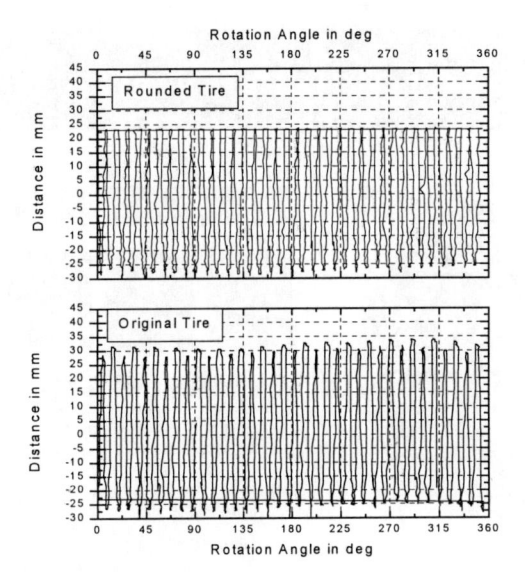

Figure 3: Noncircularity of the tire before and after rounding. Tire pressure during rounding: 200 kPa, tire pressure during measuring: 60 kPa. Tread surface is equal to positive distance.

CONCLUSION

The very small (even to tire diameter) noncircularity of a tire and the difference in hight of one tread bar to its neighbour enlarge the acceleration of tread bar gearing up to 25 % in the half tread bar gearing resonance frequency. The resulting vibration reduce the ride comfort and may have influence to ride safty of tractors with spring mounted axles. To get a good simulation modell of tractor tires such effects have to be examined and if poosible described by mathematical methods. With higher requirements to tractor dynamics it is necessary to describe the properties of tires in more detail. Therefore the research work at the TU - Berlin tries to examine the influence of tire hollow to spring - and damping rate, the absorption capacity and the effects of stimulating vibrations. With regard to that not only test stands are in use, but also field tests of tires and vehicles are applied.

REFERENCES

1. Kising, A., „Dynamische Eigenschaften von Traktor - Reifen", Dissertation, VDI Verlag Reihe 14 Nr. 40, Düsseldorf, 1988
2. Siefkes, T., „Die Dynamik in der Kontaktfläche von Reifen und Fahrbahn und ihr Einfluß auf das Verschleißverhalten von Traktor - Triebradreifen", Dissertation, VDI Verlag Rehe 14 Nr. 67, Düsseldorf, 1994
3. Yeh, Ch. - K., „Experimentelle Untersuchungen über Unwuchten und Unrundheiten von Ackerschlepperreifen", Dissertation, Selbstverlag TU - Berlin, 1992

Tractor - Tires

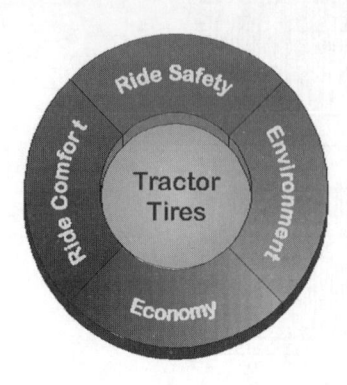

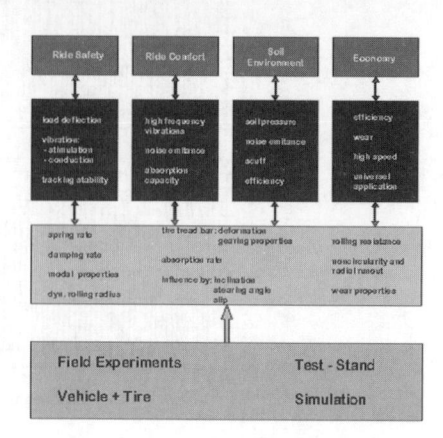

Flat - Track Test Stand

Features:
Tire diameter up to 2 m
Tire width up to 800 mm
Tire load up to 5 t
Inclination -5° to +5°
Steering angle -5° to +5°
Ride velocity up to 40 km/h
Slip variable

Flat - Track Test Stand

Features:
Tire diameter up to 2 m
Tire width up to 600 mm
Tire load up to 2,5 t
Ride velocity up to 40 km/h
Dying out facility
Stimulation of vibration in the food print

Tire - Test Rocker

Features:
Tire diameter up to 2 m
Tire width up to 1000 mm
Tire load up to 3 t
Double tires
Stimulation of vibration to the axle
Stimulation of lateral vibration

Four - Piston Vibration Test Stand

Features:
Vehicles up to 6 t
Track width up to 2,5 m
Wheel base up to 3 m
Single triggered pistons
Driven wheels

Grinding Facility

Removing of tire non-circularity. Without changing tire tread contour by special guiding appliance of the grinding machine. Special abrasive for tire grinding.

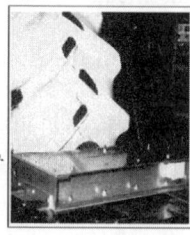

A Comparison of the Moreland and Von Schlippe-Dietrich Landing Gear Tire Shimmy Models

WILLIAM E. KRABACHER

ABSTRACT

A detailed review of two classical landing gear tire shimmy models is presented. First, derivations of each of the two models are presented along with the detailed definitions of the various tire parameters used in each model. A mathematical derivation is presented which establishes a clear mathematical connection between these two tire models. Also, a mathematical connection between the Moreland tire parameters and the Von Schlippe-Dietrich tire parameters is established.

INTRODUCTION

To the present date, two apparently different aircraft landing gear tire shimmy models have been developed. In 1941, Von Schlippe and Dietrich [1] developed what has come to be known as the stretched string model that considers the tire to be a perfectly elastic string. Two critical components of this model are the tire relaxation length and the tire half footprint length. In 1954, Moreland [2] developed what has come to be known as the tire point contact model which considers all forces and torques as acting at the center point of the tire contact patch. Two critical components of this model were the tire time constant and the tire yaw coefficient.

In the intervening years since the introduction of the Moreland tire model, there has been recurrent disputes over which of these two tire models is more predictively accurate in dealing with the problem of landing gear shimmy. It is the main objective of this paper to show that these two tire models are really quite similar and that they are essentially two different approaches to the same results.

MODEL DEFINITON

The first model to be presented is the Von Schlippe-Dietrich tire model.[1] To define the various variables and parameters in this model reference is made to Figure 1. The coordinate of the wheel axle center is y_a and the coordinate of the

---

[1] The version of the Von Schlippe Dietrich model presented here follows the presentation of Black [3].

most forward position of the tire footprint is y_f. Since the strut is deflected, the wheel is tilted at the angle ϕ and yawed about the vertical axis an angle ψ_s.

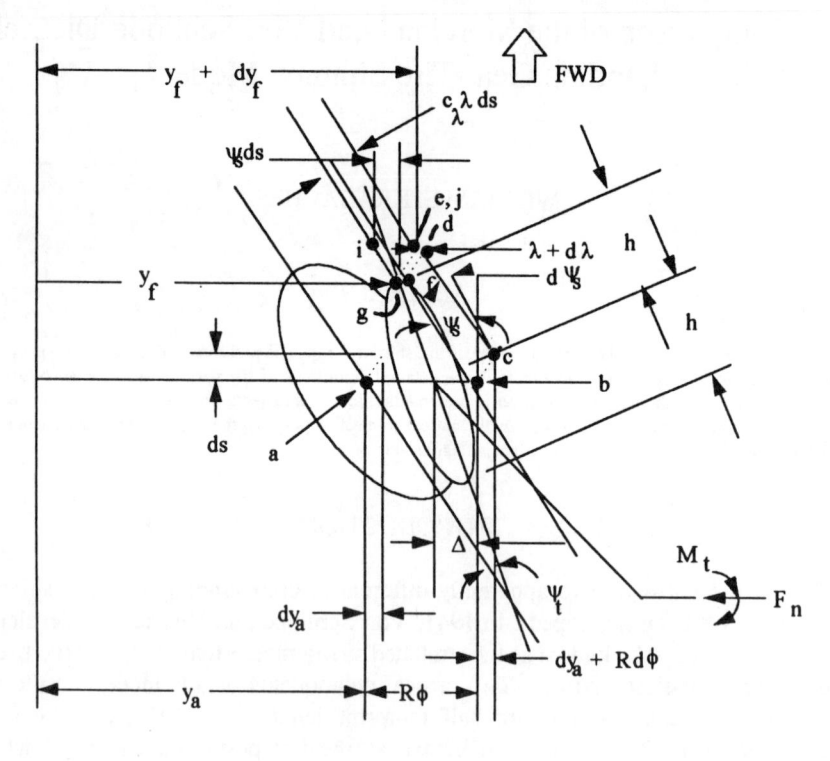

Figure 1

The parameter c_λ is essentially the reciprocal of the tire relaxation length. The variable λ is the lateral displacement of the tire outside the footprint area.

Now, from Figure 1, assume that the tire moves forward an amount ds. Then all of the other variables will change by the amounts dy_a, $d\lambda$, $c_\lambda \lambda ds$, $d\phi$, and $d\psi_s$. If one then sums the differentials around the path defined by a to b to c to d to e, one then obtains

$$y_f + dy_f = y_a + R\phi + dy_a + Rd\phi - h(\psi_s + d\psi_s) - (\lambda + d\lambda)$$

Similarly, if one sums the differentials around the path defined by a to b to f to g to i to j, one obtains

$$y_f + dy_f = y_a + R\phi - h\psi_s - \lambda - \psi_s ds + c_\lambda \lambda ds$$

Setting these last two equations equal to one another, one obtains

$$dy_a + R\phi - hd\psi_s - d\lambda = -\psi_s ds + c_\lambda \lambda ds$$

or dividing by ds one obtains

$$d\lambda/ds + c_\lambda \lambda = dy_a/ds + rd\phi/ds + \psi_s - h\, d\psi_s/ds$$

Eliminating λ from this last equation using the relation $\lambda = y_a - y_f + R\phi - h\psi_s$ one obtains

$$dy_f/ds + c_\lambda y_f = c_\lambda(y_a + R\phi) - (1 + c_\lambda h)\psi_s$$

Letting $D = d/ds$, this last equation can be written as

$$y_f = (1/(D + c_\lambda))(c_\lambda(y_a + R\phi) - (1+c_\lambda)\psi_s)$$

Due to adhesion of the tire footprint to the ground, $y_f(s = -h) = y_a + R\phi - \Delta$ where $s = -h$ is the center of the contact patch. Taking the series expansion of the left side of this last equation and making use of the fact that for a constant forward velocity $D = (1/V)D_t$

$$(1 - (h/V)D_t + (h^2/2!V^2)D_t^2 - (h^3/3!V^3)D_t^3 + ...)(c_\lambda(y_a + R\phi) - (1 + c_\lambda h)\psi_s) = (D_t/V + c_\lambda)(y_a + R\phi - \Delta)$$

This last equation is the basic equation of the Von Schlippe-Dietrich tire model. The two basic tire parameters of the Von Schlippe-Dietrich model are the tire relaxation length, L, and the tire half footprint length, h. A dynamic definition of the tire relaxation length is, giving the wheel tire combination a lateral deflection, allow the tire to roll forward. Eventually, the deflected footprint will return to its undeflected position. The length of the distance from center of the contact patch at the starting point to the point where the deflection is (1/e) of its original value is defined as the relaxation length of the tire. The definition of the tire footprint half length is simply the length along the center of the footprint from the very foremost point of contact to the center point of the contact patch.

Defining the Moreland tire model is considerably simpler than the Von Schlippe-Dietrich tire model. Moreland made the simple physical observation that whenever a lateral force is supplied to a wheel and tire there is a physical time delay before the tire responds. This time delay is the basis of his tire model. The basic equation of motion for this time delay is given by

$$cF_N = \psi + c_1 d\psi/dt$$

where c - coefficient of yaw (rad/lb)
       c_1 - Moreland tire time constant (sec)
       F_N - Normal force acting at the center of the contact patch (lbs)
       ψ - Yaw angle about the wheel vertical axis

The Moreland tire yaw coefficient is defined by placing the wheel tire combination on a pivoting arm. This combination of wheel, tire, and freely pivoting arm are placed in a fixture that will allow the combination to pivot freely when the fixture is lowered and the tire is pressed against a rotating drum. A force is applied to the axle normal to the wheel plane which will cause the trail arm of the pivot to swing out until a fixed angle is established with respect to the free trail position. The ratio of this angle to the applied force is the coefficient of yaw of the tire, c. The force is then removed and the wheel, tire, and pivot arm are free to oscillate about the free trail position. The motion will be an oscillation of decaying amplitude. The logarithmic decrement of this decaying oscillation defines the Moreland tire time constant, c_1.

THE EQUIVALENCE OF THE TWO MODELS

While these two models appear to be significantly different, they are actually equivalent. To prove this, assume that the distance of delay of the Von Schlippe-Dietrich tire model is restricted to the first two terms as follows

$(1 - (h/V)D_t)(c_\lambda(y_a + R\phi) - (1 + c_\lambda)\psi_s) = (D_t/V + c_\lambda)(y_a + R\phi - \Delta)$

Applying the following kinematic relationship $V(\psi_s - \psi_t) = -D_t(y_a + R\phi - \Delta)$ the first equation above can be written as

$\Delta = ((1 + hc_\lambda)/c_\lambda)\psi_t + (h/V)(d\Delta/dt - ((1 + c_\lambda)/c_\lambda)d\psi_s/dt))$

Since the side force is given by $F_N = K_1\Delta + C_L d\Delta/dt$, it follows by substitution

$[c_\lambda/(K1(1 + hc_\lambda))]F_N = \psi_t + (C_L/K_1)d\psi_t/dt$
$+ [c_\lambda h(K_1 + C_L D_t]/[VK_1(1 + hc_\lambda)][d\Delta/dt - ((1 + hc_\lambda)/c_\lambda)d\psi_s/dt)]$

Since V, the aircraft forward velocity, is usually measured in the units in/sec, it follows that the third term on the right side of this last equation is small. Thus, comparing this equation to the Moreland tire equation it follows that $c_\lambda/(K_1(1 + hc_\lambda))$ approximately = c and C_L/K_1 approximately = c_1.

Based upon these considerations it appears that the Moreland tire model is based upon an approximation of the Von Schlippe-Dietrich tire model. According to Black [3], it has been found that the Moreland tire model is slightly more unstable than the Von Schlippe-Dietrich model. He also indicates that the results given by the two models agree over a wide range of cases except for those cases where the shimmy frequency is high or the velocity is very low. Due to the slightly more unstable nature of the Moreland model, it follows that its use in landing gear design analysis will provide a more conservative prediction of the stability characteristics of the gear.

CONCLUSIONS

A detailed presentation of the Von Schlippe-Dietrich and Moreland tire models for landing gear shimmy was presented. A mathematical connection between these two models was established. A mathematical connection between the Von Schlippe-Dietrich and Moreland tire parameters was also established.

REFERENCES

1. Von Schlippe, B., and Dietrich, R., *"Shimmying of a Pneumatic Wheel,"* ATI 18920 October 1941. Subsequently published as NACA TM 1365.

2. Moreland, W. J., *"The Story of Shimmy,"* Journal of the Aeronautical Sciences, 1954.

3. Black, R. J., *"Application of Tire Dynamics to Aircraft Landing Gear Design Analysis,"* NASA Tire Modeling Workshop, Langley Research Center, Hampton, Virginia, September 1982.

Relaxation Length Behaviour of Tyres

J.P. MAURICE and H.B. PACEJKA

ABSTRACT

With an analytical transfer function based on a simple linear tyre model, the relaxation length can be estimated from pendulum experiments. This relaxation length is compared to that determined from frequency response functions and from step responses of a non-linear tyre model.

1. INTRODUCTION

Pendulum experiments have been conducted to validate a physical tyre model and to investigate the relaxation length behaviour of the tyre. These experiments are described and the results are compared to the responses of the non-linear tyre model at different vertical loads and average slip angles.

2. PENDULUM EXPERIMENTS

The pendulum test stand (Fig. 1) consists of a stiff frame (pendulum arm) which can rotate about a vertical hinge. Its free end moves over the top of a 2.5 m diameter steel drum. A piezo-electric Kistler measuring hub with the wheel axle is mounted on a special steering head at the free end of the pendulum arm. With this device the average slip angle of the tyre α_0 can be adjusted and fixed between -3° and 8° (although for practical reasons the maximum α_0 is restricted to 5°). The vertical load F_z is applied by slightly tilting forward the vertical hinge. Forced oscillations are applied to the wheel axle by a hydropuls cylinder.

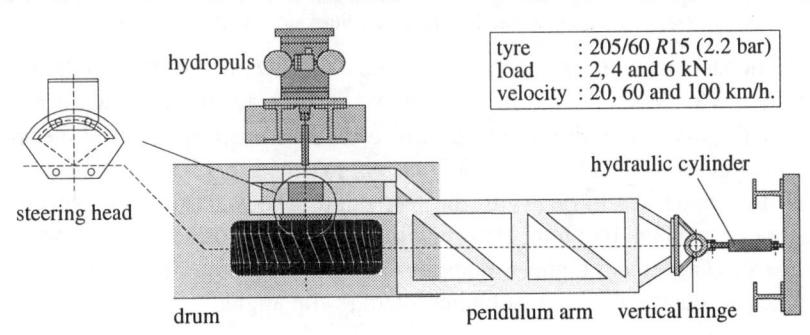

Fig. 1. Schematic top view of the pendulum test stand.

A noise signal with limited band width (0-30 Hz) is chosen as the input signal for the displacement of the hydropuls y. The frequency response functions of the lateral force F_y, the aligning moment M_z and the overturning moment M_x with respect to the lateral displacement of the hydropuls are estimated. The two contributions in the measured FRFs are tyre contact forces and inertia forces of the tyre, the wheel and a part of the measuring hub. The tyre forces can be represented by a first order relaxation length (σ) system [2]:

$$\frac{\sigma}{V}\frac{d\alpha_1}{dt} + \alpha_1 = \alpha \qquad (1)$$

where α equals the input slip angle (which is determined by the motion of the pendulum arm) and α_1 the deflection angle at the leading edge of the contact area. In case of small oscillations and taking into account the total measured mass m, the force in the measuring hub reads, with the cornering stiffness $C_{F\alpha}$:

$$F_{y,\text{hub}} = C_{F\alpha}\alpha_1 + m\ddot{y} \qquad (2)$$

Continuing from these equations, the transfer function of the lateral force in the hub with respect to the hydropuls displacement has been derived. This function resembles the measured FRFs well and is used to estimate the relaxation length and the cornering stiffness at given load, average slip angle and velocity (Fig. 2).

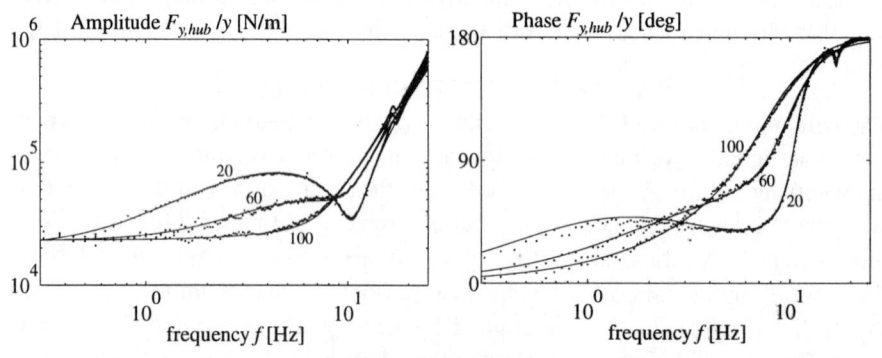

Fig. 2. Bode diagrams of experimental (dotted lines) and fitted (solid lines) FRFs at 2000 N vertical load and zero average slip angle. The three velocities are indicated in km/h.

The stationary magnitudes correspond with $C_{F\alpha}/L$, L being the length of the pendulum arm. With increasing frequency, the contact forces tend to a limit value of $C_{F\alpha}y/\sigma$. The hub force responses are determined by the total measured mass m. The small peak in the FRFs at approximately 15 Hz is caused by a wheel hop mode of the tyre with the test stand. The main problem during the experiments was the tyre temperature, which increased considerably at non-zero average slip angles and higher velocities. This causes a larger cornering stiffness and a larger relaxation length at higher average slip angles.

3. PHYSICAL TYRE MODEL

The tyre model consists of a rigid ring to represent the modes of vibration of the tyre belt with respect to the wheel in which the belt acts as a rigid body. The ring is connected to the wheel with springs (c) and dampers (k) to model the side wall and internal air pressure and has three out-of-plane degrees of freedom: in lateral direction (y_b), about the x-axis (camber angle γ_b) and about the z-axis (yaw angle

ψ_b), see Fig. 3. With the mass m_b and the moments of inertia about the x- and the z-axes I_{bx} and I_{bz}, the equations of motion of the belt relative to the wheel read:

$$m_b \ddot{y}_b + k_{by} \dot{y}_b + c_{by} y_b = F_{cy} \tag{3a}$$

$$I_{bx} \ddot{\gamma}_b + k_{b\gamma}(\dot{\gamma}_b - \Omega \psi_b) + c_{b\gamma} \gamma_b - I_{by}\Omega \dot{\psi}_b = -rF_{cy} \tag{3b}$$

$$I_{bz} \ddot{\psi}_b + k_{b\psi}(\dot{\psi}_b + \Omega \gamma_b) + c_{b\psi} \psi_b + I_{by}\Omega \dot{\gamma}_b = M_{cz} \tag{3c}$$

The gyroscopic effect of the belt is represented by the polar moment of inertia I_{by} and the angular wheel velocity Ω. The contact force F_{cy} and moment M_{cz} are generated by a brush type contact model which is attached to the lower part of the ring (with loaded radius r) by means of residual stiffnesses, which are introduced to compensate for the contributions of the higher order modes to the total deformation of the tyre. In the brush model the individual elements are followed during their passage through the contact patch. The elements may adhere or slide to the road surface, allowing large slip variations within the contact patch.

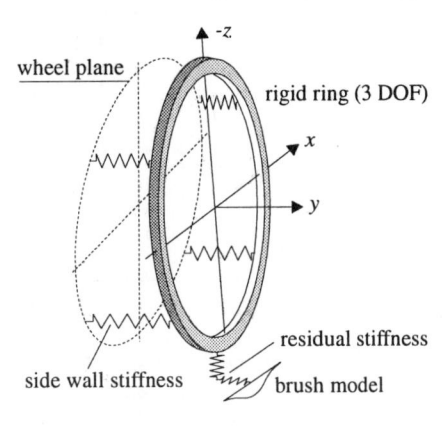

Fig. 3. Schematic view of tyre model used.

To compare the tyre model with the experiments, the measured FRFs have to be corrected for mass and inertia effects of the wheel and the moving part of the hub, as in the model the side wall forces and moments are calculated. The force and moment responses represent the experimental FRFs rather well at zero average slip. At non-zero average slip angles the tyre stiffness and friction properties change due to the rising tyre temperature, causing differences between the model and the experiments. The dynamic aspects of the tyre model have been validated by high frequency yaw oscillation experiments [1].

4. ASSESSMENT OF THE RELAXATION LENGTH

By definition the relaxation length is related to the distance needed by the tyre to reach a certain percentage of the steady state situation after a step wise change of the slip angle. To find the relaxation length at a certain level of side slip, first the steady state situation of the tyre model is determined. Next the FRF of the lateral force to lateral wheel motions is determined and the relaxation length is fitted in the same way as for the experiments. In addition, the lateral force response of the model to an incremental change of the slip angle is fitted with an exponential function. In Fig. 4 the relaxation length σ as function of the load F_z and the average slip angle α_0 as fitted from the FRFs is shown, while the results from the step responses at 4000 N are given also. Both the experiments and the tyre model show the non-linear dependency of σ on F_z and $α_0$. The relaxation length according to the model step responses corresponds with the results of the model FRFs. The differences between the experiment and the model can be explained by the lateral force characteristics in Fig. 5, as experiments and simulations show that the relaxation length variation closely resembles the variation of the local cornering stiffness $∂F_y/∂α$. Furthermore it appears that the shape of σ as function of the load changes with increasing average slip angle.

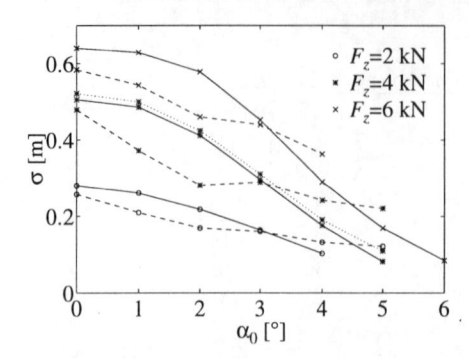

Fig. 4. Relaxation length as function of F_z and $α_0$ (V=20 km/h), according to *pendulum* FRFs (dashed), *tyre model* FRFs (solid) and *tyre model* step responses (dotted).

Fig. 5. Lateral force as function of F_z and $α_0$ (V=20 km/h), according to *pendulum* FRFs (dashed) and from the *tyre model* (solid).

The effect of the velocity is more evident in the experimental results than in the step responses with the tyre model. It may be assumed that this is mainly caused by the increasing tyre temperature in the experiments. Model calculations show that a larger gyroscopic moment increases the lag in the model response, thereby causing a larger effective relaxation length for the lateral force at higher velocities.

NOTATION

$F_{(c)y}$	lateral (contact) force	m_b	mass of tyre belt	$α$	slip angle
$M_{(c)z}$	aligning (contact) moment	$I_{b(x,y,z)}$	inertias of tyre belt	$α_0$	average slip angle
L	length of pendulum arm	$y_b, γ_b, ψ_b$	belt deflections	$α_1$	tyre deflection angle
V	velocity	$c_{b(y,γ,ψ)}$	stiffness parameters	$σ$	relaxation length
y	lateral displacement	$k_{b(y,γ,ψ)}$	damping parameters	$C_{Fα}$	cornering stiffness
m	total measured mass	r	ring radius	$Ω$	angular velocity

REFERENCES

1. Maurice, J.P., Zegelaar, P.W.A. and Pacejka, H.B., *The Influence of Belt Dynamics on Cornering and Braking Properties of Tyres*. To be presented on the 15th IAVSD Conference, Budapest, August 25-29, 1997.
2. Pacejka, H.B., *Modelling of the pneumatic tyre and its impact on vehicle dynamic behaviour*. Delft, 1988. Delft University of Technology, Faculty of Mechanical Engineering, Vehicle Research Laboratory.

LIST OF ADDRESS

2nd International Colloquium on
Tyre Models for Vehicle Dynamic Analysis
Technical University of Berlin, Germany
February 20 - 21, 1997

Lectures

PROF. I. R. DR. F. BÖHM, Technical University of Berlin, Institute of Mechanics, MS 2, Einsteinufer 5-7, 10587 Berlin, Germany

PROF. F. CHELI, Politecnico di Milano, Dipartimento di Meccanica, Piazza Leonardo da Vinci 32, 20133 Milano, Italy

M. EICHLER, Volkswagen AG, D-38436 Wolfsburg, EFM 1710, Germany

P. FANCHER, University of Michigan, UMTRI, 2901 Baxter Rd, Ann Arbor, Michigan 48109-2150, USA

F. R. FASSBENDER, University of Federal Armed Forces Hamburg, Institute of Automotive Engineering, IKK, Holstenhofweg 85, 22043 Hamburg, Germany

KONGHUI GUO, Jilin University of Technology Changchun, Jilin Province, 114 Stalin Avenue, 130025 Changchun, P. R. China

A. HIGUCHI, Delft University of Technology, Faculty of Mechanical Engineering and Marine Technology, Vehicle System Engineering, Mekelweg 2, 2628 CD Delft, The Netherlands

R. HOFER, Technical University of Wien, Institute of Mechanics 325/1, Wiedner Hauptstr. 8-10, A-1040 Wien, Austria

SON-JOO KIM, Delft University of Technology, Vehicle Research Laboratory, Fac. of Mechanical & Marine Engineering, Mekelweg 2, 2628 CD Delft, The Netherlands

DR. G. LEISTER, Daimler-Benz AG, HPC D201, 70322 Stuttgart, Germany

DR. G. MASTINU, Politecnico di Milano, Dipartimento di Meccanica, Sezione „"Veicoli Terrestri", Piazza Leonardo da Vinci, 32, 200133 Milano, Italy

MRS. DOZ. DR. N. L. NARSKAYA, et al., Moscow State Technical University, Faculty of Fundamental Sciences, Dept. of Theoretical Mechanics, 2. Baumanskaya 5, 107 005 Moskau, Rußland

PROF. E. M. NEGRUS, SIAR, Universitatea „Politehnica" Bucuresti, Fac. Transporturi, 313 Splaiul Independentei, 77206 Bucuresti, Romania

PROF. DR. CH. OERTEL, Fachhochschule Anhalt, FB Maschinenbau, Bernburger Str. 52-56, 06366 Köthen, Germany

H. OLDENETTEL Continental AG, Jädekamp 30, 30419 Hannover, Germany

J. J. M. VAN OOSTEN, TNO Road-Vehicles Research Institute, Schoemakerstraat 97, PO Box 6033, 2600 JA Delft, The Netherlands

PROF. DR. H. B. PACEJKA, Delft University of Technology, Faculty of Mechanical Engineering and Marine Technology, Vehicle Research Laboratory, Transport Technology Group, Mekelweg 2, 2628 CD Delft, The Netherlands

PROF. R. S. SHARP, Cranfield University, School of Mechanical Engineering, Cranfield, Wharley End, Bedford MK 43 OAL, United Kingdom

DR. SHUNICHI YAMAZAKI, Japan Automobile Research Institute, Senior Researcher, Research Division 2, Karima, Tsukuba-City, Ibaraki 305, Japan

D. ZACHOW, Technical University of Berlin, Institute of Mechanics, MS 2, Einsteinufer 5-7, 10587 Berlin, Germany

PETER W. A. ZEGELAAR, Delft University of Technology, Faculty of Mechanical Engineering and Marine Technology, Transport Technology Group, Vehicle Research Laboratory, Mekelweg 2, 2628 CD Delft, The Netherlands

Poster

R. WADE ALLEN, Systems Technology, Inc., Facsimili Transmission, 13766 Hawthorne Blvd., Hawthorne, CA 90250, USA

M. COCOSILA, Polytechnic University of Bucharest, Faculty of Transports-Cat. AR, Splaiul Independentei 313, Bucuresti, Romania

F. GAILLIEGUE, GIE SARA, Projet SARA - Centre SAMM, Chemin de la Malmaison, 91578 Biévvres Cedex, France

LIST OF ADDRESSES

PROF. DR. H. GÖHLICH, Technical University of Berlin, Institute of Agricultural Engineering and Hydraulics, LT 1, Zoppoter Str. 35, 14199 Berlin, Germany

W. E. KRABACHER, Landing Gear Development Facility WL/FIVM, BLDG 31, 1981 Fifth St, Wrigt-Patterson AFB, Ohio, USA

J. P. MAURICE, Delft University of Technology, Vehicle Research Laboratory, Mekelweg 2, 2628 CD Delft, The Netherlands

PROF. E. M. NEGRUS, SIAR, Universitatea „Politehnica" Bucuresti, Fac. Transporturi, 313 Splaiul Independentei, 77206 Bucuresti, Romania